T0140553

Graduate Texts in Mathematics **164**

Springer
New York
Berlin
Heidelberg
Barcelona
Budapest
Hong Kong
London
Milan
Paris
Santa Clara
Singapore
Tokyo

Graduate Texts in Mathematics

1 TAKEUTI/ZARING. Introduction to Axiomatic Set Theory. 2nd ed.
2 OXTOBY. Measure and Category. 2nd ed.
3 SCHAEFER. Topological Vector Spaces.
4 HILTON/STAMMBACH. A Course in Homological Algebra.
5 MAC LANE. Categories for the Working Mathematician.
6 HUGHES/PIPER. Projective Planes.
7 SERRE. A Course in Arithmetic.
8 TAKEUTI/ZARING. Axiomatic Set Theory.
9 HUMPHREYS. Introduction to Lie Algebras and Representation Theory.
10 COHEN. A Course in Simple Homotopy Theory.
11 CONWAY. Functions of One Complex Variable I. 2nd ed.
12 BEALS. Advanced Mathematical Analysis.
13 ANDERSON/FULLER. Rings and Categories of Modules. 2nd ed.
14 GOLUBITSKY/GUILLEMIN. Stable Mappings and Their Singularities.
15 BERBERIAN. Lectures in Functional Analysis and Operator Theory.
16 WINTER. The Structure of Fields.
17 ROSENBLATT. Random Processes. 2nd ed.
18 HALMOS. Measure Theory.
19 HALMOS. A Hilbert Space Problem Book. 2nd ed.
20 HUSEMOLLER. Fibre Bundles. 3rd ed.
21 HUMPHREYS. Linear Algebraic Groups.
22 BARNES/MACK. An Algebraic Introduction to Mathematical Logic.
23 GREUB. Linear Algebra. 4th ed.
24 HOLMES. Geometric Functional Analysis and Its Applications.
25 HEWITT/STROMBERG. Real and Abstract Analysis.
26 MANES. Algebraic Theories.
27 KELLEY. General Topology.
28 ZARISKI/SAMUEL. Commutative Algebra. Vol.I.
29 ZARISKI/SAMUEL. Commutative Algebra. Vol.II.
30 JACOBSON. Lectures in Abstract Algebra I. Basic Concepts.
31 JACOBSON. Lectures in Abstract Algebra II. Linear Algebra.
32 JACOBSON. Lectures in Abstract Algebra III. Theory of Fields and Galois Theory.
33 HIRSCH. Differential Topology.
34 SPITZER. Principles of Random Walk. 2nd ed.

35 WERMER. Banach Algebras and Several Complex Variables. 2nd ed.
36 KELLEY/NAMIOKA ET AL. Linear Topological Spaces.
37 MONK. Mathematical Logic.
38 GRAUERT/FRITZSCHE. Several Complex Variables.
39 ARVESON. An Invitation to C^*-Algebras.
40 KEMENY/SNELL/KNAPP. Denumerable Markov Chains. 2nd ed.
41 APOSTOL. Modular Functions and Dirichlet Series in Number Theory. 2nd ed.
42 SERRE. Linear Representations of Finite Groups.
43 GILLMAN/JERISON. Rings of Continuous Functions.
44 KENDIG. Elementary Algebraic Geometry.
45 LOÈVE. Probability Theory I. 4th ed.
46 LOÈVE. Probability Theory II. 4th ed.
47 MOISE. Geometric Topology in Dimensions 2 and 3.
48 SACHS/WU. General Relativity for Mathematicians.
49 GRUENBERG/WEIR. Linear Geometry. 2nd ed.
50 EDWARDS. Fermat's Last Theorem.
51 KLINGENBERG. A Course in Differential Geometry.
52 HARTSHORNE. Algebraic Geometry.
53 MANIN. A Course in Mathematical Logic.
54 GRAVER/WATKINS. Combinatorics with Emphasis on the Theory of Graphs.
55 BROWN/PEARCY. Introduction to Operator Theory I: Elements of Functional Analysis.
56 MASSEY. Algebraic Topology: An Introduction.
57 CROWELL/FOX. Introduction to Knot Theory.
58 KOBLITZ. p-adic Numbers, p-adic Analysis, and Zeta-Functions. 2nd ed.
59 LANG. Cyclotomic Fields.
60 ARNOLD. Mathematical Methods in Classical Mechanics. 2nd ed.
61 WHITEHEAD. Elements of Homotopy Theory.
62 KARGAPOLOV/MERLZJAKOV. Fundamentals of the Theory of Groups.
63 BOLLOBAS. Graph Theory.
64 EDWARDS. Fourier Series. Vol. I. 2nd ed.
65 WELLS. Differential Analysis on Complex Manifolds. 2nd ed.

continued after index

Melvyn B. Nathanson

Additive Number Theory

The Classical Bases

Springer

Melvyn B. Nathanson
Department of Mathematics
Lehman College of the
  City University of New York
250 Bedford Park Boulevard West
Bronx, NY 10468-1589 USA

Mathematics Subject Classifications (1991): 11-01, 11P05, 11P32

Library of Congress Cataloging-in-Publication Data
Nathanson, Melvyn B. (Melvyn Bernard), 1944–
    Additive number theory:the classical bases/Melvyn B.
  Nathanson.
        p.   cm. — (Graduate texts in mathematics;164)
    Includes bibliographical references and index.
    ISBN 0-387-94656-X (hardcover:alk. paper)
    1. Number theory.   I. Title.   II. Series.
  QA241.N347   1996
  512′.72–dc20   96-11745

Printed on acid-free paper.

Production managed by Hal Henglein; manufacturing supervised by Jeffrey Taub.
Camera-ready copy prepared from the author's LaTeX files.
Printed and bound by R.R. Donnelley & Sons, Harrisonburg, VA.
Printed in the United States of America.

9 8 7 6 5 4 3 2 1

ISBN 0-387-94656-X Springer-Verlag New York Berlin Heidelberg   SPIN 10490794

To Marjorie

To Marjorie

Preface

[Hilbert's] style has not the terseness of many of our modern authors
in mathematics, which is based on the assumption that printer's labor
and paper are costly but the reader's effort and time are not.

H. Weyl [143]

The purpose of this book is to describe the classical problems in additive number theory and to introduce the circle method and the sieve method, which are the basic analytical and combinatorial tools used to attack these problems. This book is intended for students who want to learn additive number theory, not for experts who already know it. For this reason, proofs include many "unnecessary" and "obvious" steps; this is by design.

The archetypical theorem in additive number theory is due to Lagrange: Every nonnegative integer is the sum of four squares. In general, the set A of nonnegative integers is called an *additive basis of order h* if every nonnegative integer can be written as the sum of h not necessarily distinct elements of A. Lagrange's theorem is the statement that the squares are a basis of order four. The set A is called a *basis of finite order* if A is a basis of order h for some positive integer h. Additive number theory is in large part the study of bases of finite order. The classical bases are the squares, cubes, and higher powers; the polygonal numbers; and the prime numbers. The classical questions associated with these bases are Waring's problem and the Goldbach conjecture.

Waring's problem is to prove that, for every $k \geq 2$, the nonnegative kth powers form a basis of finite order. We prove several results connected with Waring's problem, including Hilbert's theorem that every nonnegative integer is the sum of a bounded number of kth powers, and the Hardy–Littlewood asymptotic formula for the number of representations of an integer as the sum of s positive kth powers.

Goldbach conjectured that every even positive integer is the sum of at most two prime numbers. We prove three of the most important results on the Goldbach conjecture: Shnirel'man's theorem that the primes are a basis of finite order, Vinogradov's theorem that every sufficiently large odd number is the sum of three primes, and Chen's theorem that every sufficently large even integer is the sum of a prime and a number that is a product of at most two primes.

Many unsolved problems remain. The Goldbach conjecture has not been proved. There is no proof of the conjecture that every sufficiently large integer is the sum of four nonnegative cubes, nor can we obtain a good upper bound for the least number s of nonnegative kth powers such that every sufficiently large integer is the sum of s kth powers. It is possible that neither the circle method nor the sieve method is powerful enough to solve these problems and that completely new mathematical ideas will be necessary, but certainly there will be no progress without an understanding of the classical methods.

The prerequisites for this book are undergraduate courses in number theory and real analysis. The appendix contains some theorems about arithmetic functions that are not necessarily part of a first course in elementary number theory. In a few places (for example, Linnik's theorem on sums of seven cubes, Vinogradov's theorem on sums of three primes, and Chen's theorem on sums of a prime and an almost prime), we use results about the distribution of prime numbers in arithmetic progressions. These results can be found in Davenport's *Multiplicative Number Theory* [19].

Additive number theory is a deep and beautiful part of mathematics, but for too long it has been obscure and mysterious, the domain of a small number of specialists, who have often been specialists only in their own small part of additive number theory. This is the first of several books on additive number theory. I hope that these books will demonstrate the richness and coherence of the subject and that they will encourage renewed interest in the field.

I have taught additive number theory at Southern Illinois University at Carbondale, Rutgers University—New Brunswick, and the City University of New York Graduate Center, and I am grateful to the students and colleagues who participated in my graduate courses and seminars. I also wish to thank Henryk Iwaniec, from whom I learned the linear sieve and the proof of Chen's theorem.

This work was supported in part by grants from the PSC-CUNY Research Award Program and the National Security Agency Mathematical Sciences Program.

I would very much like to receive comments or corrections from readers of this book. My e-mail addresses are nathansn@alpha.lehman.cuny.edu and nathanson@worldnet.att.net. A list of errata will be available on my homepage at http://www.lehman.cuny.edu or http://math.lehman.cuny.edu/nathanson.

Melvyn B. Nathanson
Maplewood, New Jersey
May 1, 1996

Contents

Preface   vii

Notation and conventions   xiii

I   Waring's problem

1   Sums of polygons   3
  1.1   Polygonal numbers   4
  1.2   Lagrange's theorem   5
  1.3   Quadratic forms   7
  1.4   Ternary quadratic forms   12
  1.5   Sums of three squares   17
  1.6   Thin sets of squares   24
  1.7   The polygonal number theorem   27
  1.8   Notes   33
  1.9   Exercises   34

2   Waring's problem for cubes   37
  2.1   Sums of cubes   37
  2.2   The Wieferich–Kempner theorem   38
  2.3   Linnik's theorem   44
  2.4   Sums of two cubes   49
  2.5   Notes   71
  2.6   Exercises   72

3   The Hilbert–Waring theorem   75
  3.1   Polynomial identities and a conjecture of Hurwitz . . .   75
  3.2   Hermite polynomials and Hilbert's identity   77
  3.3   A proof by induction   86
  3.4   Notes   94

    3.5   Exercises 94

4 Weyl's inequality   **97**
    4.1   Tools 97
    4.2   Difference operators 99
    4.3   Easier Waring's problem 102
    4.4   Fractional parts 103
    4.5   Weyl's inequality and Hua's lemma 111
    4.6   Notes 118
    4.7   Exercises 118

5 The Hardy–Littlewood asymptotic formula   **121**
    5.1   The circle method 121
    5.2   Waring's problem for $k = 1$ 124
    5.3   The Hardy–Littlewood decomposition 125
    5.4   The minor arcs 127
    5.5   The major arcs 129
    5.6   The singular integral 133
    5.7   The singular series 137
    5.8   Conclusion 146
    5.9   Notes 147
    5.10 Exercises 147

II   The Goldbach conjecture

6 Elementary estimates for primes   **151**
    6.1   Euclid's theorem 151
    6.2   Chebyshev's theorem 153
    6.3   Mertens's theorems 158
    6.4   Brun's method and twin primes 167
    6.5   Notes 173
    6.6   Exercises 174

7 The Shnirel'man–Goldbach theorem   **177**
    7.1   The Goldbach conjecture 177
    7.2   The Selberg sieve 178
    7.3   Applications of the sieve 186
    7.4   Shnirel'man density 191
    7.5   The Shnirel'man–Goldbach theorem 195
    7.6   Romanov's theorem 199
    7.7   Covering congruences 204
    7.8   Notes 208
    7.9   Exercises 208

8 Sums of three primes   **211**
  8.1 Vinogradov's theorem 211
  8.2 The singular series 212
  8.3 Decomposition into major and minor arcs 213
  8.4 The integral over the major arcs 215
  8.5 An exponential sum over primes 220
  8.6 Proof of the asymptotic formula 227
  8.7 Notes 230
  8.8 Exercise 230

9 The linear sieve   **231**
  9.1 A general sieve 231
  9.2 Construction of a combinatorial sieve 238
  9.3 Approximations 244
  9.4 The Jurkat–Richert theorem 251
  9.5 Differential-difference equations 259
  9.6 Notes 267
  9.7 Exercises 267

10 Chen's theorem   **271**
  10.1 Primes and almost primes 271
  10.2 Weights 272
  10.3 Prolegomena to sieving 275
  10.4 A lower bound for $S(A, \mathcal{P}, z)$ 279
  10.5 An upper bound for $S(A_q, \mathcal{P}, z)$ 281
  10.6 An upper bound for $S(B, \mathcal{P}, y)$ 286
  10.7 A bilinear form inequality 292
  10.8 Conclusion 297
  10.9 Notes 298

III Appendix

Arithmetic functions   **301**
  A.1 The ring of arithmetic functions 301
  A.2 Sums and integrals 303
  A.3 Multiplicative functions 308
  A.4 The divisor function 310
  A.5 The Euler φ–function 314
  A.6 The Möbius function 317
  A.7 Ramanujan sums 320
  A.8 Infinite products 323
  A.9 Notes 327
  A.10 Exercises 327

Bibliography   **331**

Index   **341**

Notation and conventions

Theorems, lemmas, and corollaries are numbered consecutively in each chapter and in the Appendix. For example, Lemma 2.1 is the first lemma in Chapter 2 and Theorem A.2 is the second theorem in the Appendix.

The lowercase letter p denotes a prime number.

We adhere to the usual convention that the *empty sum* (the sum containing no terms) is equal to zero and the *empty product* is equal to one.

Let f be any real or complex-valued function, and let g be a positive function. The functions f and g can be functions of a real variable x or arithmetic functions defined only on the positive integers. We write

$$f = O(g)$$

or

$$f \ll g$$

or

$$g \gg f$$

if there exists a constant $c > 0$ such that

$$|f(x)| \le cg(x)$$

for all x in the domain of f. The constant c is called the *implied constant* . We write

$$f \ll_{a,b,\ldots} g$$

if there exists a constant $c > 0$ that depends on $a, b, \ldots$ such that

$$|f(x)| \le cg(x)$$

for all x in the domain of f. We write

$$f = o(g)$$

if

$$\lim_{x \to \infty} \frac{f(x)}{g(x)} = 0.$$

The function f *is asymptotic to* g, denoted

$$f \sim g,$$

if

$$\lim_{x \to \infty} \frac{f(x)}{g(x)} = 1.$$

The real-valued function f is *increasing* on the interval I if $f(x_1) \leq f(x_2)$ for all $x_1, x_2 \in I$ with $x_1 < x_2$. Similarly, the real-valued function f is *decreasing* on the interval I if $f(x_1) \geq f(x_2)$ for all $x_1, x_2 \in I$ with $x_1 < x_2$. The function f is *monotonic* on the interval I if it is either increasing on I or decreasing on I.

We use the following notation for exponential functions:

$$\exp(x) = e^x$$

and

$$e(x) = \exp(2\pi i x) = e^{2\pi i x}.$$

The following notation is standard:

$\mathbf{Z}$	the integers $0, \pm 1, \pm 2, \ldots$
$\mathbf{R}$	the real numbers
$\mathbf{R}^n$	n-dimensional Euclidean space
$\mathbf{Z}^n$	the integer lattice in $\mathbf{R}^n$
$\mathbf{C}$	the complex numbers
$|z|$	the absolute value of the complex number z
$\Re z$	the real part of the complex number z
$\Im z$	the imaginary part of the complex number z
$[x]$	the integer part of the real number x, that is, the integer uniquely determined by the inequality $[x] \leq x < [x] + 1$.
$\{x\}$	the fractional part of the real number x, that is, $\{x\} = x - [x] \in [0, 1)$.
$\|x\|$	the distance from the real number x to the nearest integer, that is, $\|x\| = \min\{|x - n| : n \in \mathbf{Z}\} = \min(\{x\}, 1 - \{x\}) \in [0, 1/2]$.
$(a_1, \ldots, a_n)$	the greatest common divisor of the integers $a_1, \ldots, a_n$
$[a_1, \ldots, a_n]$	the least common multiple of the integers $a_1, \ldots, a_n$
$|X|$	the cardinality of the set X
hA	the h-fold sumset, consisting of all sums of h elements of A

Part I

Waring's problem

1
Sums of polygons

Imo propositionem pulcherrimam et maxime generalem nos primi de-
teximus: nempe omnem numerum vel esse triangulum vex ex duobus
aut tribus triangulis compositum: esse quadratum vel ex duobus aut
tribus aut quatuorquadratis compositum: esse pentagonum vel ex duobus, tribus, quatuor aut quinque pentagonis compositum; et sic deinceps in infinitum, in hexagonis, heptagonis polygonis quibuslibet,
enuntianda videlicet pro numero angulorum generali et mirabili pro-
postione. Ejus autem demonstrationem, quae ex multis variis et abstru-
sissimis numerorum mysteriis derivatur, hic apponere non licet. . . .[1]

P. Fermat [39, page 303]

---

[1]I have discovered a most beautiful theorem of the greatest generality: Every number
is a triangular number or the sum of two or three triangular numbers; every number is a
square or the sum of two, three, or four squares; every number is a pentagonal number or
the sum of two, three, four, or five pentagonal numbers; and so on for hexagonal numbers,
heptagonal numbers, and all other polygonal numbers. The precise statement of this very
beautiful and general theorem depends on the number of the angles. The theorem is based
on the most diverse and abstruse mysteries of numbers, but I am not able to include the
proof here. . . .

1.1   Polygonal numbers

Polygonal numbers are nonnegative integers constructed geometrically from the regular polygons. The triangular numbers, or triangles, count the number of points in the triangular array

The sequence of triangles is 0, 1, 3, 6, 10, 15, ...
   Similarly, the square numbers count the number of points in the square array

The sequence of squares is 0, 1, 4, 9, 16, 25, ...
   The pentagonal numbers count the number of points in the pentagonal array

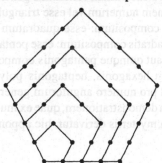

The sequence of pentagonal numbers is 0, 1, 5, 12, 22, 35, ... There is a similar sequence of m-gonal numbers corresponding to every regular polygon with m sides.
   Algebraically, for every $m \geq 1$, the kth polygonal number of order $m+2$, denoted $p_m(k)$, is the sum of the first k terms of the arithmetic progression with initial value 1 and difference m, that is,

$$p_m(k) = 1 + (m+1) + (2m+1) + \cdots + ((k-1)m+1)$$
$$= \frac{mk(k-1)}{2} + k.$$

This is a quadratic polynomial in k. The triangular numbers are the numbers

$$p_1(k) = \frac{k(k+1)}{2},$$

the squares are the numbers

$$p_2(k) = k^2,$$

the pentagonal numbers are the numbers

$$p_3(k) = \frac{k(3k-1)}{2},$$

and so on. This notation is awkward but traditional.

The epigraph to this chapter is one of the famous notes that Fermat wrote in the margin of his copy of Diophantus's *Arithmetica*. Fermat claims that, for every $m \geq 1$, every nonnegative integer can be written as the sum of $m + 2$ polygonal numbers of order $m + 2$. This was proved by Cauchy in 1813. The goal of this chapter is to prove Cauchy's polygonal number theorem. We shall also prove the related result of Legendre that, for every $m \geq 3$, every sufficiently large integer is the sum of five polygonal numbers of order $m + 2$.

1.2 Lagrange's theorem

We first prove the polygonal number theorem for squares. This theorem of Lagrange is the most important result in additive number theory.

Theorem 1.1 (Lagrange) *Every nonnegative integer is the sum of four squares.*

Proof. It is easy to check the formal polynomial identity

$$(x_1^2 + x_2^2 + x_3^2 + x_4^2)(y_1^2 + y_2^2 + y_3^2 + y_4^2) = z_1^2 + z_2^2 + z_3^2 + z_4^2, \tag{1.1}$$

where

$$\left.\begin{array}{rcl} z_1 &=& x_1y_1 + x_2y_2 + x_3y_3 + x_4y_4 \\ z_2 &=& x_1y_2 - x_2y_1 - x_3y_4 + x_4y_3 \\ z_3 &=& x_1y_3 - x_3y_1 + x_2y_4 - x_4y_2 \\ z_4 &=& x_1y_4 - x_4y_1 - x_2y_3 + x_3y_2 \end{array}\right\} \tag{1.2}$$

This implies that if two numbers are both sums of four squares, then their product is also the sum of four squares. Every nonnegative integer is the product of primes, so it suffices to prove that every prime number is the sum of four squares. Since $2 = 1^2 + 1^2 + 0^2 + 0^2$, we consider only odd primes p.

The set of squares

$$\{a^2 \mid a = 0, 1, \ldots, (p-1)/2\}$$

represents $(p + 1)/2$ distinct congruence classes modulo p. Similarly, the set of integers

$$\{-b^2 - 1 \mid b = 0, 1, \ldots, (p-1)/2\}$$

represents $(p + 1)/2$ distinct congruence classes modulo p. Since there are only p different congruence classes modulo p, by the pigeonhole principle there must exist integers a and b such that $0 \le a, b \le (p - 1)/2$ and

$$a^2 \equiv -b^2 - 1 \pmod{p},$$

that is,

$$a^2 + b^2 + 1 \equiv 0 \pmod{p}.$$

Let $a^2 + b^2 + 1 = np$. Then

$$p \le np = a^2 + b^2 + 1^2 + 0^2 \le 2 \left(\frac{p-1}{2} \right)^2 + 1 < \frac{p^2}{2} + 1 < p^2,$$

and so

$$1 \le n < p.$$

Let m be the least positive integer such that mp is the sum of four squares. Then there exist integers x_1, x_2, x_3, x_4 such that

$$mp = x_1^2 + x_2^2 + x_3^2 + x_4^2$$

and

$$1 \le m \le n < p.$$

We must show that $m = 1$.

Suppose not. Then $1 < m < p$. Choose integers y_i such that

$$y_i \equiv x_i \pmod{m}$$

and

$$-m/2 < y_i \le m/2$$

for $i = 1, \ldots, 4$. Then

$$y_1^2 + y_2^2 + y_3^2 + y_4^2 \equiv x_1^2 + x_2^2 + x_3^2 + x_4^2 = mp \equiv 0 \pmod{m}$$

and

$$mr = y_1^2 + y_2^2 + y_3^2 + y_4^2$$

for some nonnegative integer r. If $r = 0$, then $y_i = 0$ for all i and each x_i^2 is divisible by m^2. It follows that mp is divisible by m^2, and so p is divisible by m. This is impossible, since p is prime and $1 < m < p$. Therefore, $r \ge 1$ and

$$mr = y_1^2 + y_2^2 + y_3^2 + y_4^2 \le 4(m/2)^2 = m^2.$$

Moreover, $r = m$ if and only if m is even and $y_i = m/2$ for all i. In this case, $x_i \equiv m/2 \pmod{m}$ for all i, and so $x_i^2 \equiv (m/2)^2 \pmod{m^2}$ and

$$mp = x_1^2 + x_2^2 + x_3^2 + x_4^2 \equiv 4(m/2)^2 = m^2 \equiv 0 \pmod{m^2}.$$

This implies that p is divisible by m, which is absurd. Therefore,

$$1 \le r < m.$$

Applying the polynomial identity (1.1), we obtain

$$m^2 rp = (mp)(mr)$$
$$= (x_1^2 + x_2^2 + x_3^2 + x_4^2)(y_1^2 + y_2^2 + y_3^2 + y_4^2)$$
$$= z_1^2 + z_2^2 + z_3^2 + z_4^2,$$

where the z_i are defined by equations (1.2). Since $x_i \equiv y_i \pmod{m}$, these equations imply that $z_i \equiv 0 \pmod{m}$ for $i = 1, \ldots, 4$. Let $w_i = z_i/m$. Then $w_1, \ldots, w_4$ are integers and

$$rp = w_1^2 + w_2^2 + w_3^2 + w_4^2,$$

which contradicts the minimality of m. Therefore, $m = 1$ and the prime p is the sum of four squares. This completes the proof of Lagrange's theorem.

A set of integers is called a *basis of order h* if every nonnegative integer can be written as the sum of h not necessarily distinct elements of the set. A set of integers is called a *basis of finite order* if the set is a basis of order h for some h. Lagrange's theorem states that the set of squares is a basis of order four. Since 7 cannot be written as the sum of three squares, it follows that the squares do not form a basis of order three. The central problem in additive number theory is to determine if a given set of integers is a basis of finite order. Lagrange's theorem gives the first example of a natural and important set of integers that is a basis. In this sense, it is the archetypical theorem in additive number theory. Everything in this book is a generalization of Lagrange's theorem. We shall prove that the polygonal numbers, the cubes and higher powers, and the primes are all bases of finite order. These are the classical bases in additive number theory.

1.3 Quadratic forms

Let $A = (a_{i,j})$ be an $m \times n$ matrix with integer coefficients. In this chapter, we shall only consider matrices with integer coefficients. Let A^T denote the transpose of the matrix A, that is, $A^T = \left(a_{i,j}^T \right)$ is the $n \times m$ matrix such that

$$a_{i,j}^T = a_{j,i}$$

for $i = 1, \ldots, n$ and $j = 1, \ldots, m$. Then $(A^T)^T = A$ for every $m \times n$ matrix A, and $(AB)^T = B^T A^T$ for any pair of matrices A and B such that the number of columns of A is equal to the number of rows of B.

Let $M_n(\mathbf{Z})$ be the ring of $n \times n$ matrices. A matrix $A \in M_n(\mathbf{Z})$ is *symmetric* if $A^T = A$. If A is a symmetric matrix and U is any matrix in $M_n(\mathbf{Z})$, then $U^T A U$ is also symmetric, since

$$(U^T A U)^T = U^T A^T (U^T)^T = U^T A U.$$

Let $SL_n(\mathbf{Z})$ denote the group of $n \times n$ matrices of determinant 1. This group acts on the ring $M_n(\mathbf{Z})$ as follows: If $A \in M_n(\mathbf{Z})$ and $U \in SL_n(\mathbf{Z})$, we define

$$A \cdot U = U^T A U.$$

This is a group action, since

$$A \cdot (UV) = (UV)^T A(UV) = V^T(U^T AU)V = (U^T AU) \cdot V = (A \cdot U) \cdot V.$$

We say that two matrices A and B in $M_n(\mathbf{Z})$ are *equivalent*, denoted

$$A \sim B,$$

if A and B lie in the same orbit of the group action, that is, if $B = A \cdot U = U^T AU$ for some $U \in SL_n(\mathbf{Z})$. It is easy to check that this is an equivalence relation. Since $\det(U) = 1$ for all $U \in SL_n(\mathbf{Z})$, it follows that

$$\det(A \cdot U) = \det(U^T AU) = \det(U^T)\det(A)\det(U) = \det(A)$$

for all $A \in M_n(\mathbf{Z})$, and so the group action preserves determinants. Also, if A is symmetric, then $A \cdot U$ is also symmetric. Thus, for any integer d, the group action partitions the set of symmetric $n \times n$ matrices of determinant d into equivalence classes.

To every $n \times n$ symmetric matrix $A = (a_{i,j})$ we associate the *quadratic form* F_A defined by

$$F_A(x_1, \ldots, x_n) = \sum_{i=1}^{n} \sum_{j=1}^{n} a_{i,j} x_i x_j.$$

This is a homogeneous function of degree two in the n variables $x_1, \ldots, x_n$. For example, if I_n is the $n \times n$ identity matrix, then the associated quadratic form is

$$F_{I_n}(x_1, \ldots, x_n) = x_1^2 + x_2^2 + \cdots + x_n^2.$$

Let x denote the $n \times 1$ matrix (or column vector)

$$x = \begin{pmatrix} x_1 \\ \vdots \\ x_n \end{pmatrix}.$$

We can write the quadratic form in matrix notation as follows:

$$F_A(x_1, \ldots, x_n) = x^T Ax.$$

The *discriminant* of the quadratic form F_A is the determinant of the matrix A. Let A and B be $n \times n$ symmetric matrices, and let F_A and F_B be their corresponding quadratic forms. We say that these forms are *equivalent*, denoted

$$F_A \sim F_B,$$

if the matrices are equivalent, that is, if $A \sim B$. Equivalence of quadratic forms is an equivalence relation, and equivalent quadratic forms have the same discriminant.

The quadratic form F_A represents the integer N if there exist integers $x_1, \ldots, x_n$ such that

$$F_A(x_1, \ldots, x_n) = N.$$

If $F_A \sim F_B$, then $A \sim B$ and there exists a matrix $U \in SL_n(\mathbf{Z})$ such that $A = B \cdot U = U^T B U$. It follows that

$$F_A(x) = x^T A x = x^T U^T B U x = (Ux)^T B(Ux) = F_B(Ux).$$

Thus, if the quadratic form F_A represents the integer N, then every form equivalent to F_A also represents N. Since equivalence of quadratic forms is an equivalence relation, it follows that any two quadratic forms in the same equivalence class represent exactly the same set of integers. Lagrange's theorem implies that, for $n \geq 4$, any form equivalent to the form $x_1^2 + \cdots + x_n^2$ represents all nonnegative integers.

The quadratic form F_A is called *positive-definite* if $F_A(x_1, \ldots, x_n) \geq 1$ for all $(x_1, \ldots, x_n) \neq (0, \ldots, 0)$. Every form equivalent to a positive-definite quadratic form is positive-definite.

A quadratic form in two variables is called a *binary quadratic form*. A quadratic form in three variables is called a *ternary quadratic form*. For binary and ternary quadratic forms, we shall prove that there is only one equivalence class of positive-definite forms of discriminant 1. We begin with binary forms.

Lemma 1.1 *Let*

$$A = \begin{pmatrix} a_{1,1} & a_{1,2} \\ a_{1,2} & a_{2,2} \end{pmatrix}$$

be a 2×2 symmetric matrix, and let

$$F_A(x_1, x_2) = a_{1,1}x_1^2 + 2a_{1,2}x_1x_2 + a_{2,2}x_2^2$$

be the associated quadratic form. The binary quadratic form F_A is positive-definite if and only if

$$a_{1,1} \geq 1$$

and the discriminant d satisfies

$$d = \det(A) = a_{1,1}a_{2,2} - a_{1,2}^2 \geq 1.$$

Proof. If the form F_A is positive-definite, then

$$F_A(1, 0) = a_{1,1} \geq 1$$

and

$$\begin{aligned} F_A(-a_{1,2}, a_{1,1}) &= a_{1,1}a_{1,2}^2 - 2a_{1,1}a_{1,2}^2 + a_{1,1}^2a_{2,2} \\ &= a_{1,1}\left(a_{1,1}a_{2,2} - a_{1,2}^2\right) \\ &= a_{1,1}d \geq 1, \end{aligned}$$

and so $d \geq 1$. Conversely, if $a_{1,1} \geq 1$ and $d \geq 1$, then

$$a_{1,1} F_A(x_1, x_2) = (a_{1,1}x_1 + a_{1,2}x_2)^2 + dx_2^2 \geq 0,$$

and $F_A(x_1, x_2) = 0$ if and only if $(x_1, x_2) = (0, 0)$. This completes the proof.

Lemma 1.2 *Every equivalence class of positive-definite binary quadratic forms of discriminant d contains at least one form*

$$F_A(x_1, x_2) = a_{1,1}x_1^2 + 2a_{1,2}x_1x_2 + a_{2,2}x_2^2$$

for which

$$2|a_{1,2}| \leq a_{1,1} \leq \frac{2}{\sqrt{3}}\sqrt{d}.$$

Proof. Let $F_B(x_1, x_2) = b_{1,1}x_1^2 + 2b_{1,2}x_1x_2 + b_{2,2}x_2^2$ be a positive-definite quadratic form, where

$$B = \begin{pmatrix} b_{1,1} & b_{1,2} \\ b_{1,2} & b_{2,2} \end{pmatrix}$$

is the 2×2 symmetric matrix associated with F. Let $a_{1,1}$ be the smallest positive integer represented by F. Then there exist integers r_1, r_2 such that

$$F(r_1, r_2) = a_{1,1}.$$

If the positive integer h divides both r_1 and r_2, then, by the homogeneity of the form and the minimality of $a_{1,1}$, we have

$$a_{1,1} \leq F(r_1/h, r_2/h) = \frac{F(r_1, r_2)}{h^2} = \frac{a_{1,1}}{h^2} \leq a_{1,1},$$

and so $h = 1$. Therefore, $(r_1, r_2) = 1$ and there exist integers s_1 and s_2 such that

$$1 = r_1 s_2 - r_2 s_1 = r_1(s_2 + r_2 t) - r_2(s_1 + r_1 t)$$

for all integers t. Then

$$U = \begin{pmatrix} r_1 & s_1 + r_1 t \\ r_2 & s_2 + r_2 t \end{pmatrix} \in SL_2(\mathbf{Z})$$

for all $t \in \mathbf{Z}$. Let

$$\begin{aligned}
A &= U^T B U \\
&= \begin{pmatrix} F(r_1, r_2) & a'_{1,2} + F(r_1, r_2)t \\ a'_{1,2} + F(r_1, r_2)t & F(s_1 + r_1 t, s_2 + r_2 t) \end{pmatrix} \\
&= \begin{pmatrix} a_{1,1} & a_{1,2} \\ a_{1,2} & a_{2,2} \end{pmatrix},
\end{aligned}$$

where

$$a'_{1,2} = b_{1,1}r_1s_1 + b_{1,2}(r_1s_2 + r_2s_1) + b_{2,2}r_2s_2$$
$$a_{1,2} = a'_{1,2} + a_{1,1}t$$
$$a_{2,2} = F(s_1 + r_1t, s_2 + r_2t) \geq a_{1,1}$$

since $(s_1 + r_1t, s_2 + r_2t) \neq (0,0)$ for all $t \in \mathbf{Z}$, and $a_{1,1}$ is the smallest positive number represented by the form F. Since $\{a'_{1,2} + a_{1,1}t : t \in \mathbf{Z}\}$ is a congruence class modulo $a_{1,1}$, we can choose t so that

$$|a_{1,2}| = |a'_{1,2} + a_{1,1}t| \leq \frac{a_{1,1}}{2}.$$

Then $A \sim B$, and the form F_B is equivalent to the form $F_A(x_1, x_2) = a_{1,1}x_1^2 + 2a_{1,2}x_1x_2 + a_{2,2}x_2^2$, where

$$2|a_{1,2}| \leq a_{1,1} \leq a_{2,2}.$$

If d is the discriminant of the form, then

$$d = a_{1,1}a_{2,2} - a_{1,2}^2,$$

and the inequality

$$a_{1,1}^2 \leq a_{1,1}a_{2,2} = d + a_{1,2}^2 \leq d + \frac{a_{1,1}^2}{4}$$

implies that

$$\frac{3a_{1,1}^2}{4} \leq d$$

or, equivalently,

$$a_{1,1} \leq \frac{2}{\sqrt{3}}\sqrt{d}.$$

This completes the proof.

Theorem 1.2 *Every positive-definite binary quadratic form of discriminant 1 is equivalent to the form $x_1^2 + x_2^2$.*

Proof. Let F be a positive-definite binary quadratic form of discriminant 1. By Lemma 1.2, the form F is equivalent to a form $a_{1,1}x_1^2 + 2a_{1,2}x_1x_2 + a_{2,2}x_2^2$ for which

$$2|a_{1,2}| \leq a_{1,1} \leq \frac{2}{\sqrt{3}} < 2.$$

Since $a_{1,1} \geq 1$, we must have $a_{1,1} = 1$. This implies that $a_{1,2} = 0$. Since the discriminant is 1, we have

$$a_{2,2} = a_{1,1}a_{2,2} - a_{1,2}^2 = 1.$$

Thus, the form F is equivalent to $x_1^2 + x_2^2$. This completes the proof.

1.4   Ternary quadratic forms

We shall now prove an analogous result for positive-definite ternary quadratic forms.

Lemma 1.3 *Let*

$$A = \begin{pmatrix} a_{1,1} & a_{1,2} & a_{1,3} \\ a_{1,2} & a_{2,2} & a_{2,3} \\ a_{1,3} & a_{2,3} & a_{3,3} \end{pmatrix}$$

be a 3×3 symmetric matrix, and let F_A be the corresponding ternary quadratic form. Let d be the discriminant of F_A. Then

$$a_{1,1} F_A(x_1, x_2, x_3) = (a_{1,1}x_1 + a_{1,2}x_2 + a_{1,3}x_3)^2 + G_{A^*}(x_2, x_3), \qquad (1.3)$$

where G_{A^*} is the binary quadratic form corresponding to the matrix

$$A^* = \begin{pmatrix} a_{1,1}a_{2,2} - a_{1,2}^2 & a_{1,1}a_{2,3} - a_{1,2}a_{1,3} \\ a_{1,1}a_{2,3} - a_{1,2}a_{1,3} & a_{1,1}a_{3,3} - a_{1,3}^2 \end{pmatrix} \qquad (1.4)$$

and G_{A^*} has discriminant $a_{1,1}d$. If F_A is positive-definite, then G_{A^*} is positive-definite. Moreover, the form F_A is positive-definite if and only if the following three determinants are positive:

$$a_{1,1} = \det(a_{1,1}) \geq 1,$$

$$d' = \det \begin{pmatrix} a_{1,1} & a_{1,2} \\ a_{1,2} & a_{2,2} \end{pmatrix} \geq 1,$$

and

$$d = \det(A) \geq 1.$$

Proof. We obtain identities (1.3) and (1.4) as well as the discriminant of G_{A^*} by straightforward calculation.

If F_A is positive-definite, then

$$F_A(1, 0, 0) = a_{1,1} \geq 1.$$

If $G_{A^*}(x_2, x_3) \leq 0$ for some integers x_2, x_3, then $G_{A^*}(a_{1,1}x_2, a_{1,1}x_3) = a_{1,1}^2 G_{A^*}(x_2, x_3) \leq 0$. Let $x_1 = -(a_{1,2}x_2 + a_{1,3}x_3)$. Then

$$a_{1,1}x_1 + a_{1,2}a_{1,1}x_2 + a_{1,3}a_{1,1}x_3 = 0,$$

and so

$$\begin{aligned} a_{1,1} F_A(x_1, a_{1,1}x_2, a_{1,1}x_3) \\ = (a_{1,1}x_1 + a_{1,2}a_{1,1}x_2 + a_{1,3}a_{1,1}x_3)^2 + G_{A^*}(a_{1,1}x_2, a_{1,1}x_3) \\ = G_{A^*}(a_{1,1}x_2, a_{1,1}x_3) \\ = a_{1,1}^2 G_{A^*}(x_2, x_3) \\ \leq 0. \end{aligned}$$

Since F_A is positive-definite, it follows that $x_2 = x_3 = 0$, and so the binary form G_{A^*} is also positive-definite. By Lemma 1.1, the leading coefficient of G_{A^*} is positive, that is,

$$d' = a_{1,1}a_{2,2} - a_{1,2}^2 \geq 1,$$

and also the discriminant of G_{A^*} is positive, hence

$$d = \det(A) \geq 1.$$

This proves that if F_A is positive-definite, then the integers $a_{1,1}, d'$, and d are positive.

Conversely, if these three numbers are positive, then Lemma 1.1 implies that the binary form G_{A^*} is positive-definite. If $F_A(x_1, x_2, x_3) = 0$, then it follows from identity (1.3) that

$$G_{A^*}(x_2, x_3) = 0$$

and

$$a_{1,1}x_1 + a_{1,2}x_2 + a_{1,3}x_3 = 0.$$

The first equation implies that $x_2 = x_3 = 0$, and the second equation implies that $x_1 = 0$. Therefore, the form F_A is positive-definite.

Lemma 1.4 *Let $B = (b_{i,j})$ be a 3×3 symmetric matrix such that the ternary quadratic form F_B is positive-definite. Let G_{B^*} be the unique positive-definite binary quadratic form such that*

$$b_{1,1}F_B(y_1, y_2, y_3) = (b_{1,1}y_1 + b_{1,2}y_2 + b_{1,3}y_3)^2 + G_{B^*}(y_2, y_3).$$

For any matrix $V^* = (v_{i,j}^*) \in SL_2(\mathbf{Z})$, let

$$A^* = (V^*)^T B^* V^* \tag{1.5}$$

and let G_{A^*} be the positive-definite binary quadratic form corresponding to the symmetric matrix A^* and equivalent to the form G_{B^*}. For any integers r and s, let

$$V_{r,s} = (v_{i,j}) = \begin{pmatrix} 1 & r & s \\ 0 & v_{1,1}^* & v_{1,2}^* \\ 0 & v_{2,1}^* & v_{2,2}^* \end{pmatrix} \in SL_3(\mathbf{Z}) \tag{1.6}$$

and

$$A_{r,s} = V_{r,s}^T B V_{r,s} = (a_{i,j}). \tag{1.7}$$

Let $F_{A_{r,s}}$ be the corresponding ternary quadratic form. Then $a_{1,1} = b_{1,1}$ and

$$a_{1,1}F_{A_{r,s}}(x_1, x_2, x_3) = (a_{1,1}x_1 + a_{1,2}x_2 + a_{1,3}x_3)^2 + G_{A^*}(x_2, x_3),$$

where the matrix A^* defined by (1.5) is independent of r and s.

Proof. Since $v_{1,1} = 1$ and $v_{2,1} = v_{3,1} = 0$, it follows from the matrix equation (1.7) that

$$a_{1,j} = \sum_{k=1}^{3} \sum_{i=1}^{3} v_{1,k}^{T} b_{k,i} v_{i,j} = \sum_{k=1}^{3} \sum_{i=1}^{3} v_{k,1} b_{k,i} v_{i,j} = \sum_{i=1}^{3} b_{1,i} v_{i,j}$$

and so $a_{1,1} = b_{1,1}$. Let

$$x = \begin{pmatrix} x_1 \\ x_2 \\ x_3 \end{pmatrix} \qquad \text{and} \qquad V_{r,s} x = y = \begin{pmatrix} y_1 \\ y_2 \\ y_3 \end{pmatrix},$$

so

$$y_i = \sum_{j=1}^{3} v_{i,j} x_j.$$

In particular,

$$y_2 = v2, \ 1x_1 + v_{2,2}x_2 + v_{2,3}x_3 = v_{1,1}^{*}x_2 + v_{1,2}^{*}x_3$$

$$y_3 = v3, \ 1x_1 + v_{3,2}x_2 + v_{3,3}x_3 = v_{2,1}^{*}x_2 + v_{2,2}^{*}x_3.$$

Let

$$y^{*} = \begin{pmatrix} y_2 \\ y_3 \end{pmatrix} \qquad \text{and} \qquad x^{*} = \begin{pmatrix} x_2 \\ x_3 \end{pmatrix}.$$

Then

$$V^{*}x^{*} = y^{*}.$$

It follows that

$$G_{B^{*}}(y_2, y_3) = G_{B^{*}}(V^{*}x^{*}) = G_{A^{*}}(x_2, x_3).$$

Moreover,

$$b_{1,1}y_1 + b_{1,2}y_2 + b_{1,3}y_3 = \sum_{i=1}^{3} b_{1,i} \sum_{j=1}^{3} v_{i,j}x_j$$

$$= \sum_{j=1}^{3} \left(\sum_{i=1}^{3} b_{1,i} v_{i,j} \right) x_j$$

$$= a_{1,1}x_1 + a_{1,2}x_2 + a_{1,3}x_3.$$

Since

$$F_{A_{r,s}}(x_1, x_2, x_3) = x^{T} A_{r,s} x = (V_{r,s}x)^{T} B(V_{r,s}x) = y^{T} By = F_{B}(y_1, y_2, y_3),$$

it follows that

$$(a_{1,1}x_1 + a_{1,2}x_2 + a_{1,3}x_3)^2 + G_{A_{r,s}^{*}}(x_2, x_3)$$

$$= a_{1,1} F_{A_{r,s}}(x_1, x_2, x_3)$$

$$= b_{1,1} F_{A_{r,s}}(x_1, x_2, x_3)$$

$$= b_{1,1} F_{B}(y_1, y_2, y_3)$$

$$= (b_{1,1}y_1 + b_{1,2}y_2 + b_{1,3}y_3)^2 + G_{B^{*}}(y_2, y_3)$$

$$= (a_{1,1}x_1 + a_{1,2}x_2 + a_{1,3}x_3)^2 + G_{A^{*}}(x_2, x_3),$$

and so

$$G_{A^*}(x_2, x_3) = G_{A^*_{r,s}}(x_2, x_3)$$

for all integers r and s. This completes the proof.

Lemma 1.5 *Let $u_{1,1}, u_{2,1},$ and $u_{3,1}$ be integers such that*

$$(u_{1,1}, u_{2,1}, u_{3,1}) = 1.$$

Then there exist six integers $u_{i,j}$ for $i = 1, 2, 3$ and $j = 2, 3$ such that the matrix $U = (u_{i,j}) \in SL_3(\mathbb{Z})$, that is, $\det(U) = 1$.

Proof. Let $(u_{1,1}, u_{2,1}) = a$. Choose integers $u_{1,2}$ and $u_{2,2}$ such that

$$u_{1,1}u_{2,2} - u_{2,1}u_{1,2} = a.$$

Since $(a, u_{3,1}) = (u_{1,1}, u_{2,1}, u_{3,1}) = 1$, we can choose integers $u_{3,3}$ and b such that

$$au_{3,3} - bu_{3,1} = 1.$$

Let

$$u_{1,3} = \frac{u_{1,1}b}{a},$$

$$u_{2,3} = \frac{u_{2,1}b}{a},$$

$$u_{3,2} = 0.$$

Then the matrix

$$U = (u_{i,j}) = \begin{pmatrix} u_{1,1} & u_{1,2} & \left(\frac{u_{1,1}}{a}\right)b \\ u_{2,1} & u_{2,2} & \left(\frac{u_{2,1}}{a}\right)b \\ u_{3,1} & 0 & u_{3,3} \end{pmatrix}$$

has integer coefficients and determinant 1. This completes the proof.

Lemma 1.6 *Every equivalence class of positive-definite ternary quadratic forms of discriminant d contains at least one form $\sum_{i,j=1}^{3} a_{i,j}x_i x_j$ for which*

$$2 \max (|a_{1,2}|, |a_{1,3}|) \le a_{1,1} \le \frac{4}{3}\sqrt[3]{d}.$$

Proof. Let F be a positive-definite ternary quadratic form of determinant d, and let C be the corresponding 3×3 symmetric matrix. Let $a_{1,1}$ be the smallest positive integer represented by F. Then there exist integers $u_{1,1}, u_{2,1},$ and $u_{3,1}$ such that

$$F(u_{1,1}, u_{2,1}, u_{3,1}) = a_{1,1}.$$

If $(u_{1,1}, u_{2,1}, u_{3,1}) = h$, then the form F also represents $a_{1,1}/h^2$, and so, by the minimality of $a_{1,1}$, we have $(u_{1,1}, u_{2,1}, u_{3,1}) = 1$. By Lemma 1.5, there exist integers $u_{i,j}$ for $i = 1, 2, 3$ and $j = 2, 3$ such that the matrix $U = (u_{i,j}) \in SL_3(\mathbb{Z})$. Let

$$B = U^T C U = (b_{i,j}).$$

Then F is equivalent to the form F_B, and

$$b_{1,1} = a_{1,1}$$

is also the smallest integer represented by F_B. By Lemma 1.3,

$$a_{1,1}F_B(x_1, x_2, x_3) = (b_{1,1}x_1 + b_{1,2}x_2 + b_{1,3}x_3)^2 + G_{B^*}(x_2, x_3),$$

where $G_{B^*}(x_2, x_3)$ is a positive-definite binary quadratic form of determinant $a_{1,1}d$. By Lemma 1.2, the form $G_{B^*}(x_2, x_3)$ is equivalent to a binary form

$$G_{A^*}(x_2, x_3) = a_{1,1}^* x_2^2 + a_{1,2}^* x_2 x_3 + a_{2,2}^* x_3^2$$

such that

$$a_{1,1}^* \leq \frac{2}{\sqrt{3}}\sqrt{a_{1,1}d}.$$

Choose $V^* \in SL_2(\mathbf{Z})$ such that $A^* = (V^*)^T B^* V^*$. Let $r, s \in \mathbf{Z}$, and let $V_{r,s} \in SL_3(\mathbf{Z})$ be the matrix defined by (1.6) in Lemma 1.4. Let

$$A = V_{r,s}^T B V_{r,s} = (a_{i,j}). \tag{1.8}$$

Note that the integer in the upper left corner of the matrix is still $a_{1,1}$, the smallest positive integer represented by any form in the equivalence class of F, and that, by Lemma 1.3,

$$a_{1,1}^* = a_{1,1}a_{2,2} - a_{1,2}^2.$$

Finally, it follows from (1.8) that

$$a_{1,2} = a_{1,1}r + b_{1,2}v_{1,1}^* + b_{1,3}v_{2,1}^*$$

and

$$a_{1,3} = a_{1,1}s + b_{1,2}v_{1,2}^* + b_{1,3}v_{2,2}^*.$$

Therefore, we can choose r such that

$$|a_{1,2}| \leq \frac{a_{1,1}}{2}$$

and choose s such that

$$|a_{1,3}| \leq \frac{a_{1,1}}{2}.$$

Since

$$a_{1,1} \leq F_A(0, 1, 0) = a_{2,2},$$

we have

$$a_{1,1}^2 \leq a_{1,1}a_{2,2}$$

$$= a_{1,1}a_{2,2} - a_{1,2}^2 + a_{1,2}^2$$

$$= a_{1,1}^* + a_{1,2}^2$$

$$\leq \frac{2}{\sqrt{3}}\sqrt{a_{1,1}d} + \frac{a_{1,1}^2}{4}.$$

This implies that

$$a_{1,1}^2 \leq \left(\frac{2}{\sqrt{3}}\right)^3 \sqrt{a_{1,1}d}$$

or, equivalently,

$$a_{1,1} \leq \frac{4}{3}\sqrt[3]{d}.$$

This completes the proof.

Theorem 1.3 *Every positive-definite ternary quadratic form of discriminant 1 is equivalent to the form* $x_1^2 + x_2^2 + x_3^2$.

Proof. Let F be a positive-definite ternary quadratic form of discriminant 1. By Lemma 1.6, the form F is equivalent to a form $F_A = \sum a_{i,j} x_i x_j$ for which

$$0 \leq 2 \max\left(|a_{1,2}|, |a_{1,3}|\right) \leq a_{1,1} \leq \frac{4}{3}.$$

This implies that $a_{1,2} = a_{1,3} = 0$. Since $d \neq 0$, it follows that $a_{1,1} \neq 0$ and so $a_{1,1} = 1$. Therefore,

$$A = \begin{pmatrix} 1 & 0 & 0 \\ 0 & a_{2,2} & a_{2,3} \\ 0 & a_{2,3} & a_{3,3} \end{pmatrix},$$

where the 2×2 matrix

$$A^* = \begin{pmatrix} a_{2,2} & a_{2,3} \\ a_{2,3} & a_{3,3} \end{pmatrix}$$

has determinant 1. By Theorem 1.2, there exists a matrix

$$U^* = \begin{pmatrix} u_{2,2} & u_{2,3} \\ u_{2,3} & u_{3,3} \end{pmatrix} \in SL_2(\mathbf{Z})$$

such that $(U^*)^T A^* U^*$ is the 2×2 identity matrix I_2. Let

$$U = \begin{pmatrix} 1 & 0 & 0 \\ 0 & u_{2,2} & u_{2,3} \\ 0 & u_{2,3} & u_{3,3} \end{pmatrix}.$$

Then $U^T A U$ is the 3×3 identity matrix I_3. This completes the proof.

1.5   Sums of three squares

In this section, we determine the integers that can be written as the sum of three squares. The proof uses the fact that a number is the sum of three squares if and only if it can be represented by some positive-definite ternary quadratic form of discriminant 1, together with two important theorems of elementary number theory: Gauss's law of quadratic reciprocity and Dirichlet's theorem on primes in arithmetic progressions.

The statement that a is a quadratic residue modulo m means that there exist integers x and y such that $x^2 - a = ym$. If p is prime and $(a, p) = 1$, then the Legendre symbol $\left(\frac{a}{p}\right)$ is defined by $\left(\frac{a}{p}\right) = 1$ if a is a quadratic residue modulo p and $\left(\frac{a}{p}\right) = -1$ if a is not a quadratic residue modulo p. By quadratic reciprocity, if p and q are distinct odd primes, then $\left(\frac{p}{q}\right) = \left(\frac{q}{p}\right)$ if $p \equiv 1 \pmod 4$ or $q \equiv 1$ (mod 4), and $\left(\frac{p}{q}\right) = -\left(\frac{q}{p}\right)$ if $p \equiv q \equiv 3 \pmod 4$. Also, $\left(\frac{-1}{p}\right) = 1$ if and only if $p \equiv 1 \pmod 4$, and $\left(\frac{2}{p}\right) = 1$ if and only if $p \equiv 1$ or $7 \pmod 8$.

Lemma 1.7 *Let $n \geq 2$. If there exists a positive integer d' such that $-d'$ is a quadratic residue modulo $d'n - 1$, then n can be represented as the sum of three squares.*

Proof. If $-d'$ is a quadratic residue modulo $d'n - 1$, then there exist integers $a_{1,2}$ and $a_{1,1}$ such that

$$a_{1,2}^2 + d' = a_{1,1}(d'n - 1) = a_{1,1}a_{2,2},$$

where

$$a_{2,2} = d'n - 1 \geq 2d' - 1 \geq 1$$

and so

$$a_{1,1} \geq 1.$$

Equivalently,

$$d' = a_{1,1}a_{2,2} - a_{1,2}^2.$$

The symmetric matrix

$$A = \begin{pmatrix} a_{1,1} & a_{1,2} & 1 \\ a_{1,2} & a_{2,2} & 0 \\ 1 & 0 & n \end{pmatrix}$$

has determinant

$$\det(A) = (a_{1,1}a_{2,2} - a_{1,2}^2)n - a_{2,2} = d'n - a_{2,2} = 1.$$

By Lemma 1.3, the quadratic form F_A corresponding to the matrix A is positive. Moreover, F_A has discriminant 1 and represents n, since $F_A(0, 0, 1) = n$. By Theorem 1.3, the form $x_1^2 + x_2^2 + x_3^2$ must also represent n. This completes the proof.

Lemma 1.8 *If n is a positive integer and $n \equiv 2 \pmod 4$, then n can be represented as the sum of three squares.*

Proof. Since $(4n, n - 1) = 1$, it follows from Dirichlet's theorem that the arithmetic progression $\{4nj + n - 1 : j = 1, 2, \ldots\}$ contains infinitely many primes. Choose $j \geq 1$ such that

$$p = 4nj + n - 1 = (4j + 1)n - 1$$

is prime. Let $d' = 4j + 1$. Since $n \equiv 2 \pmod 4$, we have

$$p = d'n - 1 \equiv 1 \pmod 4.$$

By Lemma 1.7, it suffices to prove that $-d'$ is a quadratic residue modulo p. Let

$$d' = \prod_{q_i \mid d'} q_i^{k_i},$$

where the q_i are the distinct primes dividing d'. Then

$$p = d'n - 1 \equiv -1 \pmod{q_i}$$

for all i, and

$$d' \equiv \prod_{\substack{q_i \mid d' \\ q_i \equiv 3 \ (\mathrm{mod}\ 4)}} (-1)^{k_i} \equiv 1 \pmod 4.$$

Therefore,

$$\prod_{\substack{q_i \mid d' \\ q_i \equiv 3 \ (\mathrm{mod}\ 4)}} (-1)^{k_i} = 1.$$

By quadratic reciprocity we have

$$\left(\frac{-1}{p}\right) = 1$$

since $p \equiv 1 \pmod 4$, and

$$
\begin{aligned}
\left(\frac{-d'}{p}\right) &= \left(\frac{-1}{p}\right)\left(\frac{d'}{p}\right) \\
&= \left(\frac{d'}{p}\right) \\
&= \prod_{q_i \mid d'} \left(\frac{q_i}{p}\right)^{k_i} \\
&= \prod_{q_i \mid d'} \left(\frac{p}{q_i}\right)^{k_i} \\
&= \prod_{q_i \mid d'} \left(\frac{-1}{q_i}\right)^{k_i} \\
&= \prod_{\substack{q_i \mid d' \\ q_i \equiv 3 \ (\mathrm{mod}\ 4)}} (-1)^{k_i} \\
&= 1.
\end{aligned}
$$

This completes the proof.

Lemma 1.9 *If n is a positive integer such that $n \equiv 1, 3,$ or 5 (mod 8), then n can be represented as the sum of three squares.*

Proof. Clearly, 1 is a sum of three nonnegative squares. Let $n \geq 2$. Let

$$c = \begin{cases} 3 & \text{if } n \equiv 1 \pmod 8 \\ 1 & \text{if } n \equiv 3 \pmod 8 \\ 3 & \text{if } n \equiv 5 \pmod 8. \end{cases}$$

If $n \equiv 1$ or 3 (mod 8), then

$$\frac{cn - 1}{2} \equiv 1 \pmod 4.$$

If $n \equiv 5$ (mod 8), then

$$\frac{cn - 1}{2} \equiv 3 \pmod 4.$$

In all three cases,

$$\left(4n, \frac{cn - 1}{2}\right) = 1.$$

By Dirichlet's theorem, there exists a prime number p of the form

$$p = 4nj + \frac{cn - 1}{2}$$

for some positive integer j. Let

$$d' = 8j + c.$$

Then

$$2p = (8j + c)n - 1 = d'n - 1.$$

By Lemma 1.7, it suffices to prove that $-d'$ is a quadratic residue modulo $2p$.
   If $-d'$ is a quadratic residue modulo p, then there exists an integer x_0 such that

$$(x_0 + p)^2 + d' \equiv x_0^2 + d' \equiv 0 \pmod p.$$

Let $x = x_0$ if x_0 is odd, and let $x = x_0 + p$ if x_0 is even. Then x is odd and $x^2 + d'$ is even. Since

$$x^2 + d' \equiv 0 \pmod 2$$

and

$$x^2 + d' \equiv 0 \pmod p,$$

it follows that

$$x^2 + d' \equiv 0 \pmod{2p}.$$

Therefore, it suffices to prove that $-d'$ is a quadratic residue modulo p.

Let

$$d' = \prod_{q_i \mid d'} q_i^{k_i}$$

be the factorization of the odd integer d' into a product of powers of distinct odd primes q_i. Since

$$2p \equiv -1 \pmod{d'},$$

it follows that

$$2p \equiv -1 \pmod{q_i}$$

and

$$(p, q_i) = 1$$

for every prime q_i that divides d'.

If $n \equiv 1$ or $3 \pmod 8$, then $p \equiv 1 \pmod 4$ and

$$\left(\frac{-d'}{p}\right) = \left(\frac{-1}{p}\right)\left(\frac{d'}{p}\right)$$

$$= \left(\frac{d'}{p}\right)$$

$$= \prod_{q_i \mid d'} \left(\frac{q_i}{p}\right)^{k_i}$$

$$= \prod_{q_i \mid d'} \left(\frac{p}{q_i}\right)^{k_i}.$$

If $n \equiv 5 \pmod 8$, then $p \equiv 3 \pmod 4$ and $d' \equiv 3 \pmod 8$. From the factorization of d', we obtain

$$d' = \prod_{\substack{q_i \mid d' \\ q_i \equiv 1 \ (\mathrm{mod}\ 4)}} q_i^{k_i} \prod_{\substack{q_i \mid d' \\ q_i \equiv 3 \ (\mathrm{mod}\ 4)}} q_i^{k_i}$$

$$\equiv \prod_{\substack{q_i \mid d' \\ q_i \equiv 3 \ (\mathrm{mod}\ 4)}} (-1)^{k_i} \pmod 4$$

$$\equiv -1 \pmod 4$$

and so

$$\prod_{\substack{q_i \mid d' \\ q_i \equiv 3 \ (\mathrm{mod}\ 4)}} (-1)^{k_i} = -1.$$

It follows from quadratic reciprocity that

$$\left(\frac{-d'}{p}\right) = \left(\frac{-1}{p}\right)\left(\frac{d'}{p}\right)$$

$$= -\left(\frac{d'}{p}\right)$$

$$= - \prod_{\substack{q_i | d' \\ q_i \equiv 1 \ (\text{mod } 4)}} \left(\frac{q_i}{p}\right)^{k_i} \prod_{\substack{q_i | d' \\ q_i \equiv 3 \ (\text{mod } 4)}} \left(\frac{q_i}{p}\right)^{k_i}$$

$$= - \prod_{\substack{q_i | d' \\ q_i \equiv 1 \ (\text{mod } 4)}} \left(\frac{p}{q_i}\right)^{k_i} \prod_{\substack{q_i | d' \\ q_i \equiv 3 \ (\text{mod } 4)}} \left(\frac{p}{q_i}\right)^{k_i} \prod_{\substack{q_i | d' \\ q_i \equiv 3 \ (\text{mod } 4)}} (-1)^{k_i}$$

$$= \prod_{\substack{q_i | d' \\ q_i \equiv 1 \ (\text{mod } 4)}} \left(\frac{p}{q_i}\right)^{k_i} \prod_{\substack{q_i | d' \\ q_i \equiv 3 \ (\text{mod } 4)}} \left(\frac{p}{q_i}\right)^{k_i}$$

$$= \prod_{q_i | d'} \left(\frac{p}{q_i}\right)^{k_i}.$$

In both cases,

$$\left(\frac{-d'}{p}\right) = \prod_{q_i | d'} \left(\frac{p}{q_i}\right)^{k_i}$$

$$= \prod_{q_i | d'} \left(\frac{2}{q_i}\right)^{k_i} \left(\frac{2p}{q_i}\right)^{k_i}$$

$$= \prod_{q_i | d'} \left(\frac{2}{q_i}\right)^{k_i} \prod_{q_i | d'} \left(\frac{-1}{q_i}\right)^{k_i}$$

$$= \prod_{\substack{q_i | d' \\ q_i \equiv 3,5 \ (\text{mod } 8)}} (-1)^{k_i} \prod_{\substack{q_i | d' \\ q_i \equiv 3,7 \ (\text{mod } 8)}} (-1)^{k_i}$$

$$= \prod_{\substack{q_i | d' \\ q_i \equiv 5,7 \ (\text{mod } 8)}} (-1)^{k_i}.$$

Therefore, $-d'$ is a quadratic residue modulo $2p = d'n - 1$ if

$$\sum_{\substack{q_i | d' \\ q_i \equiv 5,7 \ (\text{mod } 8)}} k_i \equiv 0 \pmod 2.$$

This is what we shall prove. We have

$$d' = \prod_{\substack{q_i | d' \\ q_i \equiv 1 \ (\text{mod } 8)}} q_i^{k_i} \prod_{\substack{q_i | d' \\ q_i \equiv 3 \ (\text{mod } 8)}} q_i^{k_i} \prod_{\substack{q_i | d' \\ q_i \equiv 5 \ (\text{mod } 8)}} q_i^{k_i} \prod_{\substack{q_i | d' \\ q_i \equiv 7 \ (\text{mod } 8)}} q_i^{k_i}$$

$$\equiv \prod_{\substack{q_i | d' \\ q_i \equiv 3 \ (\text{mod } 8)}} 3^{k_i} \prod_{\substack{q_i | d' \\ q_i \equiv 5 \ (\text{mod } 8)}} (-3)^{k_i} \prod_{\substack{q_i | d' \\ q_i \equiv 7 \ (\text{mod } 8)}} (-1)^{k_i} \pmod 8$$

$$\equiv \prod_{\substack{q_i | d' \\ q_i \equiv 3,5 \ (\text{mod } 8)}} 3^{k_i} \prod_{\substack{q_i | d' \\ q_i \equiv 5,7 \ (\text{mod } 8)}} (-1)^{k_i} \pmod 8.$$

If $n \equiv 1$ or $5 \pmod 8$, then $c = 3$ and

$$d' = 8j + 3 \equiv 3 \pmod 8.$$

This implies that

$$\sum_{\substack{q_i \mid d' \\ q_i \equiv 3,5 \ (\text{mod } 8)}} k_i \equiv 1 \pmod{2}$$

and

$$\sum_{\substack{q_i \mid d' \\ q_i \equiv 5,7 \ (\text{mod } 8)}} k_i \equiv 0 \pmod{2}.$$

If $n \equiv 3 \pmod 8$, then $c = 1$ and

$$d' = 8j + 1 \equiv 1 \pmod 8.$$

It follows that

$$\sum_{\substack{q_i \mid d' \\ q_i \equiv 3,5 \ (\text{mod } 8)}} k_i \equiv 0 \pmod{2}$$

and

$$\sum_{\substack{q_i \mid d' \\ q_i \equiv 5,7 \ (\text{mod } 8)}} k_i \equiv 0 \pmod{2}.$$

This completes the proof.

Theorem 1.4 (Gauss) *A positive integer N can be represented as the sum of three squares if and only if N is not of the form*

$$N = 4^a(8k + 7).$$

Proof. Since

$$x^2 \equiv 0, 1, \text{ or } 4 \pmod 8$$

for every integer x, it follows that a sum of three squares can never be congruent to 7 modulo 8. If the integer $4m$ is the sum of three squares, then there exist integers x_1, x_2, x_3 such that

$$4m = x_1^2 + x_2^2 + x_3^2.$$

This is possible only if x_1, x_2, x_3 are all even, and so

$$m = \left(\frac{x_1}{2}\right)^2 + \left(\frac{x_2}{2}\right)^2 + \left(\frac{x_3}{2}\right)^2.$$

Therefore, $4^a m$ is the sum of three squares if and only if m is the sum of three squares. This proves that no integer of the form $4^a(8k+7)$ can be the sum of three squares.

Every positive integer N can be written uniquely in the form $N = 4^a m$, where $m \equiv 2 \pmod 4$ or $m \equiv 1, 3, 5,$ or $7 \pmod 8$. By Lemma 1.8 and Lemma 1.9, the positive integer N is the sum of three squares unless $m \equiv 7 \pmod 8$. This completes the proof.

Theorem 1.5 *If N is a positive integer such that $N \equiv 3 \pmod 8$, then N is the sum of three odd squares.*

Proof. Recall that $x^2 \equiv 0, 1,$ or 4 (mod 8) for every integer x. If $N \equiv 3$ (mod 8) is a sum of three squares, then each of the squares must be congruent to 1 modulo 8, and so each of the squares must be odd. This completes the proof.

1.6   Thin sets of squares

If A is a finite set of nonnegative integers such that every integer from 0 to N can be written as the sum of h elements of A, with repetitions allowed, then A is called a *basis of order h for N*. A simple counting argument shows that if A is a basis of order h for N, then A cannot be too small.

Theorem 1.6 *Let $h \geq 2$. There exists a positive constant $c = c(h)$ such that, if A is a basis of order h for N, then*

$$|A| > cN^{1/h}.$$

Proof. Let $|A| = k$. If A is a basis of order h for N, then each of the integers $0, 1, \ldots, N$ is a sum of h elements of A, with repetitions allowed. The number of combinations of h elements, with repetitions allowed, of a set of cardinality k is the binomial coefficient $\binom{k+h-1}{h}$. Therefore,

$$N + 1 \leq \binom{k + h - 1}{h} = \frac{k(k + 1) \cdots (k + h - 1)}{h!} \leq \frac{c'k^h}{h!}$$

for some constant $c' > 0$ and all k, and so

$$|A| = k > \left(\frac{h!N}{c'}\right)^{1/h} = cN^{1/h}.$$

This completes the proof.

Since the squares form a basis of order 4, it follows that for every $N \geq 0$ the set Q_N of all squares up to N is a basis of order 4 for N. Moreover,

$$|Q_N| = 1 + [N^{1/2}] > N^{1/2}.$$

This is much larger than $cN^{1/4}$, which is a lower bound for the thinnest possible basis of order 4. It is natural to ask if for every N there exists a set A_N of squares that is a basis of order 4 for N and satisfies

$$\lim_{N \to \infty} \frac{A_N}{N^{1/2}} = 0.$$

The answer is provided by the following theorem.

Theorem 1.7 (Choi–Erdős–Nathanson) *For every $N \geq 2$, there exists a set A_N of squares such that A_N is a basis of order 4 for N and*

$$|A_N| \leq \left(\frac{4}{\log 2}\right) N^{1/3} \log N.$$

Proof. The sets $A_2 = A_3 = \{0, 1\}$ and $A_4 = A_5 = \{0, 1, 4\}$ satisfy the requirements of the theorem. Therefore, we can assume that $N \geq 6$.

We begin with a simple remark. By Theorem 1.4, if ℓ is a nonnegative integer and $\ell \equiv 1$ or $2 \pmod 4$, then ℓ is the sum of three squares. Since the square of an even integer is $0 \pmod 4$ and the square of an odd integer is $1 \pmod 4$, it follows that if $m \not\equiv 0 \pmod 4$ and a is any positive integer such that $a^2 \leq m$, then either $m - a^2$ is the sum of three squares or $m - (a - 1)^2$ is the sum of three squares.

For $N \geq 6$, we let $A_N^{(1)}$ consist of the squares of all nonnegative integers up to $2N^{1/3}$. Then

$$|A_N^{(1)}| \leq 2N^{1/3} + 1.$$

Let $A_N^{(2)}$ consist of the squares of all integers of the form

$$\left[k^{1/2}N^{1/3}\right] \quad \text{or} \quad \left[k^{1/2}N^{1/3}\right] - 1,$$

where

$$4 \leq k \leq N^{1/3}.$$

Then

$$|A_N^{(2)}| \leq 2(N^{1/3} - 3) = 2N^{1/3} - 6.$$

Let

$$A_N^{(0)} = A_N^{(1)} \cup A_N^{(2)}.$$

Then

$$|A_N^{(0)}| < 4N^{1/3}.$$

Since $A_N^{(0)}$ contains all the squares up to $4N^{2/3}$, it follows from Lagrange's theorem that every nonnegative integer up to $4N^{2/3}$ is the sum of four squares belonging to $A_N^{(0)}$.

Let m be an integer such that

$$4N^{2/3} < m \leq N$$

and

$$m \not\equiv 0 \pmod 4.$$

We shall prove that there exists an integer $a_0 \in A_N^{(2)}$ such that

$$0 \leq m - a_0^2 \leq 4N^{2/3}$$

and $m - a_0^2$ is the sum of three squares. Since

$$4 < \frac{m}{N^{2/3}} \leq N^{1/3},$$

it follows that

$$4 \leq k = \left[\frac{m}{N^{2/3}}\right] \leq N^{1/3}.$$

Let

$$a = \left[k^{1/2}N^{1/3}\right].$$

Then $a^2 \in A_N^{(2)}, (a-1)^2 \in A_N^{(2)},$

$$a^2 \leq kN^{2/3} \leq m < (k+1)N^{2/3},$$

and

$$a > k^{1/2}N^{1/3} - 1.$$

It follows from our initial remark that either $m - a^2$ or $m - (a-1)^2$ is the sum of three squares. Choose $a_0^2 \in \{(a-1)^2, a^2\} \subseteq A_N^{(2)}$ such that $m - a_0^2$ is a sum of three squares. Since $4 < 3N^{1/6}$ for $N \geq 6$, we have

$$
\begin{aligned}
0 \leq m - a^2 \\
\leq m - a_0^2 \\
\leq m - (a-1)^2 \\
< (k+1)N^{2/3} - (k^{1/2}N^{1/3} - 2)^2 \\
< (k+1)N^{2/3} - kN^{2/3} + 4k^{1/2}N^{1/3} \\
= N^{2/3} + 4k^{1/2}N^{1/3} \\
\leq N^{2/3} + 4N^{1/2} \\
< 4N^{2/3},
\end{aligned}
$$

and so $m - a_0^2$ is the sum of three squares belonging to $A_N^{(1)}$. Therefore, if $0 \leq m \leq N$ and $m \not\equiv 0 \pmod 4$, then m is the sum of four squares belonging to $A_N^{(0)}$.

Let

$$A_N = \left\{(2^i a)^2 : 0 \leq i \leq \frac{\log N}{\log 4} \quad \text{and} \quad a \in A_N^{(0)}\right\}.$$

Then A_N is a set of squares and

$$|A_N| \leq \left(\frac{\log N}{\log 4} + 1\right)|A_N^{(0)}| < \left(\frac{2\log N}{\log 4}\right)4N^{1/3} = \left(\frac{4}{\log 2}\right)N^{1/3}\log N.$$

Let $n \in [0, N]$. If $n \not\equiv 0 \pmod 4$, then n is the sum of four squares belonging to $A_N^{(0)} \subseteq A_N$. If $n \equiv 0 \pmod 4$, then $n = 4^i m$, where $m \not\equiv 0 \pmod 4$ and $0 \leq i \leq \log N / \log 4$. Then

$$m = a_1^2 + a_2^2 + a_3^2 + a_4^2,$$

where $a_1, a_2, a_3, a_4 \in A_N^{(0)}$, and so

$$n = 4^i m = (2^i a_1)^2 + (2^i a_2)^2 + (2^i a_3)^2 + (2^i a_4)^2$$

is a sum of four squares belonging to A_N. This completes the proof.

1.7   The polygonal number theorem

We begin by proving Gauss's theorem that the triangles form a basis of order three. Equivalently, as Gauss wrote in his journal on July 10, 1796,

$$\text{E\Upsilon PHKA!}\quad \text{num} = \Delta + \Delta + \Delta.$$

Theorem 1.8 (Gauss) *Every nonnegative integer is the sum of three triangles.*

Proof. The triangular numbers are integers of the form $k(k + 1)/2$. Let $N \geq 1$. By Theorem 1.5, the integer $8N + 3$ is the sum of three odd squares, and so there exist nonnegative integers k_1, k_2, k_3 such that

$$8N + 3 = (2k_1 + 1)^2 + (2k_2 + 1)^2 + (2k_3 + 1)^2$$
$$= 4(k_1^2 + k_1 + k_2^2 + k_2 + k_3^2 + k_3) + 3.$$

Therefore,

$$N = \frac{k_1(k_1 + 1)}{2} + \frac{k_2(k_2 + 1)}{2} + \frac{k_3(k_3 + 1)}{2}.$$

This completes the proof.

Lagrange's theorem (Theorem 1.1) is the polygonal number theorem for squares, and Gauss's theorem is the polygonal number theorem for triangles. We shall now prove the theorem for polygonal numbers of order $m + 2$ for all $m \geq 3$. It is easy to check the polygonal number theorem for small values of N/m. Recall that the kth polygonal number of order $m + 2$ is

$$p_m(k) = \frac{mk(k - 1)}{2} + k.$$

The first six polygonal numbers are

$$p_m(0) = 0$$
$$p_m(1) = 1$$
$$p_m(2) = m + 2$$
$$p_m(3) = 3m + 3$$
$$p_m(4) = 6m + 4$$
$$p_m(5) = 10m + 5.$$

If $k_1, \ldots, k_s$ are positive integers, then, for $r = 0, 1, \ldots, m + 2 - s$, the numbers of the form

$$p_m(k_1) + p_m(k_2) + \cdots + p_m(k_s) + r p_m(1) \tag{1.9}$$

are an interval of $m + 3 - s$ consecutive integers, each of which is a sum of exactly $m + 2$ polygonal numbers. Here is a short table of representations of integers as sums of $m + 2$ polygonal numbers of order $m + 2$. The first column expresses the integer as a sum of polygonal numbers in the form (1.9), and the next two columns give the smallest and largest integers that the expression represents.

$rp_m(1)$	0	$m+2$
$p_m(2)+rp_m(1)$	$m+2$	$2m+3$
$2p_m(2)+rp_m(1)$	$2m+4$	$3m+4$
$p_m(3)+rp_m(1)$	$3m+3$	$4m+4$
$p_m(3)+p_m(2)+rp_m(1)$	$4m+5$	$5m+5$
$4p_m(2)+rp_m(1)$	$4m+8$	$5m+6$
$p_m(3)+2p_m(2)+rp_m(1)$	$5m+7$	$6m+4$
$p_m(4)+rp_m(1)$	$6m+4$	$7m+5$
$p_m(4)+p_m(2)+rp_m(1)$	$7m+6$	$8m+6$
$2p_m(3)+p_m(2)$	$7m+8$	$8m+7$
$p_m(4)+2p_m(2)+rp_m(1)$	$8m+8$	$9m+7$
$p_m(4)+p_m(3)+rp_m(1)$	$9m+7$	$10m+7$
$p_m(5)+rp_m(1)$	$10m+5$	$11m+6$
$p_m(5)+p_m(2)+rp_m(1)$	$11m+7$	$12m+7$

This table gives explicit polygonal number representations for all integers up to $12m+7$. It is not difficult to extend this computation. Pepin [95] and Dickson [23] published tables of representations of N as a sum of $m+2$ polygonal numbers of order $m+2$ for all $m \geq 3$ and $N \leq 120m$. Therefore, it suffices to prove the polygonal number theorem for $N > 120m$.

We need the following lemmas.

Lemma 1.10 *Let $m \geq 3$ and $N \geq 2m$. Let L denote the length of the interval*

$$I = \left(\frac{1}{2} + \sqrt{\frac{6N}{m} - 3}, \quad \frac{2}{3} + \sqrt{\frac{8N}{m} - 8} \right).$$

Then

$$L > 4 \quad \text{if } N \geq 108m$$

and

$$L > \ell m \quad \text{if } \ell \geq 3 \text{ and } N \geq 7\ell^2 m^3.$$

Proof. This is a straightforward computation. Let

$$x = N/m \geq 2$$

and

$$\ell_0 = \ell - \frac{1}{6}.$$

We see that

$$L = \sqrt{8x-8} - \sqrt{6x-3} + \frac{1}{6} > \ell$$

if and only if

$$\sqrt{8x-8} > \sqrt{6x-3} + \ell_0,$$

or, after squaring both sides and rearranging,

$$2x - \ell_0^2 - 5 > 2\ell_0\sqrt{6x - 3}.$$

Squaring and rearranging again, we obtain

$$4x\left(x - (7\ell_0^2 + 5)\right) + (\ell_0^2 + 5)^2 + 12\ell_0^2 > 0.$$

This inequality certainly holds if

$$x \geq 7\ell_0^2 + 5 = 7\left(\ell - \frac{1}{6}\right)^2 + 5.$$

Therefore,

$$L > \ell \quad \text{if} \quad \frac{N}{m} \geq 7\left(\ell - \frac{1}{6}\right)^2 + 5.$$

Since

$$7\left(4 - \frac{1}{6}\right)^2 + 5 = 107.86\ldots,$$

it follows that $L > 4$ if $N \geq 108m$. Since

$$7\ell^2 > 7\left(\ell - \frac{1}{6}\right)^2 + 5$$

for $\ell \geq 3$, it follows that $L > \ell$ if $\ell \geq 3$ and $N/m \geq 7\ell^2$. Therefore, if $\ell \geq 3$ and $N \geq 7\ell^2 m^3$, then $L > \ell m$. This completes the proof.

Lemma 1.11 *Let $m \geq 3$ and $N \geq 2m$. Let $a, b,$ and r be nonnegative integers such that*

$$0 \leq r < m$$

and

$$N = \frac{m}{2}(a - b) + b + r. \tag{1.10}$$

Consider the open interval

$$I = \left(\frac{1}{2} + \sqrt{\frac{6N}{m} - 3}, \ \frac{2}{3} + \sqrt{\frac{8N}{m} - 8}\right).$$

If

$$b \in I,$$

then

$$b^2 < 4a \tag{1.11}$$

and

$$3a < b^2 + 2b + 4. \tag{1.12}$$

Proof. From equation (1.10), we have

$$a = \left(1 - \frac{2}{m}\right) b + 2 \left(\frac{N-r}{m}\right).$$

By the quadratic formula,

$$b^2 - 4a = b^2 - 4\left(1 - \frac{2}{m}\right) b - 8\left(\frac{N-r}{m}\right) < 0$$

if

$$0 \le b < 2\left(1 - \frac{2}{m}\right) + \sqrt{4\left(1 - \frac{2}{m}\right)^2 + 8\left(\frac{N-r}{m}\right)}.$$

If $b \in I$, then

$$0 < b < \frac{2}{3} + \sqrt{\frac{8N}{m} - 8}$$

$$< 2\left(1 - \frac{2}{m}\right) + \sqrt{8\left(\frac{N-r}{m}\right)}$$

$$< 2\left(1 - \frac{2}{m}\right) + \sqrt{4\left(1 - \frac{2}{m}\right)^2 + 8\left(\frac{N-r}{m}\right)}.$$

This proves (1.11).

Again by the quadratic formula,

$$b^2 + 2b + 4 - 3a = b^2 - \left(1 - \frac{6}{m}\right) b - \left(6\left(\frac{N-r}{m}\right) - 4\right) > 0$$

if

$$b > \left(\frac{1}{2} - \frac{3}{m}\right) + \sqrt{\left(\frac{1}{2} - \frac{3}{m}\right)^2 + 6\left(\frac{N-r}{m}\right) - 4}.$$

If $b \in I$, then

$$b > \frac{1}{2} + \sqrt{\frac{6N}{m} - 3}$$

$$> \left(\frac{1}{2} - \frac{3}{m}\right) + \sqrt{\left(\frac{1}{2} - \frac{3}{m}\right)^2 + \frac{6N}{m} - 4}$$

$$> \left(\frac{1}{2} - \frac{3}{m}\right) + \sqrt{\left(\frac{1}{2} - \frac{3}{m}\right)^2 + 6\left(\frac{N-r}{m}\right) - 4}.$$

This proves inequality (1.12).

The following result is sometimes called *Cauchy's lemma*.

Lemma 1.12 *Let a and b be odd positive integers such that*

$$b^2 < 4a$$

and

$$3a < b^2 + 2b + 4.$$

Then there exist nonnegative integers s, t, u, v such that

$$a = s^2 + t^2 + u^2 + v^2 \tag{1.13}$$

and

$$b = s + t + u + v. \tag{1.14}$$

Proof. Since a and b are odd, it follows that $4a - b^2 \equiv 3 \pmod 8$. By Theorem 1.5, there exist odd positive integers $x \geq y \geq z$ such that

$$4a - b^2 = x^2 + y^2 + z^2.$$

We can choose the sign of $\pm z$ so that $b + x + y \pm z \equiv 0 \pmod 4$. Define integers s, t, u, v as follows:

$$s = \frac{b + x + y \pm z}{4}$$

$$t = \frac{b + x}{2} - s = \frac{b + x - y \mp z}{4}$$

$$u = \frac{b + y}{2} - s = \frac{b - x + y \mp z}{4}$$

$$v = \frac{b \pm z}{2} - s = \frac{b - x - y \pm z}{4}.$$

These numbers satisfy equations (1.13) and (1.14) and

$$s \geq t \geq u \geq v.$$

We must show that $v \geq 0$. By Exercise 8, the maximum value of $x + y + z$ subject to the constraint $x^2 + y^2 + z^2 = 4a - b^2$ is $\sqrt{12a - 3b^2}$. Also, the inequality $3a < b^2 + 2b + 4$ implies that $\sqrt{12a - 3b^2} < b + 4$. Therefore,

$$x + y + z \leq \sqrt{12a - 3b^2} < b + 4,$$

and so

$$v \geq \frac{b - x - y - z}{4} > -1.$$

Since v is an integer, we must have $v \geq 0$. This completes the proof.

The following result is a strong form of Cauchy's polygonal number theorem.

Theorem 1.9 (Cauchy) *If $m \geq 4$ and $N \geq 108m$, then N can be written as the sum of $m + 1$ polygonal numbers of order $m + 2$, at most four of which are different from 0 or 1. If $N \geq 324$, then N can be written as the sum of five pentagonal numbers, at least one of which is 0 or 1.*

Proof. By Lemma 1.10, the length of the interval

$$I = \left(\frac{1}{2} + \sqrt{\frac{6N}{m} - 3}, \quad \frac{2}{3} + \sqrt{\frac{8N}{m} - 8} \right)$$

is greater than 4 since $N \geq 108m$, and so I contains four consecutive integers and, consequently, two consecutive odd numbers b_1 and b_2. If $m \geq 4$, the set of numbers of the form $b + r$, where $b \in \{b_1, b_2\}$ and $r \in \{0, 1, \ldots, m - 3\}$, contains a complete set of representatives of the congruence classes modulo m, and so we can choose $b \in \{b_1, b_2\} \subseteq I$ and $r \in \{0, 1, \ldots, m - 3\}$ such that

$$N \equiv b + r \pmod{m}.$$

Then

$$a = 2\left(\frac{N - b - r}{m}\right) + b = \left(1 - \frac{2}{m}\right)b + 2\left(\frac{N - r}{m}\right) \qquad (1.15)$$

is an odd positive integer, and

$$N = \frac{m}{2}(a - b) + b + r.$$

By Lemma 1.11, since $b \in I$, we have

$$b^2 < 4a$$

and

$$3a < b^2 + 2b + 4.$$

By Lemma 1.12, there exist nonnegative integers s, t, u, v such that

$$a = s^2 + t^2 + u^2 + v^2$$

and

$$b = s + t + u + v.$$

Therefore,

$$N = \frac{m}{2}(a - b) + b + r$$
$$= \frac{m}{2}\left(s^2 - s + t^2 - t + u^2 - u + v^2 - v\right) + (s + t + u + v) + r$$
$$= p_m(s) + p_m(t) + p_m(u) + p_m(v) + r.$$

Since $0 \leq r \leq m - 3$ and since 0 and 1 are polygonal numbers of order $m + 2$ for every m, we obtain Cauchy's theorem for $m \geq 4$, that is, for polygonal numbers of order at least six. To obtain the result for pentagonal numbers, that is, for $m = 3$, we consider numbers of the form $b_1 + r$ and $b_2 + r$, where b_1, b_2 are consecutive odd integers in the interval I, and $r = 0$ or 1.

Theorem 1.10 (Legendre) *Let $m \geq 3$ and $N \geq 28m^3$. If m is odd, then N is the sum of four polygonal numbers of order $m + 2$. If m is even, then N is the sum of five polygonal numbers of order $m + 2$, at least one of which is 0 or 1.*

Proof. By Lemma 1.10, the length of the interval I is greater than $2m$, so I contains m consecutive odd numbers. If m is odd, these form a complete set of representatives of the congruence classes modulo m, so $N \equiv b$ (mod m) for some odd integer $b \in I$. Let $r = 0$ and define a by formula (1.15). Then

$$N = \frac{m}{2}(a - b) + b,$$

and it follows from Lemma 1.11 and Lemma 1.12 that N is the sum of four polygonal numbers of order $m + 2$.

If m is even and N is odd, then $N \equiv b$ (mod m) for some odd integer $b \in I$ and N is the sum of four polygonal numbers of order $m + 2$. If m is even and N is even, then $N - 1 \equiv b$ (mod m) for some odd integer $b \in I$ and N is the sum of five polygonal numbers of order $m + 2$, one of which is $p_m(1) = 1$. This completes the proof.

A set of integers is called an *asymptotic basis of order h* if every sufficiently large integer can be written as the sum of h not necessarily distinct elements of the set. Legendre's theorem shows that if $m \geq 3$ and m is odd, then the polygonal numbers of order $m + 2$ form an asymptotic basis of order 4, and if $m \geq 4$ and m is even, then the polygonal numbers of order $m + 2$ form an asymptotic basis of order 5.

1.8 Notes

Polygonal numbers go back at least as far as Pythagoras. They are discussed at length by Diophantus in his book *Arithmetica* and in a separate essay *On polygonal numbers*. An excellent reference is *Diophantus of Alexandria: A Study in the History of Greek Algebra*, by T. L. Heath [53]. Dickson's *History of the Theory of Numbers* [22, Vol. II, Ch.1] provides a detailed history of polygonal numbers and sums of squares.

There are many different proofs of Lagrange's theorem that every nonnegative integer is the sum of four squares. For a proof using the geometry of numbers, see Nathanson [93]. There is a vast literature concerned with the number of representations of an integer as the sum of s squares. Extensive treatments of these matters can be found in the monographs of Grosswald [43], Knopp [74], and Rademacher [98]. Liouville discovered an important and powerful elementary method that produces many of the same results (see Dickson [22, Vol. II, Ch. 11] or Uspensky and Heaslet [122]).

Legendre and Gauss determined the numbers that can be represented as the sum of three squares. See Dickson [22, Vol. II] for historical references. In this chapter, I followed the beautiful exposition of Landau [78]. There is also a nice proof by

Weil [140] that every positive integer congruent to 3   (mod 8) is the sum of three odd squares.

Cauchy [9] published the first proof of the polygonal number theorem. Legendre's theorem that the polygonal numbers of order m form an asymptotic basis of order 4 or 5 appears in [80, Vol. 2, pp. 331–356]. In this chapter I gave a simple proof of Nathanson [91, 92], which is based on Pepin [95].

Theorem 1.7 is due to Choi, Erdős, and Nathanson [13]. Using a probabilistic result of Erdős and Nathanson [36], Zöllner [152] has proved the existence of a basis of order 4 for N consisting of $\ll N^{1/4+\varepsilon}$ squares. It is not known if the ε can be removed from this inequality. Nathanson [89], Spencer [118], Wirsing [145], and Zöllner [151] proved the existence of "thin" subsets of the squares that are bases of order 4 for the set of all nonnegative integers.

1.9   Exercises

1. Let $m \geq 2$. Show that the polygonal numbers of order $m + 2$ can be written in terms of the triangular numbers as follows:

$$p_m(k) = mp_1(k) + k$$

for all $k \geq 0$.

2. (Nicomachus, 100 A.D.) Prove that the sum of two consecutive triangular numbers is a square. Prove that the sum of the nth square and the $(n-1)$-st triangular number is the nth pentagonal number.

3. Let $v(2)$ be the smallest number such that every integer N can be written in the form

$$N = \pm x_1^2 \pm \cdots \pm x_{v(2)}^2.$$

Prove that $v(2) = 3$. This is called the *easier Waring's problem for squares*. Hint: Use the identities

$$2x + 1 = (x + 1)^2 - x^2$$

and

$$2x = (x + 1)^2 - x^2 - 1^2.$$

4. Prove that if m is the sum of two squares and n is the sum of two squares, then mn is the sum of two squares. Hint: Use the polynomial identity

$$(x_1^2 + x^2)(y_1^2 + y_2^2) = (x_1y_1 + x_2y_2)^2 + (x_1y_2 - x_2y_1)^2.$$

5. (Nathanson [88]) Prove that there does not exist a polynomial identity of the form

$$(x_1^2 + x^2 + x_3^2)(y_1^2 + y_2^2 + y_3^2) = z_1^2 + z_2^2 + z_3^2,$$

where z_1, z_2, z_3 are polynomials in $x_1, x_2, x_3, y_1, y_2, y_3$ with integral coefficients.

6. Prove that Theorem 1.4 implies Lagrange's theorem (Theorem 1.1).

7. Prove that the set of triangular numbers is not a basis of order 2.

8. Let $S^2 = \{(x, y, z) \in \mathbf{R}^3 : x^2 + y^2 + z^2 = 1\}$. Prove that

$$\{x + y + z : (x, y, z) \in S^2\} = [-\sqrt{3}, \sqrt{3}].$$

9. Let

$$F_A(x_1, \ldots, x_n) = \sum_{i,j=1}^{n} a_{i,j} x_i x_j$$

and

$$F_B(x_1, \ldots, x_n) = \sum_{i,j=1}^{n} b_{i,j} x_i x_j$$

be quadratic forms in n variables such that

$$F_A(x_1, \ldots, x_n) = F_B(x_1, \ldots, x_n)$$

for all $x_1, \ldots, x_n \in \mathbf{Z}$. Prove that $a_{i,j} = b_{i,j}$ for all $i.j = 1, \ldots, n$.

10. Let A be an $n \times n$ symmetric matrix, and let F_A be the corresponding quadratic form. Let

$$U = (u_{i,j})$$

and

$$B = U^T A U = (b_{i,j}).$$

Prove that

$$b_{j,j} = F_A(u_{1,j}, u_{2,j}, \ldots, u_{n,j})$$

for $j = 1, \ldots, n$.

11. For $N \geq 1$, let $k = \left[\sqrt{N}\right]$ and

$$A = \{0, 1, \ldots, k - 1\} \cup \{k, 2k, \ldots, (k-1)k\}.$$

Show that A is a basis of order 2 for N such that

$$|A| \leq 2\sqrt{N} + 1.$$

12. Let $h \geq 2, k \geq 2$, and

$$A = \{0\} \cup \bigcup_{i=0}^{h-1} \{a_i k^i : a_i = 1, \ldots, k - 1\}.$$

Prove that A is a basis of order h for $k^h - 1$ and

$$|A| \leq h(k-1) + 1.$$

13. (Raikov [99], Stöhr [119]) Let $h \geq 2$ and $N \geq 2^h$. Let A be the set constructed in the preceding exercise with

$$k = \left[N^{1/h}\right] + 1.$$

Prove that A is a basis of order h for N such that

$$|A| \leq hN^{1/h} + 1.$$

2
Waring's problem for cubes

Omnis integer numerus vel est cubus; vel e duobus, tribus, 4,5,6,7,8,
vel novem cubus compositus: est etiam quadratoquadratus; vel e duobus, tribus &c. usque ad novemdecim compositus &sic deinceps.[1]

E. Waring [138]

2.1 Sums of cubes

In his book *Meditationes Algebraicae*, published in 1770, Edward Waring stated
without proof that every nonnegative integer is the sum of four squares, nine cubes,
19 fourth powers, and so on. Waring's problem is to prove that, for every $k \geq 2$,
the set of nonnegative kth powers is a basis of finite order.

Waring's problem for cubes is to prove that every nonnegative integer is the
sum of a bounded number of nonnegative cubes. The least such number is denoted
$g(3)$. Wieferich and Kempner proved that $g(3) = 9$, and so the cubes are a basis
of order nine. This is clearly best possible, since there are integers, such as 23 and
239, that cannot be written as sums of eight cubes.

Immediately after Wieferich published his theorem, Landau observed that, in
fact, only finitely many positive integers actually require nine cubes, that is, every

---

[1]Every positive integer is either a cube or the sum of 2,3,4,5,6,7,8, or 9 cubes; similarly,
every integer is either a fourth power, or the sum of 2, 3, . . . , or 19 fourth powers; and so
on.

sufficiently large integer is the sum of eight cubes. Indeed, 23 and 239 are the *only* positive integers that cannot be written as sums of eight nonnegative cubes. A set of integers is called an *asymptotic basis of order h* if every sufficiently large integer can be written as the sum of exactly h elements of the set. Thus, Landau's theorem states that the cubes are an asymptotic basis of order eight. Later, Linnik proved that only finitely many integers require eight cubes, so every sufficiently large integer is the sum of seven cubes, that is, the cubes are an asymptotic basis of order seven. On the other hand, an examination of congruences modulo 9 shows that there are infinitely many positive integers that cannot be written as sums of three cubes.

Let $G(3)$ denote the smallest integer h such that the cubes are an asymptotic basis of order h, that is, such that every sufficiently large positive integer can be written as the sum of h nonnegative cubes. Then

$$4 \leq G(3) \leq 7.$$

To determine the exact value of $G(3)$ is a major unsolved problem of additive number theory. It is known that almost all positive integers are sums of four cubes, and it is possible that $G(3) = 4$.

The principal results of this chapter are the theorems of Wieferich–Kempner and of Linnik. Because of the mystery surrounding sums of few cubes, we also include a section about sums of two cubes. We shall prove that there are integers with arbitrarily many representations as the sum of two nonnegative cubes, but that almost all numbers that can be written in at least one way as the sum of two nonnegative cubes have essentially only one such representation.

2.2 The Wieferich–Kempner theorem

The proof that $g(3) = 9$ requires four lemmas.

Lemma 2.1 *Let A and m be nonnegative integers such that $m \leq A^2$ and m can be written as the sum of three squares. Then*

$$6A(A^2 + m)$$

is a sum of six nonnegative cubes.

Proof. Let m_1, m_2, m_3 be nonnegative integers such that

$$m = m_1^2 + m_2^2 + m_3^2.$$

Then

$$0 \leq m_i \leq \sqrt{m} \leq A$$

for $i = 1, 2, 3$, and

$$6A(A^2 + m) = 6A(A^2 + m_1^2 + m_2^2 + m_3^2) = \sum_{i=1}^{3} \left((A + m_i)^3 + (A - m_i)^3 \right).$$

This completes the proof.

Lemma 2.2 *Let* $t \geq 1$. *For every odd integer* w, *there is an odd integer* b *such that*

$$w \equiv b^3 \quad (\bmod\ 2^t).$$

Proof. If b is odd and $w \equiv b^3 \pmod{2^t}$, then w is odd. Let b_1 and b_2 be odd integers such that

$$b_1^3 \equiv b_2^3 \quad (\bmod\ 2^t).$$

Then 2^t divides

$$b_2^3 - b_1^3 = (b_2 - b_1)(b_2^2 + b_2 b_1 + b_1^2).$$

Since $b_2^2 + b_2 b_1 + b_1^2$ is odd, it follows that 2^t divides $b_2 - b_1$, that is,

$$b_1 \equiv b_2 \quad (\bmod\ 2^t).$$

This means that if b_1 and b_2 are odd integers such that

$$0 < b_1 < b_2 < 2^t,$$

then

$$b_1^3 \not\equiv b_2^3 \quad (\bmod\ 2^t),$$

and so every odd integer is congruent to a cube modulo 2^t. This completes the proof.

Lemma 2.3 *If*

$$r \geq 10648 = 22^3,$$

then there exists an integer $d \in [0, 22]$ *and an integer* m *that is a sum of three squares such that*

$$r = d^3 + 6m.$$

Proof. If the nonnegative integer m is not the sum of three squares, then there exist nonnegative integers s and t such that

$$m = 4^s(8t + 7),$$

and so

$$6m = 6 \cdot 4^s(8t + 7) \equiv \begin{cases} 0 & (\bmod\ 96) & \text{if } s \geq 2 \\ 72 & (\bmod\ 96) & \text{if } s = 1 \\ 42 & (\bmod\ 96) & \text{if } s = 0 \text{ and } t \text{ is even} \\ 90 & (\bmod\ 96) & \text{if } s = 0 \text{ and } t \text{ is odd.} \end{cases}$$

It follows that if m is a positive integer and

$$6m \equiv h \quad (\bmod\ 96)$$

for some

$$h \in \mathcal{H} = \{6, 12, 18, 24, 30, 36, 48, 54, 60, 66, 78, 84\},$$

then m is the sum of three squares. The following table lists, for various $h \in \mathcal{H}$ and

$$d \in \mathcal{D} = \{0, 1, 2, 3, 4, 5, 6, 7, 8, 9, 10, 11, 13, 14, 15, 17, 18, 22\},$$

the least nonnegative residue in the congruence class

$$d^3 + h \quad (\text{mod } 96).$$

The elements of $\mathcal{H}$ are listed in the top row, and the elements of $\mathcal{D}$ are listed in the column on the left.

	6	12	18	24	30	36	48	54	60	66	78	84
0	6	12	18	24	30	36	48	54	60	66	78	84
1	7	13	19	25	31	37	49	55	61	67	79	85
2	14	20	26	32	38	44	56	62	68	74	86	92
3	33	39	45	51	57	63	75	81	87	93	9	15
4	70	76	82	88	94	4	16	22	28	34	46	52
5	35	41	47	53	59	65	77	83	89	95	11	17
6	42	72	90									
7	73	91	43									
8	50	80	2									
9	69	21	27									
10	58	64	10									
11	5	23	71									
13	1											
14	8											
15	3											
17	29											
18	0											
22	40											

Every congruence class modulo 96 appears in this table. Since $0 \leq d \leq 22$ for all $d \in \mathcal{D}$, it follows that if $r \geq 22^3$, then there exists an integer $d \in \mathcal{D}$ such that $r - d^3$ is nonnegative and $r - d^3 \equiv h \pmod{96}$ for some $h \in \mathcal{H}$. Therefore, $r - d^3 = 6m$, where m is the sum of three squares. This completes the proof.

Lemma 2.4 *If* $1 \leq N \leq 40,000$, *then*

(i) N is a sum of nine nonnegative cubes;

(ii) if N $\neq$ 23 or 239, then N is a sum of eight nonnegative cubes;

(iii) *if $N \neq 23$ or 239 and if N is not one of the following fifteen numbers:*

$$
\begin{array}{ccccc}
15 & 22 & 50 & 114 & 167 \\
175 & 186 & 212 & 231 & 238 \\
303 & 364 & 420 & 428 & 454
\end{array}
$$

then N is a sum of seven nonnegative cubes;

(iv) *if $N > 8042$, then N is a sum of six nonnegative cubes.*

Proof. Let $s(N)$ denote the least integer h such that N is the sum of h nonnegative cubes. Von Sterneck computed $s(N)$ for all N up to 40,000. The four statements in the lemma are obtained by examining von Sterneck's list of values of $s(N)$. Using a computer, one can quickly verify (and extend) von Sterneck's list (see Exercise 8).

Theorem 2.1 (Wieferich–Kempner)   *Every nonnegative integer is the sum of nine nonnegative cubes.*

Proof. We shall first prove the theorem for integers

$$N > 8^{10}.$$

Let

$$n = \left[N^{1/3}\right].$$

Then

$$2^{10} \leq n \leq 2 \cdot 8^{k+1}.$$

There exists an integer $k \geq 3$ such that

$$8 \cdot 8^{3k} < N \leq 8 \cdot 8^{3(k+1)}.$$

Let

$$N_i = N - i^3.$$

For $i = 1, \ldots, n$ we have

$$d_i = N_{i-1} - N_i = i^3 - (i-1)^3 = 3i^2 - 3i + 1$$

$$< 3i^2 \leq 3N^{2/3} \leq \frac{3 \cdot 8^{2k+3}}{2}.$$

Choose i so that

$$N_{i+1} < 8 \cdot 8^{3k} \leq N_i.$$

Then $i \geq 1$. Since $k \geq 3$, we have

$$
\begin{aligned}
N_n &= N - n^3 \\
&\leq (n+1)^3 - n^3 - 1 \\
&= 3n^2 + 3n \\
&< 6n^2 \\
&\leq 3 \cdot 8^{2k+3} \\
&\leq 8 \cdot 8^{3k}.
\end{aligned}
$$

Therefore, $i \leq n - 1$. It follows that

$$N_i < N_{i-1} = (N_{i-1} - N_i) + (N_i - N_{i+1}) + N_{i+1}$$
$$= d_i + d_{i+1} + N_{i+1}$$
$$< 3 \cdot 8^{2k+3} + 8 \cdot 8^{3k}$$
$$\leq 11 \cdot 8^{3k}.$$

Since $N_{i-1} - N_i = d_i$ is odd, exactly one of the integers N_i and N_{i-1} is odd. Choose $a \in \{i - 1, i\}$ such that $N_a = N - a^3$ is odd. By Lemma 2.2, there is an odd integer $b \in [1, 8^k - 1]$ such that

$$N - a^3 \equiv b^3 \pmod{8^k}.$$

Then

$$7 \cdot 8^{3k} = 8 \cdot 8^{3k} - 8^{3k} < N - a^3 - b^3 < N_a < 11 \cdot 8^{3k}$$

and

$$N - a^3 - b^3 = 8^k q,$$

where

$$7 \cdot 8^{2k} < q < 11 \cdot 8^{2k}.$$

Let

$$r = q - 6 \cdot 8^{2k}.$$

Then

$$22^3 < 8^6 \leq 8^{2k} < r < 5 \cdot 8^{2k}.$$

It follows from Lemma 2.3 that r can be written in the form

$$r = d^3 + 6m,$$

where $0 \leq d \leq 22$ and m is a sum of three squares. Let

$$A = 8^k.$$

Then

$$m \leq \frac{r}{6} < \frac{5 \cdot 8^{2k}}{6} < A^2.$$

Let

$$c = 2^k d.$$

Then

$$N = a^3 + b^3 + 8^k q$$
$$= a^3 + b^3 + 8^k (6 \cdot 8^{2k} + r)$$
$$= a^3 + b^3 + 8^k (6 \cdot 8^{2k} + d^3 + 6m)$$
$$= a^3 + b^3 + (2^k d)^3 + 8^k (6 \cdot 8^{2k} + 6m)$$
$$= a^3 + b^3 + c^3 + 6A(A^2 + m).$$

By Lemma 2.1, $6A(A^2 + m)$ is a sum of six nonnegative cubes, so N is the sum of nine nonnegative cubes.

Now let
$$40,000 < N \leq 8^{10}.$$

Then
$$a = \left[(N - 10,000)^{1/3}\right] > 30,000^{1/3} > 31,$$

so
$$d = (a + 1)^3 - a^3 = 3a^2 + 3a + 1 < 4a^2 < 4N^{2/3}.$$

Therefore,
$$N - (a + 1)^3 < 10,000 \leq N - a^3 = N - (a + 1)^3 + d < 10,000 + 4N^{2/3}.$$

If $N - a^3 \leq 40,000$, then $N - a^3$ is a sum of six nonnegative cubes by Lemma 2.4. If $N - a^3 > 40,000$, then we choose the integer
$$b = \left[(N - a^3 - 10,000)^{1/3}\right] > 31,$$

and obtain
$$N - a^3 - (b + 1)^3 < 10,000 \leq N - a^3 - b^3 < 10,000 + 4(N - a^3)^{2/3}.$$

If $N - a^3 - b^3 \leq 40,000$, then $N - a^3 - b^3$ is a sum of six nonnegative cubes by Lemma 2.4. If $N - a^3 - b^3 > 40,000$, then we choose the integer
$$c = \left[(N - a^3 - b^3 - 10,000)^{1/3}\right] > 31$$

and obtain
$$
\begin{aligned}
N - a^3 &- b^3 - (c + 1)^3 \\
&< 10,000 \\
&\leq N - a^3 - b^3 - c^3 \\
&< 10,000 + 4\left(N - a^3 - b^3\right)^{2/3} \\
&< 10,000 + 4\left(10,000 + 4\left(10,000 + 4N^{2/3}\right)^{2/3}\right)^{2/3} \\
&\leq 10,000 + 4\left(10,000 + 4\left(10,000 + 4\left(8^{10}\right)^{2/3}\right)^{2/3}\right)^{2/3} \\
&< 20,000.
\end{aligned}
$$

Thus, if $40,000 < N < 8^{10}$, then there exist three nonnegative integers a, b, and c such that
$$10,000 < N - a^3 - b^3 - c^3 \leq 40,000.$$

By Lemma 2.4, $N - a^3 - b^3 - c^3$ is the sum of six nonnegative cubes. This completes the proof.

2.3   Linnik's theorem

Let $G(3)$ denote the smallest integer s such that every sufficiently large integer is the sum of s nonnegative cubes.

Theorem 2.2 *If $N \equiv \pm 4$ (mod 9), then N is not the sum of three integral cubes. In particular,*
$$G(3) \geq 4.$$

Proof. Since every integer, positive or negative, is congruent to $0, 1$, or -1 modulo 9, it follows that every sum of three cubes belongs to one of the seven congruence classes, $0, \pm 1, \pm 2, \pm 3$ (mod 9). Therefore, if $N \equiv \pm 4$ (mod 9), then N cannot be the sum of three cubes, so $G(3) \geq 4$.

Lemma 2.5 *Let n be a positive integer. If there exist distinct primes p, q, r such that*

$$p \equiv q \equiv r \equiv -1 \quad (\text{mod } 6), \tag{2.1}$$
$$r < q < 1.02r, \tag{2.2}$$
$$\tfrac{3}{4} p^3 q^{18} < n < p^3 q^{18}, \tag{2.3}$$
$$4n \equiv p^3 r^{18} \quad (\text{mod } q^6), \tag{2.4}$$
$$2n \equiv p^3 q^{18} \quad (\text{mod } r^6), \tag{2.5}$$
$$n \equiv 3p \quad (\text{mod } 6p), \tag{2.6}$$

then n is the sum of six positive integral cubes.

Proof. It follows from (2.2) and (2.3) that

$$p^3(4q^{18} + 2r^{18}) < 6p^3 q^{18}$$
$$< 8n$$
$$< 8p^3 q^{18}$$
$$< p^3(4q^{18} + 4(1.02r)^{18})$$
$$< p^3(4q^{18} + 8r^{18}).$$

Thus,

$$p^3(4q^{18} + 2r^{18}) < 8n < p^3(4q^{18} + 8r^{18}). \tag{2.7}$$

Congruences (2.6), (2.4), and (2.5) imply that

$$8n \equiv 2p^3 r^{18} \equiv p^3(4q^{18} + 2r^{18}) + 18pq^6 r^6 \quad (\text{mod } q^6),$$
$$8n \equiv 4p^3 q^{18} \equiv p^3(4q^{18} + 2r^{18}) + 18pq^6 r^6 \quad (\text{mod } r^6),$$
$$8n \equiv 0 \equiv p^3(4q^{18} + 2r^{18}) + 18pq^6 r^6 \quad (\text{mod } p),$$

so

$$8n \equiv p^3(4q^{18} + 2r^{18}) + 18pq^6 r^6 \quad (\text{mod } pq^6 r^6). \tag{2.8}$$

It follows from (2.1) and (2.6) that

$$n \equiv 3p \equiv -3 \equiv 3 \quad (\text{mod } 6),$$

so

$$8n \equiv 24 \quad (\text{mod } 48). \tag{2.9}$$

By (2.1), the primes p, q, r are odd; hence

$$p^2 \equiv q^2 \equiv r^2 \equiv 1 \quad (\text{mod } 8)$$

and

$$p^3 \left(2q^{18} + r^{18}\right) + 9pq^6r^6 \equiv (2+1)p + p \equiv 4p \equiv 4 \quad (\text{mod } 8).$$

Therefore,

$$p^3(4q^{18} + 2r^{18}) + 18pq^6r^6 \equiv 8 \quad (\text{mod } 16).$$

Similarly, since $p \equiv q \equiv r \equiv -1 \quad (\text{mod } 3)$, we have

$$p^3(4q^{18} + 2r^{18}) + 18pq^6r^6 \equiv 0 \quad (\text{mod } 3)$$

so

$$p^3(4q^{18} + 2r^{18}) + 18pq^6r^6 \equiv 24 \quad (\text{mod } 48). \tag{2.10}$$

Since $(pqr, 48) = 1$, we can combine (2.8), (2.9), and (2.10) to obtain

$$8n \equiv p^3(4q^{18} + 2r^{18}) + 18pq^6r^6 \quad (\text{mod } 48pq^6r^6).$$

Therefore, there exists an integer u such that

$$8n = p^3(4q^{18} + 2r^{18}) + 18pq^6r^6 + 48pq^6r^6u$$
$$= p^3(4q^{18} + 2r^{18}) + 6pq^6r^6(8u + 3).$$

It follows from (2.7) that

$$0 < 6pq^6r^6(8u + 3) < 6p^3r^{18},$$

so

$$0 < 8u + 3 < p^2q^{-6}r^{12}.$$

By Theorem 1.5,

$$8u + 3 = x^2 + y^2 + z^2,$$

where x, y, z are odd positive integers less than $pq^{-3}r^6$, that is,

$$\max\{q^3x, q^3y, q^3z\} < pr^6. \tag{2.11}$$

Therefore,

$$8n = p^3(4q^{18} + 2r^{18}) + 6pq^6r^6(x^2 + y^2 + z^2)$$
$$= (pq^6 + r^3x)^3 + (pq^6 - r^3x)^3 + (pq^6 + r^3y)^3$$
$$+ (pq^6 - r^3y)^3 + (pr^6 + q^3z)^3 + (pr^6 - q^3z)^3.$$

Since each of the six integers p, q, r, x, y, z is odd, it follows that each of the six cubes in the preceding expression is even. Moreover, each of these cubes is positive, since, by (2.2) and (2.11),

$$0 < r^3 x < q^3 x < pr^6 < pq^6,$$

$$0 < r^3 y < q^3 y < pr^6 < pq^6,$$

and

$$0 < q^3 z < pr^6.$$

Therefore,

$$n = \left(\frac{pq^6 + r^3 x}{2}\right)^3 + \left(\frac{pq^6 - r^3 x}{2}\right)^3 + \left(\frac{pq^6 + r^3 y}{2}\right)^3$$
$$+ \left(\frac{pq^6 - r^3 y}{2}\right)^3 + \left(\frac{pr^6 + q^3 z}{2}\right)^3 + \left(\frac{pr^6 - q^3 z}{2}\right)^3$$

is a sum of six positive cubes.

Theorem 2.3 (Linnik) *Every sufficiently large integer is the sum of seven positive cubes, that is,*

$$G(3) \leq 7.$$

Proof. Let k and ℓ be integers such that $k \geq 1$ and $(k, \ell) = 1$. We define the Chebyshev function for the arithmetic progression ℓ modulo k by

$$\vartheta(x; k, \ell) = \sum_{\substack{p \leq x \\ p \equiv \ell \ (\mathrm{mod}\ k)}} \log p.$$

The Siegel-Walfisz theorem states that for any $A > 0$ and for all $x > 1$,

$$\vartheta(x; k, \ell) = \frac{x}{\varphi(k)} + O\left(\frac{x}{(\log x)^A}\right), \tag{2.12}$$

where $\varphi(k)$ is the Euler φ-function, and the implied constant depends only on A. It follows that, for any $\delta > 0$,

$$\vartheta((1 + \delta)x; k, \ell) - \vartheta(x; k, \ell) = \frac{\delta x}{\varphi(k)} + O\left(\frac{x}{(\log x)^A}\right).$$

Let $k = 6$, $\ell = -1$, $\delta = 1/50$, and $x = (50/51)(\log N)^2$. For any integer $N > 2$,

$$\sum_{\substack{(50/51)(\log N)^2 < p \leq (\log N)^2 \\ p \equiv -1 \ (\mathrm{mod}\ 6)}} \log p$$

$$= \vartheta((\log N)^2; 6, -1) - \vartheta((50/51)(\log N)^2; 6, -1)$$

$$= \frac{(\log N)^2}{102} + O\left(\frac{(\log N)^2}{(\log \log N)^A}\right).$$

Since

$$\sum_{\substack{p|N \\ p \equiv -1 \ (\mathrm{mod}\ 6)}} \log p \le \sum_{p|N} \log p \le \log N,$$

it follows that, for N sufficiently large, there must exist at least two prime numbers, q and r, such that

$$q \equiv r \equiv -1 \quad (\mathrm{mod}\ 6),$$

$$(q, N) = (r, N) = 1,$$

and

$$\frac{50}{51}(\log N)^2 < r < q < (\log N)^2 < \frac{51r}{50} = 1.02r.$$

The multiplicative group of congruence classes relatively prime to q^6 is cyclic of order $\varphi(q^6) = q^5(q - 1)$. Since $q \equiv -1 \quad (\mathrm{mod}\ 6)$, it follows that $(\varphi(q^6), 3) = 1$, so every integer relatively prime to q^6 is a cubic residue modulo q^6. Similarly, every integer relatively prime to r^6 is a cubic residue modulo r^6. Since

$$(2Nr, q) = (2Nq, r) = 1,$$

there exist integers u and v such that

$$(u, q) = (v, r) = 1,$$

$$4N \equiv u^3 r^{18} \quad (\mathrm{mod}\ q^6),$$

and

$$2N \equiv v^3 q^{18} \quad (\mathrm{mod}\ r^6).$$

The numbers 6, q^6, and r^6 are pairwise relatively prime. By the Chinese remainder theorem, there exists an integer ℓ such that

$$\ell \equiv u \quad (\mathrm{mod}\ q^6),$$

$$\ell \equiv v \quad (\mathrm{mod}\ r^6),$$

$$\ell \equiv -1 \quad (\mathrm{mod}\ 6).$$

Then

$$4N \equiv \ell^3 r^{18} \quad (\mathrm{mod}\ q^6)$$

and

$$2N \equiv \ell^3 q^{18} \quad (\mathrm{mod}\ r^6).$$

Let

$$k = 6q^6 r^6.$$

Then

$$(k, \ell) = (6q^6 r^6, \ell) = 1.$$

Let

$$x = N^{1/3} q^{-6}.$$

Since $q < (\log N)^2$, we have, for N sufficiently large,

$$\log x = \frac{1}{3} \log N - 6 \log q > \frac{1}{3} \log N - 12 \log \log N > \frac{1}{4} \log N$$

and

$$k = 6q^6 r^6 < 6(\log N)^{24} < 6(4 \log x)^{24} \ll (\log x)^{24}.$$

By the Siegel-Walfisz theorem with $A = 25$ and $\delta = 1/50$,

$$\vartheta((51/50)x; k, \ell) - \vartheta(x; k, \ell) = \frac{x}{50\varphi(k)} + O\left(\frac{x}{(\log x)^{25}}\right)$$

$$\geq \frac{x}{50k} + O\left(\frac{x}{(\log x)^{25}}\right)$$

$$\gg \frac{x}{(\log x)^{24}} + O\left(\frac{x}{(\log x)^{25}}\right)$$

$$> 0.$$

Therefore, if N is sufficiently large, there exists a prime p such that

$$x < p < \frac{51x}{50} = 1.02x$$

and

$$p \equiv \ell \pmod{6q^6 r^6}.$$

The primes p, q, r are distinct because $(qr, \ell) = 1$. Since $p \equiv -1 \pmod{3}$, every integer is a cubic residue modulo $6p$, and there exists an integer s such that

$$s^3 \equiv N - 3p \pmod{6p}.$$

By the Chinese remainder theorem, there exists t such that

$$t^3 \equiv N - 3p \pmod{6p},$$

$$t \equiv 0 \pmod{q^2 r^2},$$

and

$$1 \leq t \leq 6pq^2 r^2.$$

Let

$$n = N - t^3.$$

Then

$$4n = 4N - 4t^3 \equiv 4N \equiv \ell^3 r^3 \equiv p^3 r^{18} \pmod{q^6},$$

$$2n = 2N - 2t^3 \equiv 2N \equiv \ell^3 q^{18} \equiv p^3 q^{18} \pmod{r^6},$$

$$n = N - t^3 \equiv 3p \pmod{6p}.$$

Finally,

$$n = N - t^3 < N = x^3 q^{18} < p^3 q^{18}$$

and

$$n = N - t^3$$
$$\geq x^3 q^{18} - 216 p^3 q^6 r^6$$
$$> (1.02)^{-3} p^3 q^{18} - 216 p^3 q^{12}$$
$$= \frac{3}{4} p^3 q^{18} + \left(\left((1.02)^{-3} - \frac{3}{4} \right) q^6 - 216 \right) p^3 q^{12}$$
$$> \frac{3}{4} p^3 q^{18}$$

for N sufficiently large. Thus, the integer $n = N - t^3$ and the primes p, q, r satisfy conditions (2.1)–(2.5) of Lemma 2.5, so $N - t^3$ is a sum of six positive cubes. Since t is positive, we see that N is a sum of seven positive cubes. This proves Linnik's theorem.

2.4 Sums of two cubes

The subject of this book is additive bases. The generic theorem states that a certain classical sequence of integers, such as the cubes, has the property that every non-negative integer, or every sufficiently large integer, can be written as the sum of a bounded number of terms of the sequence. In this section, we diverge from this theme to study sums of two cubes. [2] This is important for several reasons. First, it is part of the unsolved problem of determining $G(3)$, the order of the set of cubes as an asymptotic basis and, in particular, the conjecture that every sufficiently large integer is the sum of four cubes. Second, the equation

$$N = x^3 + y^3 \tag{2.13}$$

is an elliptic curve. If $r_{3,2}(N)$ denotes the number of representations of the integer N as the sum of two positive cubes, then $r_{3,2}(N)$ counts the number of integral points with positive coordinates that lie on this curve. Counting the number of integral points on a curve is a deep and difficult problem in arithmetic geometry, and the study of sums of two cubes is an important special case.

If $N = x^3 + y^3$ and $x \neq y$, then $N = y^3 + x^3$ is another representation of N as a sum of two cubes. We call two representations

$$N = x_1^3 + y_1^3 = x_2^3 + y_2^3$$

essentially distinct if $\{x_1, y_1\} \neq \{x_2, y_2\}$. Note that N has two essentially distinct representations if and only if $r_{3,2}(N) \geq 3$.

---

[2] This section can be omitted on the first reading.

Here are some examples. The smallest number that has two essentially distinct representations as the sum of two positive cubes is 1729. The representations are

$$1729 = 1^3 + 12^3 = 9^3 + 10^3.$$

These give four positive integral points on the curve

$$1729 = x^3 + y^3,$$

so

$$r_{3,2}(1729) = 4.$$

The smallest number that has three essentially distinct representations as the sum of two positive cubes is 87,539,319. The representations are

$$87539319 = 167^3 + 436^3$$
$$= 228^3 + 423^3$$
$$= 255^3 + 414^3.$$

The cubes in these equations are not relatively prime, because

$$(228, 423) = (255, 414) = 3.$$

The smallest number that has three essentially distinct representations as the sum of two relatively prime positive cubes is 15,170,835,645. The representations are

$$15,170,835,645 = 2468^3 + 517^3$$
$$= 2456^3 + 709^3$$
$$= 2152^3 + 1733^3.$$

The smallest number that has four essentially distinct representations as the sum of two positive cubes is 6,963,472,309,248. The representations are

$$6,963,472,309,248 = 2421^3 + 19,083^3$$
$$= 5436^3 + 18,948^3$$
$$= 10,200^3 + 18,072^3$$
$$= 13,322^3 + 16,630^3.$$

It is an unsolved problem to find an integer N that has four essentially distinct representations as the sum of two positive cubes that are relatively prime.

In this section, we shall prove three theorems on sums of two cubes. The first is Fermat's result that there are integers with arbitrarily many representations as the sum of two positive cubes, that is,

$$\limsup_{N \to \infty} r_{3,2}(N) = \infty.$$

Next we shall prove a theorem of Erdős and Mahler. Let $C_2(n)$ be the number of integers up to n that can be represented as the sum of two positive cubes. Since the number of positive cubes up to n is $n^{1/3}$, it follows that $C_2(n)$ is at most $n^{2/3}$. Erdős and Mahler proved that this is the correct order of magnitude for $C_2(n)$, that is,

$$C_2(n) = \sum_{\substack{N \leq n \\ r_{3,2}(N) \geq 1}} 1 \gg n^{2/3}.$$

However, numbers with two or more essentially distinct representations as sums of two cubes are rare. Erdős observed that the number $C_2^*(n)$ of integers up to n that have at least two essentially distinct representations as the sum of two cubes is $o(n^{2/3})$. More precisely, we shall prove a theorem of Hooley that states that

$$C_2^*(n) \ll n^{(5/9)+\varepsilon}.$$

This implies that almost every integer that can be written as the sum of two positive cubes has an essentially unique representation in this form.

Theorem 2.4 (Fermat) *For every $k \geq 1$, there exists an integer N and k pairwise disjoint sets of positive integers $\{x_i, y_i\}$ such that*

$$N = x_i^3 + y_i^3$$

for $i = 1, \ldots, k$. Equivalently,

$$\limsup_{N \to \infty} r_{3,2}(N) = \infty.$$

Proof. The functions

$$f(x, y) = \frac{x(x^3 + 2y^3)}{x^3 - y^3}$$

and

$$g(x, y) = \frac{y(2x^3 + y^3)}{x^3 - y^3}$$

satisfy the polynomial identity

$$f(x, y)^3 - g(x, y)^3 = x^3 + y^3.$$

If

$$F(u, v) = \frac{u(u^3 - 2v^3)}{u^3 + v^3} = f(u, -v)$$

and

$$G(u, v) = \frac{v(2u^3 - v^3)}{u^3 + v^3} = -g(u, -v),$$

then

$$F(u, v)^3 + G(u, v)^3 = f(u, -v)^3 - g(u, -v)^3 = u^3 + (-v)^3 = u^3 - v^3.$$

Let

$$0 < \varepsilon < \frac{1}{4}.$$

Let x_1 and y_1 be positive rational numbers such that

$$0 < \frac{y_1}{x_1} < \varepsilon.$$

We define

$$u = f(x_1, y_1),$$
$$v = g(x_1, y_1).$$

Then u and v are positive rational numbers such that

$$u^3 - v^3 = x_1^3 + y_1^3 > 0.$$

Moreover,

$$\frac{u}{v} = \frac{x_1(x_1^3 + 2y_1^3)}{y_1(2x_1^3 + y_1^3)} = \frac{x_1}{2y_1}\left(\frac{1 + 2\rho^3}{1 + \rho^3/2}\right),$$

where $\rho = y_1/x_1 \in (0, 1/4)$. Since

$$1 < \frac{1 + 2\rho^3}{1 + \rho^3/2} = 1 + \frac{3\rho^3}{2 + \rho^3} < 1 + \frac{3\rho^3}{2},$$

it follows that

$$0 < \frac{u}{v} - \frac{x_1}{2y_1} < \frac{3x_1\rho^3}{4y_1} = \frac{3x_1}{4y_1}\left(\frac{y_1}{x_1}\right)^3 = \frac{3}{4}\left(\frac{y_1}{x_1}\right)^2 < \frac{3\varepsilon^2}{4}$$

and

$$\frac{u}{v} > \frac{x_1}{2y_1} > \frac{1}{2\varepsilon} > 2. \qquad (2.14)$$

Next, we define

$$x_2 = F(u, v),$$
$$y_2 = G(u, v).$$

Since $u > 2v$, it follows from the definition of the functions $F(u, v)$ and $G(u, v)$ that x_2 and y_2 are positive rational numbers. Moreover,

$$x_2^3 + y_2^3 = u^3 - v^3 = x_1^3 + y_1^3.$$

Let $\sigma = v/u$. Then

$$0 < \sigma < 2\varepsilon < 1/2$$

by (2.14) and

$$\frac{x_2}{y_2} = \frac{u(u^3 - 2v^3)}{v(2u^3 - v^3)}$$

$$= \frac{u}{2v}\left(\frac{1 - 2\sigma^3}{1 - \sigma^3/2}\right)$$

$$= \frac{u}{2v}\left(1 - \frac{3\sigma^3}{2 - \sigma^3}\right)$$

$$= \frac{u}{2v} - \frac{3u\sigma^2}{2v}\left(\frac{\sigma}{2 - \sigma^3}\right)$$

$$= \frac{u}{2v} - \frac{3v}{2u}\left(\frac{\sigma}{2 - \sigma^3}\right).$$

Since

$$0 < \frac{\sigma}{2 - \sigma^3} < \sigma < \frac{1}{2},$$

it follows that

$$0 < \frac{u}{2v} - \frac{x_2}{y_2} = \frac{3v}{2u}\left(\frac{\sigma}{2 - \sigma^3}\right) \le \frac{3v}{4u} < \frac{3\varepsilon}{2}.$$

Thus,

$$\left|\frac{x_2}{y_2} - \frac{x_1}{4y_1}\right| \le \left|\frac{x_2}{y_2} - \frac{u}{2v}\right| + \frac{1}{2}\left|\frac{u}{v} - \frac{x_1}{2y_1}\right| < \frac{3\varepsilon}{2} + \frac{3\varepsilon^2}{8} < 2\varepsilon,$$

and so

$$\frac{x_2}{y_2} > \frac{x_1}{4y_1} - 2\varepsilon > \frac{1}{4\varepsilon} - 2\varepsilon > \frac{1}{8\varepsilon} > 0.$$

This proves that if x_1 and y_1 are positive rational numbers such that

$$0 < \frac{y_1}{x_1} < \varepsilon < 1/4,$$

then there exist positive rational numbers x_2 and y_2 such that

$$x_2^3 + y_2^3 = x_1^3 + y_1^3,$$

$$0 < \frac{y_2}{x_2} < 8\varepsilon,$$

and

$$\left|\frac{4x_2}{y_2} - \frac{x_1}{y_1}\right| < 8\varepsilon.$$

If $8\varepsilon < 1/4$, then there exist positive rational numbers x_3 and x_4 such that

$$x_3^3 + y_3^3 = x_2^3 + y_2^3,$$

$$0 < \frac{y_3}{x_3} < 8^2\varepsilon,$$

and

$$\left| \frac{4x_3}{y_3} - \frac{x_2}{y_2} \right| < 8^2\varepsilon.$$

Similarly, if $k \geq 2$ and

$$0 < 8^{k-2}\varepsilon < \frac{1}{4},$$

then there exist positive rational numbers $x_1, y_1, x_2, y_2, \ldots, x_k, y_k$ such that

$$x_1^3 + y_1^3 = x_2^3 + y_2^3 = \cdots = x_k^3 + y_k^3,$$

$$0 < \frac{y_i}{x_i} < 8^{i-1}\varepsilon \qquad \text{for } i = 1, \ldots, k,$$

and

$$\left| \frac{4x_{i+1}}{y_{i+1}} - \frac{x_i}{y_i} \right| < 8^i\varepsilon \qquad \text{for } i = 1, \ldots, k-1.$$

Let $\varepsilon = 8^{-k}$. We shall prove that the k sets $\{x_i, y_i\}$ are pairwise disjoint. Since

$$\left| \frac{4^j x_{i+j}}{y_{i+j}} - \frac{4^{j-1} x_{i+j-1}}{y_{i+j-1}} \right| < 4^{j-1} \cdot 8^{i+j-1}\varepsilon = 8^i \cdot 32^{j-1}\varepsilon$$

for $j = 1, \ldots, k - i$, it follows that

$$\left| \frac{4^\ell x_{i+\ell}}{y_{i+\ell}} - \frac{x_i}{y_i} \right| \leq \sum_{j=1}^{\ell} \left| \frac{4^j x_{i+j}}{y_{i+j}} - \frac{4^{j-1} x_{i+j-1}}{y_{i+j-1}} \right|$$

$$\leq 8^i \varepsilon \sum_{j=1}^{\ell} 32^{j-1}$$

$$< 8^i 32^\ell \varepsilon$$

for $1 \leq i < i + \ell \leq k$. If $x_i = x_{i+\ell}$ and $y_i = y_{i+\ell}$ for some $\ell \geq 1$, then

$$\frac{x_{i+\ell}}{y_{i+\ell}} = \frac{x_i}{y_i}$$

and

$$\frac{3x_i}{y_i} \leq (4^\ell - 1)\frac{x_i}{y_i} = \left| \frac{4^\ell x_{i+\ell}}{y_{i+\ell}} - \frac{x_i}{y_i} \right| < 8^i 32^\ell \varepsilon.$$

It follows that

$$3 \leq 8^i 32^\ell \varepsilon \left(\frac{y_i}{x_i} \right)$$

$$< 8^{2i-1} 32^\ell \varepsilon^2$$

$$< 8^{2k} \varepsilon^2$$

$$= 1,$$

which is absurd. Therefore, $\{x_1, y_1\}, \ldots, \{x_k, y_k\}$ are k pairwise disjoint sets of positive rational numbers. Let d be a common denominator for the $2k$ numbers x_1, $\ldots, x_k, y_1, \ldots, y_k$, and let $N = (dx_1)^3 + (dy_1)^3$. Then $\{dx_1, dy_1\}, \ldots, \{dx_k, dy_k\}$ are pairwise disjoint sets of positive integers, and

$$(dx_1)^3 + (dy_1)^3 = (dx_2)^3 + (dy_2)^3 = \cdots = (dx_k)^3 + (dy_k)^3 = N,$$

that is, $r_{3,2}(N) \geq k$. This proves Fermat's theorem.

Next, we shall prove the Erdős–Mahler theorem. This requires four elementary lemmas.

Lemma 2.6 *Let a and b be positive integers such that*

$$a < b.$$

Let $r(a, b)$ denote the number of pairs (x, y) of integers such that

$$x^3 + (a - x)^3 = y^3 + (b - y)^3 \tag{2.15}$$

and

$$0 < x < \frac{a}{2} \quad \text{and} \quad 0 < y < \frac{b}{2}. \tag{2.16}$$

Then

$$r(a, b) < 5a^{2/3}.$$

Proof. The function

$$f_a(x) = x^3 + (a - x)^3 = 3ax^2 - 3a^2x + a^3$$

is strictly decreasing for $0 \leq x \leq a/2$. Let $r = r(a, b) \geq 1$. Let $(x_1, y_1), \ldots, (x_r, y_r)$ be the distinct solutions of equation (2.15) that satisfy inequalities (2.16), and let

$$0 < x_1 < \cdots < x_r < \frac{a}{2}.$$

Then

$$\frac{b^3}{4} = f_b\left(\frac{b}{2}\right) < f_b(y_1) = f_a(x_1) < f_a(0) = a^3,$$

and so

$$a < b < 4^{1/3}a < 2a. \tag{2.17}$$

For $i = 1, \ldots, r - 1$ we have

$$f_b(y_{i+1}) = f_a(x_{i+1}) < f_a(x_i) = f_b(y_i),$$

and so

$$0 < y_1 < \cdots < y_r < \frac{b}{2}.$$

Moreover, the point (x_i, y_i) is a solution of equation (2.15) if and only if (x_i, y_i) lies on the hyperbola

$$a\left(x - \frac{a}{2}\right)^2 - b\left(y - \frac{b}{2}\right)^2 = c,$$

where

$$c = \frac{b^3 - a^3}{12} > 0.$$

For $i = 1, \ldots, r$, let

$$u_i = \frac{a}{2} - x_i$$

and

$$v_i = \frac{b}{2} - y_i.$$

Then

$$0 < u_r < \cdots < u_1 < \frac{a}{2},$$

$$0 < v_r < \cdots < v_1 < \frac{b}{2},$$

and (u_i, v_i) is a point in the first quadrant of the uv-plane lies on the hyperbola

$$au^2 - bv^2 = c.$$

Since the hyperbola is convex downwards in the first quadrant, it follows that

$$\frac{v_{i+1} - v_i}{u_{i+1} - u_i} > \frac{v_i - v_{i-1}}{u_i - u_{i-1}}$$

for $i = 2, \ldots, r - 1$, and so the $r - 1$ fractions

$$\frac{v_{i+1} - v_i}{u_{i+1} - u_i} = \frac{y_{i+1} - y_i}{x_{i+1} - x_i}$$

are distinct for $i = 1, \ldots, r - 1$. If r_1 is the number of points (x_i, y_i) such that

$$x_{i+1} - x_i > \frac{a^{1/3}}{2},$$

then

$$\frac{a^{1/3} r_1}{2} < \frac{a}{2},$$

and so

$$r_1 < a^{2/3}.$$

Similarly, if r_2 is the number of points (x_i, y_i) such that

$$y_{i+1} - y_i > \frac{a^{1/3}}{2},$$

then

$$\frac{a^{1/3}r_2}{2} < \frac{b}{2} < a$$

by (2.17), and so

$$r_2 < 2a^{2/3}.$$

Let r_3 be the number of points (x_i, y_i) such that

$$1 \leq x_{i+1} - x_i \leq \frac{a^{1/3}}{2}$$

and

$$1 \leq y_{i+1} - y_i \leq \frac{a^{1/3}}{2}.$$

Since the fractions

$$\frac{y_{i+1} - y_i}{x_{i+1} - x_i}$$

are distinct, and the numerators and denominators are bounded by $a^{1/3}/2$, we have

$$r_3 \leq \left(\frac{a^{1/3}}{2}\right)^2 = \frac{a^{2/3}}{4}.$$

Therefore,

$$r(a, b) \leq r_1 + r_2 + r_3 + 1 < 3a^{2/3} + \frac{a^{2/3}}{4} + 1 < 5a^{2/3}.$$

This completes the proof.

Lemma 2.7 *Let x and y be positive integers, $(x, y) = 1$. If the prime $p \neq 3$ divides*

$$\frac{x^3 + y^3}{x + y},$$

then

$$p \equiv 1 \pmod 3.$$

Proof. Let $p \neq 3$ be a prime such that

$$x^2 - xy + y^2 = \frac{x^3 + y^3}{x + y} \equiv 0 \pmod p.$$

If p divides y, then p also divides x, which is impossible because $(x, y) = 1$. Therefore, $(p, y) = 1$. Since

$$(2x - y)^2 + 3y^2 \equiv 0 \pmod p,$$

it follows that -3 is a quadratic residue modulo p. Let $\left(\frac{a}{p}\right)$ be the Legendre symbol. By quadratic reciprocity, we have

$$\left(\frac{-3}{p}\right) = \left(\frac{p}{3}\right) = 1$$

if and only if $p \equiv 1 \pmod 3$. This completes the proof.

In the proof of the next lemma, we shall use some results from multiplicative number theory. Let $\pi(x; 3, 2)$ denote the number of primes $p \le x$ such that $p \equiv 2 \pmod 3$. By the prime number theorem for arithmetic progressions, $\pi(x; 3, 2) \sim x/(2 \log x)$. Moreover, there exists a constant A such that

$$\sum_{\substack{p \le x \\ p \equiv 2 \pmod 3}} \frac{1}{p} = \frac{1}{2} \log \log x + A + O\left(\frac{1}{\log x}\right).$$

This implies that

$$\sum_{\substack{x^{10/11} < p \le x \\ p \equiv 2 \pmod 3}} \frac{1}{p} = \frac{1}{2} \log \log x - \frac{1}{2} \log \log x^{10/11} + O\left(\frac{1}{\log x}\right)$$

$$= \frac{1}{2} \log \frac{11}{10} + O\left(\frac{1}{\log x}\right).$$

Lemma 2.8 *For any positive integer a, let $h(a)$ denote the largest divisor of a consisting only of primes $p \equiv 1 \pmod 3$, that is,*

$$h(a) = \prod_{\substack{p^k \| a \\ p \equiv 1 \pmod 3}} p^k. \tag{2.18}$$

Let $H(x)$ denote the number of positive integers a up to x such that $h(a) < a^{1/10}$ and a is not divisible by 3. There exists a constant $\delta_1 \in (0, 1)$ such that

$$H(x) > \delta_1 x$$

for all $x \ge 2$.

Proof. Let $H_0(x)$ denote the number of positive integers $a \le x$ of the form $a = pb$, where $p \equiv 2 \pmod 3$ is a prime such that $p > x^{10/11}$, and b is an integer not divisible by 3. An integer a has at most one representation of this form. Moreover,

$$h(a) = h(b) \le b = \frac{a}{p} < x^{1/11} < p^{1/10} \le a^{1/10}.$$

It follows that every number of the form pb is counted in $H(x)$, and so

$$H_0(x) \le H(x).$$

Also, $H_0(2) = H(2) = 1$. Let $g(x)$ denote the number of positive integers up to x not divisible by 3. Then

$$g(x) > \frac{2x}{3} - 1$$

and

$$H_0(x) = \sum_{\substack{x^{10/11} < p \le x \\ p \equiv 2 \pmod 3}} g\left(\frac{x}{p}\right)$$

$$\ge \sum_{\substack{x^{10/11} < p \le x \\ p \equiv 2 \pmod 3}} \left(\frac{2x}{3p} - 1\right)$$

$$\ge \frac{2x}{3} \sum_{\substack{x^{10/11} < p \le x \\ p \equiv 2 \pmod 3}} \frac{1}{p} - \pi(x; 3, 2)$$

$$= \frac{2x}{3} \left(\frac{1}{2} \log \frac{11}{10} + O\left(\frac{1}{\log x}\right)\right) + O\left(\frac{x}{\log x}\right)$$

$$= \frac{x}{3} \log \frac{11}{10} + +O\left(\frac{x}{\log x}\right)$$

$$\gg x.$$

This completes the proof.

Lemma 2.9 *Let $\varphi(d)$ be the Euler φ-function, and let $0 < \delta < 1$. There exists a constant $c_1 = c_1(\delta) > 0$ such that, if n is a positive integer and $t \ge \delta n$, and if*

$$a_1 < \cdots < a_t \le n$$

are any t positive integers, then

$$\sum_{i=1}^{t} \varphi(a_i) > c_1 n^2.$$

Proof. For any $p \ge 7$, we have

$$\left(1 - \frac{2}{p^2}\right)^p = \sum_{k=0}^{p} \binom{p}{k} \frac{(-2)^k}{p^{2k}}$$

$$= 1 - \frac{2}{p} + \sum_{k=2}^{p} \binom{p}{k} \frac{(-2)^k}{p^{2k}}$$

$$< 1 - \frac{2}{p} + \sum_{k=2}^{p} p^k \frac{2^k}{p^{2k}}$$

$$= 1 - \frac{2}{p} + \sum_{k=2}^{p} \left(\frac{2}{p}\right)^k$$

$$< 1 - \frac{2}{p} + \frac{4}{p(p-2)}$$
$$< 1 - \frac{1}{p}.$$

Since the infinite product

$$\prod_{p \geq 7} \left(1 - \frac{2}{p^2} \right)$$

converges, we have

$$\prod_p \left(1 - \frac{1}{p} \right)^{n/p} = \prod_{p<7} \left(1 - \frac{1}{p} \right)^{n/p} \prod_{p \geq 7} \left(1 - \frac{1}{p} \right)^{n/p}$$

$$> \prod_{p<7} \left(1 - \frac{1}{p} \right)^n \prod_{p \geq 7} \left(1 - \frac{2}{p^2} \right)^n$$

$$= c_2^n,$$

where

$$0 < c_2 < 1.$$

Since $\varphi(d) = d \prod_{p|d} \left(1 - \frac{1}{p} \right)$ and $n! > (n/e)^n$, it follows that

$$\prod_{d=1}^n \varphi(d) = \prod_{d=1}^n d \prod_{p|d} \left(1 - \frac{1}{p} \right)$$

$$= n! \prod_{p \leq n} \left(1 - \frac{1}{p} \right)^{[n/p]}$$

$$\geq n! \prod_{p \leq n} \left(1 - \frac{1}{p} \right)^{n/p}$$

$$\geq n! c_2^n$$

$$> \left(\frac{c_2 n}{e} \right)^n.$$

Choose $c_3 > 0$ so that

$$c_3^{\delta/2} < \frac{c_2}{e} < 1.$$

Let

$$m = \left[\frac{\delta n}{2} \right] \leq \frac{\delta n}{2} < m + 1.$$

Suppose that there exists a set $\mathcal{D} \subseteq [1, n]$ such that $|\mathcal{D}| = m + 1$ and $\varphi(d) \leq c_3 n$ for all $d \in \mathcal{D}$. Since $\varphi(d) \leq d \leq n$ for all $d \leq n$, we have

$$\prod_{d=1}^n \varphi(d) = \prod_{\substack{d=1 \\ d \in \mathcal{D}}}^n \varphi(d) \prod_{\substack{d=1 \\ d \notin \mathcal{D}}}^n \varphi(d)$$

$$\leq \prod_{\substack{d-1 \\ d \in D}}^{n} c_3 n \prod_{\substack{d-1 \\ d \notin D}}^{n} n$$

$$\leq (c_3 n)^{m+1} n^{n-m-1}$$
$$= c_3^{m+1} n^n$$
$$< c_3^{\delta n/2} n^n$$
$$< \left(\frac{c_2 n}{e}\right)^n,$$

which is impossible. It follows that there exist at most m integers in $[1, n]$ with $\varphi(d_i) \leq c_3 n$. In particular, among the $t \geq \delta n$ integers a_i, there must be at least

$$t - m \geq \delta n - \left[\frac{\delta n}{2}\right] \geq \frac{\delta n}{2}.$$

integers for which $\varphi(a_i) > c_3 n$, and so

$$\sum_{i=1}^{t} \varphi(a_i) > \left(\frac{\delta n}{2}\right) c_3 n = \frac{c_3 \delta}{2} n^2 = c_1 n^2,$$

where $c_1 = c_3 \delta / 2$. This completes the proof.

Theorem 2.5 (Erdős–Mahler) *Let $C_2'(n)$ denote the number of integers not exceeding n that can be written as the sum of two positive, relatively prime integral cubes. Then*

$$C_2'(n) \gg n^{2/3}.$$

Proof. Let

$$h(a) = \prod_{\substack{p^k \|a \\ p=1 \ (\mathrm{mod}\ 3)}} p^k$$

and let

$$a_1 < \cdots < a_t \leq n^{1/3}$$

be the integers in $[1, n^{1/3}]$ not divisible by 3 such that

$$h(a_i) < a_i^{1/10}.$$

Then $h(1) = h(2) = 1$ and so $a_1 = 2$. By Lemma 2.8, we have

$$t = H(n^{1/3}) > \delta_1 n^{1/3}.$$

Let x and y be positive integers such that

$$x + y = a_i \qquad \text{for some } i = 1, \ldots, t.$$

Then

$$x^3 + y^3 < (x + y)^3 = a_i^3 \leq n.$$

Moreover, $(x, y) = 1$ if and only if $(x, a_i) = (y, a_i) = 1$. Therefore, the number of pairs x, y of positive integers such that $x + y = a_i$, $x < y$, and $(x, y) = 1$ is $\varphi(a_i)/2$.

Let $r(m)$ denote the number of representations of m in the form

$$m = x^3 + y^3,$$

where x and y are relatively prime positive integers such that $(x, y) = 1$ and $x + y = a_i$ for some i. Then

$$R_1 = \sum_{m=1}^{n} r(m) = \frac{1}{2} \sum_{i=2}^{t} \varphi(a_1) > c_6 n^{2/3}$$

by Lemma 2.9.

Let R_2 be the number of ordered quadruples (x, y, u, v) of positive integers such that

$$x^3 + y^3 = u^3 + v^3,$$

$$a_i = x + y < u + v = a_j \qquad \text{for } i, j \in [1, t],$$

$$(x, y) = (u, v) = 1,$$

$$x < y \qquad \text{and} \qquad u < v.$$

Note that if $x^3 + y^3 = u^3 + v^3$, then $x + y = u + v$ if and only if $\{x, y\} = \{u, v\}$ (Exercise 7). Then

$$R_2 = \sum_{m=1}^{n} \binom{r(m)}{2}.$$

Let (x, y, u, v) be a quadruple counted in R_2. Since

$$h(a_i)\frac{a_i}{h(a_i)}\frac{x^3 + y^3}{x + y} = h(a_j)\frac{a_j}{h(a_j)}\frac{u^3 + v^3}{u + v}$$

and a_i and a_j are not divisible by 3, it follows from (2.18) that $a_i/h(a_i)$ and $a_j/h(a_j)$ are products of primes $p \equiv 2 \pmod 3$. By Lemma 2.7,

$$\left(p, \frac{x^3 + y^3}{x + y}\right) = \left(p, \frac{u^3 + v^3}{u + v}\right) = 1$$

if $p \equiv 2 \pmod 3$. Therefore,

$$\frac{a_i}{h(a_i)} = \frac{a_j}{h(a_j)}.$$

Fix the integer a_i. Since

$$0 < \left(\frac{a_i}{h(a_i)}\right) h(a_j) = a_j \le n^{1/3}$$

and

$$\frac{a_i}{h(a_i)} > a_i^{9/10},$$

it follows that

$$1 \le h(a_j) < \frac{n^{1/3}}{a_i^{9/10}}.$$

Therefore, to each a_i there correspond fewer than

$$\frac{n^{1/3}}{a_i^{9/10}}$$

different integers a_j. By Lemma 2.6, the number of quadruples (x, y, u, v) such that $x + y = a_i$ and $u + v = a_j$ is smaller than $3a_i^{2/3}$. Therefore, the number $R_{2,i}$ of quadruples (x, y, u, v) such that $x + y = a_i$ satisfies

$$R_{2,i} < 3a_i^{2/3} \frac{n^{1/3}}{a_i^{9/10}} = \frac{3n^{1/3}}{a_i^{7/30}},$$

and so

$$R_2 = \sum_{i=1}^{t} R_{2,i}$$

$$< 3 \sum_{i=1}^{t} \frac{n^{1/3}}{a_i^{7/30}}$$

$$\le 3n^{1/3} \sum_{1 \le i \le n^{1/3}} \frac{1}{i^{7/30}}$$

$$\le 3n^{1/3}(n^{1/3})^{23/30}$$

$$= 3n^{(2/3)-(7/90)}.$$

Let $C_2'(n)$ count the number of integers m up to n of the form $m = x^3 + y^3$, where x and y are relatively prime positive integers. Since

$$r \le 1 + \binom{r}{2}$$

for all integers r, we have

$$R_1 = \sum_{\substack{m=1 \\ r(m) \ge 1}}^{n} r(m) \le \sum_{\substack{m=1 \\ r(m) \ge 1}}^{n} 1 + \sum_{\substack{m=1 \\ r(m) \ge 1}}^{n} \binom{r(m)}{2} \le C_2'(n) + R_2.$$

Therefore,

$$C_2'(n) \ge R_1 - R_2 \ge n^{2/3} - n^{(2/3)-(7/90)} \gg n^{2/3}.$$

This completes the proof.

The Erdős–Mahler theorem states that many integers can be written as the sum of two positive cubes. Hooley showed that very few numbers have two essentially distinct representations in this form. To prove this, we need the following result of Vaughan–Wooley [130, Lemma 3.5] from the elementary theory of binary quadratic forms.

Lemma 2.10 *Let $\varepsilon > 0$. For any nonzero integers D and N, the number of solutions of the equation*

$$X^2 - DY^2 = N$$

with

$$\max(|X|, |Y|) \ll P$$

is

$$\ll (DNP)^\varepsilon,$$

where the implied constant depends only on ε.

Proof. See Hua [63, chapter 11] or Landau [78, part 4].

The following lemma on "completing the square" shows how to transform certain quadratic equations in two variables into Pell's equations.

Lemma 2.11 *Let a, b, c be integers such that $a \neq 0$ and $D = b^2 - 4ac \neq 0$. Let (x, y) be a solution of the equation*

$$ax^2 + bxy + cy^2 + dx + ey + f = 0. \tag{2.19}$$

Let

$$X = Dy - 2ae + bd$$

and

$$Y = 2ax + by + d.$$

Then (X, Y) is a solution of the equation

$$X^2 - DY^2 = N,$$

where

$$N = (4af - d^2)D + (2ae - bd)^2. \tag{2.20}$$

Moreover, this map sending (x, y) to (X, Y) is one-to-one.

The number $D = b^2 - 4ac$ is called the *discriminant* of equation (2.19).

Proof. Multiplying equation (2.19) by $4a$, we obtain

$$
\begin{aligned}
4a^2x^2 &+ 4abxy + 4acy^2 + 4adx + 4aey + 4af \\
&= (2ax + by)^2 - Dy^2 + 2d(2ax + by) + 2(2ae - bd)y + 4af \\
&= (2ax + by + d)^2 - Dy^2 + 2(2ae - bd)y + (4af - d^2) \\
&= Y^2 - Dy^2 + 2(2ae - bd)y + (4af - d^2) \\
&= 0,
\end{aligned}
$$

where

$$Y = 2ax + by + d.$$

Multiplying by $-D$, we obtain

$$
\begin{aligned}
D^2 y^2 &- 2(2ae - bd)Dy - DY^2 - (4af - d^2)D \\
&= (Dy - 2ae + bd)^2 - DY^2 - (4af - d^2)D - (2ae - bd)^2 \\
&= X^2 - DY^2 - \big((4af - d^2)D + (2ae - bd)^2\big) \\
&= X^2 - DY^2 - N \\
&= 0,
\end{aligned}
$$

where

$$X = Dy - 2ae + bd$$

and

$$N = (4af - d^2)D + (2ae - bd)^2.$$

The determinant of the affine map that sends (x, y) to (X, Y) is

$$
\begin{vmatrix} 0 & D \\ 2a & b \end{vmatrix} = -2aD \neq 0
$$

since $a \neq 0$ and $D \neq 0$, and so the map $(x, y) \mapsto (X, Y)$ is one-to-one. This completes the proof.

Lemma 2.12 *Let $P \geq 2$, and let a, b, c, d, e, f be integers such that*

$$\max\{|a|, \ldots, |f|\} \ll P^2.$$

Let $D = b^2 - 4ac$, and define the integer N by (2.20). Let W denote the number of solutions of the equation

$$ax^2 + bxy + cy^2 + dx + ey + f = 0$$

with $\max(|x|, |y|) \ll P$. If a, D, and N are nonzero, then

$$W \ll |P|^\varepsilon$$

for any $\varepsilon > 0$, where the implied constant depends only on ε.

Proof. By Lemma 2.11, to every solution (x, y) of the quadratic equation (2.19) there corresponds a solution of the equation

$$X^2 - DY^2 = N,$$

where

$$D = b^2 - 4ac \ll P^4$$

and

$$N = (4af - d^2)D + (2ae - bd)^2 \ll P^8.$$

Moreover,

$$X = Dy - 2ae + bd \ll P^4|y| \ll P^5$$

and

$$Y = 2ax + by + d \ll P^2(|x| + |y|) \ll P^3$$

if $\max(|x|, |y|) \ll P$. It follows from Lemma 2.10 that

$$W \ll (DNP^5)^\varepsilon \ll P^{17\varepsilon} \ll P^\varepsilon.$$

This completes the proof.

Theorem 2.6 (Hooley–Wooley) *Let $D(n)$ denote the number of integers not exceeding n that have at least two essentially distinct representations as the sum of two nonnegative integral cubes. Then*

$$D(n) \ll_\varepsilon n^{5/9+\varepsilon}.$$

Proof. If N has at least two essentially distinct representations as the sum of two nonnegative cubes, then there exist integers x_1, x_2, x_3, x_4 such that

$$x_1^3 + x_2^3 = x_3^3 + x_4^3 = N$$

and

$$0 \leq x_3 < x_1 \leq x_2 \leq x_4 \leq N^{1/3}.$$

For any number $P \geq 2$, let $S(P)$ denote the number of solutions of the equation

$$x_1^3 + x_2^3 = x_3^3 + x_4^3 \tag{2.21}$$

that satisfy

$$0 \leq x_3 < x_1 \leq x_2 < x_4 \leq P. \tag{2.22}$$

Then

$$D(n) \leq S(n^{1/3}). \tag{2.23}$$

If the integers x_1, x_2, x_3, x_4 satisfy (2.21) and (2.22), then $x_1 + x_2 \neq x_3 + x_4$ by Exercise 7, and so

$$x_1 + x_2 = x_3 + x_4 + h,$$

where

$$1 \leq |h| < 2P.$$

Let $T(P, h)$ denote the number of solutions of the simultaneous equations

$$x_1^3 + x_2^3 = x_3^3 + x_4^3$$

and

$$x_1 + x_2 = x_3 + x_4 + h$$

with

$$0 \leq x_i \leq P \qquad \text{for } i = 1, \ldots, 4.$$

Choose the integer ℓ so that

$$2^\ell \leq 2P < 2^{\ell+1}.$$

Then

$$S(P) \leq \sum_{1 \leq |h| < 2P} T(P, h)$$

$$\leq \sum_{0 \leq i \leq \ell} \sum_{2^i \leq |h| < 2^{i+1}} T(P, h)$$

$$\ll \ell \max_{0 \leq i \leq \ell} \left\{ \sum_{2^i \leq |h| < 2^{i+1}} T(P, h) \right\}$$

$$\ll \log P \max_{1 \leq H \leq 2P} \left\{ \sum_{H \leq |h| < 2H} T(P, h) \right\}.$$

Since x_3 is the smallest of the four integers x_1, x_2, x_3, x_4, we have

$$2x_4 + h > x_3 + x_4 + h = x_1 + x_2 > 0.$$

For fixed h, we can use $x_1, \ldots, x_4$ to define four positive integers u_1, u_2, u_3, and y as follows:

$$u_1 = x_1 + x_2$$
$$u_2 = x_1 - x_3$$
$$u_3 = x_2 - x_3$$
$$y = 2x_4 + h,$$

where

$$1 \leq u_i \leq 2P \qquad \text{for } i = 1, 2, 3$$

and

$$1 \leq y \leq 4P.$$

Moreover,

$$u_1 + u_2 + u_3 = 2(x_1 + x_2 - x_3) = 2(x_4 + h) = y + h$$

and

$$\begin{aligned}
h\left(3y^2 + h^2\right) &= h\left(3(2x_4 + h)^2 + h^2\right) \\
&= h(12x_4^2 + 12x_4h + 4h^2) \\
&= 4(3x_4^2h + 3x_4h^2 + h^3) \\
&= 4((x_4 + h)^3 - x_4^3) \\
&= 4((x_1 + x_2 - x_3)^3 - x_1^3 - x_2^3 + x_3^3) \\
&= 12(x_1^2x_2 + x_1x_2^2 - x_2^2x_3 + x_2x_3^2 - x_1^2x_3 + x_1x_3^2 - 2x_1x_2x_3) \\
&= 12(x_1 + x_2)(x_1 - x_3)(x_2 - x_3) \\
&= 12u_1u_2u_3.
\end{aligned}$$

Conversely, the numbers u_1, u_2, u_3, and y determine $x_1, \ldots, x_4$ uniquely. It follows that

$$T(P, h) \leq U(P, h),$$

where $U(P, h)$ denotes the number of solutions of the equations

$$u_1 + u_2 + u_3 = y + h \qquad (2.24)$$

and

$$12u_1 u_2 u_3 = h(3y^2 + h^2) \qquad (2.25)$$

in positive integers $u_i \leq 2P$ and $y \leq 4P$. If $u_i = h$ for some i, say, $u_3 = h$, then $u_1 + u_2 = h$ and

$$12u_1 u_2 = 3y^2 + h^2 = 3u_1^2 + 6u_1 u_2 + 3u_2^2 + h^2.$$

This implies that

$$3(u_1 - u_2)^2 + h^2 = 0,$$

which is impossible since $h \neq 0$. Therefore, $u_i \neq h$ for all $i = 1, 2, 3$. Let u_1, u_2, u_3, h be a solution of equations (2.24) and (2.25) counted in $U(P, h)$. Let

$$(u_3, h) = \max\{(u_i, h) : i = 1, 2, 3\},$$

where (a, b) denotes the greatest common divisor of a and b. We define

$$d_3 = (u_3, h),$$

$$d_2 = \left(u_2, \frac{h}{d_3}\right),$$

$$d_1 = \left(u_1, \frac{h}{d_2 d_3}\right).$$

Then

$$d_3 = \max\{d_1, d_2, d_3\}$$

and $d_1 d_2 d_3$ divides h. Let

$$g = \frac{h}{d_1 d_2 d_3},$$

and

$$v_i = \frac{u_i}{d_i} \qquad \text{for } i = 1, 2, 3.$$

Then

$$(v_i, g) = 1 \quad \text{and} \quad 1 \leq v_i \leq \frac{2P}{d_i} \qquad \text{for } i = 1, 2, 3. \qquad (2.26)$$

It follows from (2.25) that

$$12v_1 v_2 v_3 = g(3y^2 + h^2),$$

and so g divides 12, that is,

$$fg = 12$$

for some integer f. Therefore, $|h| = |gd_1d_2d_3| \leq 12d_3^3$, and so

$$d_3 \gg |h|^{1/3}. \tag{2.27}$$

Since $u_3 \neq h$, it follows that

$$v_3 \neq gd_1d_2. \tag{2.28}$$

We can rewrite equation (2.25) in terms of the new variables v_i, d_i, f, g. Since

$$h = gd_1d_2d_3$$

and

$$y = d_1v_1 + d_2v_2 + d_3v_3 - h,$$

we have

$$12u_1u_2u_3 = fgd_1d_2d_3v_1v_2v_3 = fhv_1v_2v_3 = h(3y^2 + h^2),$$

and so

$$fv_1v_2v_3 = 3(d_1v_1 + d_2v_2 + d_3v_3 - h)^2 + h^2. \tag{2.29}$$

If we fix the integers d_1, d_2, d_3, f, g, v_3, then equation (2.29) becomes a quadratic equation in v_1, v_2:

$$3d_1^2v_1^2 + (6d_1d_2 - fv_3)v_1v_2 + 3d_2^2v_2^2 + 6d_1(d_3v_3 - h)v_1$$
$$+6d_2(d_3v_3 - h)v_2 + 3(d_3v_3 - h)^2 + h^2 = 0. \tag{2.30}$$

The discriminant of this quadratic is

$$\begin{aligned}
D &= ((6d_1d_2 - fv_3)^2 - 36d_1^2d_2^2 \\
&= f^2v_3^2 - 12d_1d_2fv_3 \\
&= f^2v_3^2 - d_1d_2f^2gv_3 \\
&= f^2v_3(v_3 - d_1d_2g) \\
&\neq 0
\end{aligned}$$

by (2.28). Similarly, the integer N defined by (2.20) is nonzero, because

$$\begin{aligned}
N &= \left(4 \cdot 3d_1^2 \left(3(d_3v_3 - h)^2 + h^2\right) - (6d_1(d_3v_3 - h))^2\right) D \\
&\quad + \left(2 \cdot 3d_1^2 \cdot 6d_2(d_3 - v_3h) - (6d_1d_2 - fv_3) \cdot 6d_1(d_3 - v_3h)\right)^2 \\
&= 12d_1^2h^2D + (6d_1fv_3(d_3v_3 - h))^2 \\
&= 12d_1^2h^2f^2v_3(v_3 - d_1d_2g) + 36d_1^2f^2v_3^2d_3^2(v_3 - d_1d_2g))^2 \\
&= 12d_1^2d_3^2f^2v_3(v_3 - d_1d_2g)\left((d_1d_2g)^2 - 3d_1d_2gv_3 + 3v_3^2\right) \\
&= 3d_1^2d_3^2f^2v_3(v_3 - d_1d_2g)\left((d_1d_2g)^2 + 3(d_1d_2g - 2v_3)2\right) \\
&\neq 0.
\end{aligned}$$

Let $W(P, d_1, d_2, d_3, f, g, v_3)$ denote the number of solutions of equation (2.30) in integers v_1, v_2 satisfying (2.26). Since the coefficients of this quadratic equation are all $\ll P^2$, it follows from Lemma 2.12 that

$$W(P, d_1, d_2, d_3, f, g, v_3) \ll P^\varepsilon.$$

Therefore,

$$S(P) \ll \log P \max_{1 \le H \le 2P} \sum_{H \le |h| < 2H} T(P, h)$$

$$\ll \log P \max_{1 \le H \le 2P} \sum_{H \le |h| < 2H} U(P, h)$$

$$\ll \log P \max_{1 \le H \le 2P} \sum_{H \le |h| < 2H} \sum_{\substack{fg=12 \\ }} \sum_{\substack{gd_1d_2d_3=h \\ d_3 \ge \max(d_1,d_2)}}$$

$$\sum_{\substack{1 \le v_3 \le 2P/d_3 \\ v_3 \nmid gd_1d_2}} W(P, d_1, d_2, d_3, f, g, v_3)$$

$$\ll \log P \max_{1 \le H \le 2P} \sum_{H \le |h| < 2H} \sum_{\substack{fg=12}} \sum_{\substack{gd_1d_2d_3=h \\ d_3 \ge \max(d_1,d_2)}} \sum_{\substack{1 \le v_3 \le 2P/d_3 \\ v_3 \nmid gd_1d_2}} P^\varepsilon$$

$$\ll P^\varepsilon \max_{1 \le H \le 2P} \sum_{H \le |h| < 2H} \sum_{\substack{fg=12}} \sum_{\substack{gd_1d_2d_3=h \\ d_3 \ge \max(d_1,d_2)}} \frac{P^{1+\varepsilon}}{d_3}$$

$$\ll P^{1+2\varepsilon} \max_{1 \le H \le 2P} \sum_{H \le |h| < 2H} \sum_{\substack{gd_1d_2d_3=h \\ d_3 \ge \max(d_1,d_2)}} \frac{1}{d_3}.$$

Since the number of factorizations of h in the form $h = gd_1d_2d_3$ is $\ll |h|^\varepsilon$, and since

$$d_3 \gg |h|^{1/3}$$

by (2.27), we have

$$\sum_{\substack{H \le |h| < 2H}} \sum_{\substack{gd_1d_2d_3=h \\ d_3 \ge \max(d_1,d_2)}} \frac{1}{d_3} \ll \sum_{H \le h < 2H} \frac{1}{h^{1/3-\varepsilon}} \ll H^{2/3+\varepsilon},$$

and so

$$S(P) \ll P^{1+2\varepsilon} \max_{1 \le H \le 2P} H^{2/3+\varepsilon} \ll P^{5/3+3\varepsilon}.$$

Therefore, by (2.23), we have

$$D(n) \le S(n^{1/3}) \ll n^{5/9+\varepsilon}.$$

This completes the proof.

Theorem 2.7 (Erdős) *Almost all integers that can be represented as the sum of two positive cubes have essentially only one such representation.*

Proof. This follows immediately from the remark that there are greater than $cn^{2/3}$ integers that can be represented in at least one way as the sum of two nonnegative cubes, but there are no more than $c'n^{5/9+\varepsilon} = o(n^{2/3})$ integers that have two or more essentially distinct representations as the sum of two cubes.

2.5 Notes

Wieferich's proof [144] that $g(3) = 9$ appeared in *Mathematische Annalen* in 1909. In the immediately following paper in the same issue of that journal, Landau [75] proved that $G(3) \leq 8$. Dickson [24] showed that 23 and 239 are the only positive integers not representable as the sum of eight nonnegative cubes. An error in Wieferich's paper was corrected by Kempner [70]. Scholz [108] gives a nice version of the Wieferich–Kempner proof.

Linnik's proof [81] of the theorem that $G(3) \leq 7$ is difficult. Watson [139] subsequently discovered a different and much more elementary proof of this result, and it is Watson's proof that is given in this chapter. Dress [25] has a simple proof that $G(3) \leq 11$.

Vaughan [126] obtained an asymptotic formula for $r_{3,8}(n)$, the number of representations of an integer as the sum of eight cubes. It is an open problem to obtain an asymptotic formula for the number of representations of an integer as the sum of seven or fewer cubes.

It is possible that every sufficiently large integer is the sum of four nonnegative cubes. Let $E(x)$ denote the number of positive integers up to x that cannot be written as the sum of four positive cubes. Davenport [17] proved that $E_{4,3}(x) \ll x^{29/30+\varepsilon}$, and so almost all positive integers can be represented as the sum of four positive cubes. Brüdern [6] proved that

$$E_{4,3}(x) \ll x^{37/42+\varepsilon}.$$

There are interesting identities that express a linear polynomial as the sum of the cubes of four polynomials with integer coefficients. Such identities enable us to represent the integers in particular congruence classes as sums of four integral cubes. See Mordell [85, 86], Demjanenko [20], and Revoy [101] for such polynomial identities.

Theorem 2.5 was first proved by Erdős and Mahler [31, 35]. The beautiful elementary proof given in this chapter is due to Erdős [31]. Similarly, Theorem 2.6 was originally proved by Hooley [57, 58]. The elementary proof presented here is due to Wooley [149]. For an elementary discussion of elliptic curves and sums of two cubes, see Silverman [115] and Silverman and Tate [116, pages 147–151].

Waring stated in 1770 that $g(2) = 4$, $g(3) = 9$, and $g(4) = 19$. The theorem that every nonnegative integer is the sum of 19 fourth powers was finally proved in 1992 in joint work of Balasubramanian [2] and Deshouillers and Dress [21].

2.6   Exercises

1. Prove that

$$3^3 + 4^3 + 5^3 = 6^3$$

   is the only solution in integers of the equation

$$(x - 3)^3 + (x - 2)^3 + (x - 1)^3 = x^3.$$

2. Let $s(N)$ be the smallest number such that N can be written as the sum of $s(N)$ positive cubes. Compute $s(N)$ for $N = 1, \ldots, 100$.

3. Prove that $s(239) = 9$, that is, 239 cannot be written as a sum of eight nonnegative cubes.

4. Show that none of the following numbers

$$
\begin{array}{ccccc}
15 & 22 & 50 & 114 & 167 \\
175 & 186 & 212 & 231 & 238 \\
303 & 364 & 420 & 428 & 454
\end{array}
$$

   can be written as a sum of seven nonnegative cubes.

5. Show that none of the following numbers

$$79, 159, 239, 319, 399, 479, 559$$

   can be written as a sum of 18 fourth powers.

6. Let $v(3)$ denote the smallest number such that every integer can be written as the sum or difference of $v(3)$ nonnegative integral cubes.

   (a) Prove that

$$4 \le v(3) \le g(3).$$

   (b) Prove that

$$v(3) \le 5.$$

   Hint: Use the polynomial identity

$$6x = (x + 1)^3 + (x - 1)^3 - 2x^3$$

   and the fact that $x = (N - N^3)/6$ is an integer for every integer N.

   It is an unsolved problem to determine whether $v(3) = 4$ or 5. This is called the *easier Waring's problem for cubes*.

7. Let x, y, u, v be positive integers. Prove that if $x + y = u + v$ and $x^3 + y^3 = u^3 + v^3$, then $\{x, y\} = \{u, v\}$.

8. (Von Sterneck [136]) Using a computer, calculate $s(n)$ for n up to 40,000. Verify the results of Lemma 2.4.

9. (Mahler [82]) Prove that 1 has infinitely many different representations as the sum of three cubes. Hint: Establish the polynomial identity

$$(9x^4)^3 + (3x - 9x^4)^3 + (1 - 9x^3)^3 = 1. \qquad (2.31)$$

Prove that

$$(9m^4)^3 + (3mn^3 - 9m^4)^3 + (n^4 - 9m^3n)^3 = n^{12}.$$

Let $r_{3,3}(N)$ denote the number of representations of N as the sum of three nonnegative cubes. Prove that if $N = n^{12}$ for some positive integer n, then

$$r_{3,3}(N) \geq 9^{-1/3}N^{1/12}.$$

Note: This is Mahler's counterexample to Hypothesis K of Hardy and Littlewood [49].

10. (Elkies and Kaplansky [27]) Verify the following polynomial identities:

$$8(x^2 + y^2 - z^3) = (2x + 2y)^2 + (2x - 2y)^2 - (2z)^3,$$

$$2x + 1 = (x^3 - 3x^2 + x)^2 + (x^2 - x - 1)^2 - (x^2 - 2x)^3,$$

$$2(2x + 1) = (2x^3 - 2x^2 - x)^2 - (2x^3 - 4x^2 - x + 1)^2 - (2x^2 - 2x - 1)^3,$$

$$4(2x + 1) = (x^3 + x + 2)^2 + (x^2 - 2x - 1)^2 - (x^2 + 1)^3.$$

Show that every integer N, positive or negative, can be written uniquely in the form

$$N = 8^q 2^r (2m + 1),$$

where $q \geq 0$, $r \in \{0, 1, 2\}$, and $m \in \mathbf{Z}$. Prove that every integer N can be written in the form

$$N = a^2 + b^2 - c^3,$$

where a, b, c are integers.

11. Let a be a positive rational number. Consider the equations

$$a = x^3 + y^3 + z^3$$
$$a = (x + y + z)^3 - 3(y + z)(z + x)(x + y)$$
$$8a = (u + v + w)^3 - 24uvw.$$

Prove that if any one of these equations has a solution in positive rational numbers, then each of the three equations does.

12. Let a be a rational number. Let r be any rational number such that $r \neq 0$ and

$$t = \frac{a}{72r^3} \neq -1.$$

For any rational number w, let

$$u = \left(\frac{24t^2}{(t+1)^3} - 1 \right) w$$

and

$$v = \left(\frac{24t}{(t+1)^3} \right) w.$$

Prove that

$$(u + v + w)^3 - 24uvw = 8a \left(\frac{w}{r(t+1)} \right)^3.$$

Let $w = r(t+1)$. Prove that there exist rational numbers x, y, z such that

$$u = y + z$$
$$v = z + x$$
$$w = x + y$$

and

$$a = x^3 + y^3 + z^3.$$

This proves that every rational number can be written as the sum of three rational cubes.

13. Let a be a positive rational number. Show that it is possible to choose r in Exercise 12 so that

$$a = x^3 + y^3 + z^3,$$

where x, y, z are positive rational numbers. This proves that every positive rational number can be written as the sum of three positive rational cubes.

3

The Hilbert–Waring theorem

Nous ne devons pas douter que ces considérations, qui permettent ainsi d'obtenir des relations arithmétiques en les faisant sortir d'identités où figurent des intégrales définies, ne puissent un jour, quand on en aura bien compris de sens, être appliquées à des problèmes bien plus étendus que celui de Waring. [1]

H. Poincaré [96]

3.1 Polynomial identities and a conjecture of Hurwitz

Waring's problem for exponent k is to prove that the set of nonnegative integers is a basis of finite order, that is, to prove that every nonnegative integer can be written as the sum of a bounded number of kth powers. We denote by $g(k)$ the smallest number s such that every nonnegative integer is the sum of exactly s kth powers of nonnegative integers. Waring's problem is to show that $g(k)$ is finite; Hilbert proved this in 1909. The goal of this chapter is to prove the Hilbert–Waring theorem: the kth powers are a basis of finite order for every positive integer k.

We have already proved Waring's problem for exponent two (the squares) and exponent three (the cubes). Other cases of Waring's problem can be deduced from

---

[1] We should not doubt that [Hilbert's] method, which makes it possible to obtain arithmetic relations from identities involving definite integrals, might one day, when it is better understood, be applied to problems far more general than Waring's.

these results by means of polynomial identities. Here are three examples. We use the notation

$$(x_1 \pm x_2 \pm \cdots \pm x_h)^k = \sum_{\varepsilon_2,\ldots,\varepsilon_h=\pm 1} (x_1 + \varepsilon_2 x_2 + \cdots + \varepsilon_h x_h)^k.$$

Theorem 3.1 (Liouville)

$$\left(x_1^2 + x_2^2 + x_3^2 + x_4^2\right)^2 = \frac{1}{6} \sum_{1 \le i < j \le 4} (x_i + x_j)^4 + \frac{1}{6} \sum_{1 \le i < j \le 4} (x_i - x_j)^4$$

is a polynomial identity, and every nonnegative integer is the sum of 53 fourth powers, that is,

$$g(4) \le 53.$$

Proof. We begin by observing that

$$(x_1 \pm x_2)^4 = (x_1 + x_2)^4 + (x_1 - x_2)^4 = 2x_1^4 + 12x_1^2 x_2^2 + 2x_2^4,$$

and so

$$\sum_{1 \le i < j \le 4} (x_i \pm x_j)^4 = \sum_{1 \le i < j \le 4} (x_i + x_j)^4 + \sum_{1 \le i < j \le 4} (x_i - x_j)^4$$

$$= \sum_{1 \le i < j \le 4} \left(2x_i^4 + 12x_i^2 x_j^2 + 2x_j^4\right)$$

$$= 6 \sum_{i=1}^{4} x_i^4 + 12 \sum_{1 \le i < j \le 4} x_i^2 x_j^2$$

$$= 6 \left(x_1^2 + x_2^2 + x_3^2 + x_4^2\right)^2.$$

This proves Liouville's identity.

Let a be a nonnegative integer. By Lagrange's theorem, $a = x_1^2 + x_2^2 + x_3^2 + x_4^2$ is the sum of four squares, and so

$$6a^2 = 6 \left(x_1^2 + x_2^2 + x_3^2 + x_4^2\right)^2$$

$$= \sum_{1 \le i < j \le 4} (x_i + x_j)^4 + \sum_{1 \le i < j \le 4} (x_i - x_j)^4$$

is the sum of 12 fourth powers. Every nonnegative integer n can be written in the form $n = 6q + r$, where $q \ge 0$ and $0 \le r \le 5$. By Lagrange's theorem again, we have $q = a_1^2 + \cdots + a_4^2$, and so $6q = 6a_1^2 + \cdots + 6a_4^2$ is the sum of 48 fourth powers. Since r is the sum of 5 fourth powers, each of them either 0^4 or 1^4, it follows that n is the sum of 53 squares. This completes the proof.

The proofs of the following two results are similar.

Theorem 3.2 (Fleck)

$$\left(x_1^2 + x_2^2 + x_3^2 + x_4^2\right)^3$$

$$= \frac{1}{60} \sum_{1 \leq i < j < k \leq 4} \left(x_i \pm x_j \pm x_k\right)^6 + \frac{1}{30} \sum_{1 \leq i < j \leq 4} \left(x_i \pm x_j\right)^6 + \frac{3}{5} \sum_{1 \leq i \leq 4} x_i^6$$

is a polynomial identity, and every nonnegative integer is the sum of a bounded number of sixth powers.

Theorem 3.3 (Hurwitz)

$$\left(x_1^2 + x_2^2 + x_3^2 + x_4^2\right)^4$$

$$= \frac{1}{840} (x_1 \pm x_2 \pm x_3 \pm x_4)^8 + \frac{1}{5040} \sum_{1 \leq i < j < k < \leq 4} \left(2x_i \pm x_j \pm x_k\right)^8$$

$$+ \frac{1}{84} \sum_{1 \leq i < j \leq 4} \left(x_i \pm x_j\right)^8 + \frac{1}{840} \sum_{1 \leq i \leq 4} (2x_i)^6$$

is a polynomial identity, and every nonnegative integer is the sum of a bounded number of eighth powers.

Suppose that

$$\left(x_1^2 + \cdots + x_4^2\right)^k = \sum_{i=1}^{M} a_i \left(b_{i,1}x_1^2 + b_{i,2}x_2^2 + b_{i,3}x_3^2 + b_{i,4}x_4^2\right)^{2k} \quad (3.1)$$

for some positive integer M, integers $b_{i,j}$, and positive rational numbers a_i. Hurwitz observed that this polynomial identity and Lagrange's theorem immediately imply that if Waring's problem is true for exponent k, then it is also true for exponent $2k$. Hilbert subsequently proved the existence of polynomial identities of the form (3.1) for all positive integers k, and he applied it to show that the set of nonnegative integral kth powers is a basis of finite order for every exponent k. This was the first proof of Waring's problem. In the next section, we obtain Hilbert's polynomial identities.

3.2   Hermite polynomials and Hilbert's identity

For $n \geq 0$, we define the *Hermite polynomial* $H_n(x)$ by

$$H_n(x) = \left(\frac{-1}{2}\right)^n e^{x^2} \frac{d^n}{dx^n} \left(e^{-x^2}\right).$$

The first five Hermite polynomials are

$$H_0(x) = 1$$

$$H_1(x) = x$$

$$H_2(x) = x^2 - \frac{1}{2}$$

$$H_3(x) = x^3 - \frac{3}{2}x$$

$$H_4(x) = x^4 - 3x^2 + \frac{3}{4}.$$

Since

$$H_n'(x) = \left(\frac{-1}{2}\right)^n \frac{d}{dx}\left(e^{x^2}\frac{d^n}{dx^n}\left(e^{-x^2}\right)\right)$$

$$= \left(\frac{-1}{2}\right)^n (2x)e^{x^2}\frac{d^n}{dx^n}\left(e^{-x^2}\right) - 2\left(\frac{-1}{2}\right)^{n+1} e^{x^2}\frac{d^{n+1}}{dx^{n+1}}\left(e^{-x^2}\right)$$

$$= 2x\,H_n(x) - 2H_{n+1}(x),$$

the Hermite polynomials satisfy the recurrence relation

$$H_{n+1}(x) = x\,H_n(x) - \frac{1}{2}H_n'(x). \tag{3.2}$$

It follows that $H_n(x)$ is a monic polynomial of degree n with rational coefficients and that $H_n(x)$ is an even polynomial for n even and an odd polynomial for n odd.

Lemma 3.1 *The Hermite polynomial $H_n(x)$ has n distinct real zeros.*

Proof. This is by induction on n. The lemma is clearly true for $n = 0$ and $n = 1$, since $H_1(x) = x$. Let $n \geq 1$, and assume that the lemma is true for n. Then $H_n(x)$ has n distinct real zeros, and these zeros must be simple. Therefore, there exist real numbers

$$\beta_n < \cdots < \beta_2 < \beta_1$$

such that

$$H_n(\beta_j) = 0$$

and

$$H_n'(\beta_j) \neq 0$$

for $j = 1, \ldots, n$. Since $H_n(x)$ is a monic polynomial of degree n, it follows that

$$\lim_{x\to\infty} H_n(x) = \infty,$$

and so

$$H_n'(\beta_1) > 0.$$

Since the $n - 1$ distinct real zeros of the derivative $H_n'(x)$ are intertwined with the n zeros of $H_n(x)$, it follows that

$$(-1)^{j+1} H_n'(\beta_j) > 0$$

for $j = 1, \ldots, n$. The recurrence relation (3.2) implies that

$$H_{n+1}(\beta_j) = \beta_j H_n(\beta_j) - \frac{1}{2} H_n'(\beta_j) = -\frac{1}{2} H_n'(\beta_j),$$

and so

$$(-1)^j H_{n+1}(\beta_j) = \frac{(-1)^{j+1}}{2} H_n'(\beta_j) > 0$$

for $j = 1, \ldots, n$. Therefore, for $j = 2, \ldots, n$, $H_{n+1}(x)$ has a zero β_j^* in each open interval (β_j, β_{j-1}). Since $\lim_{x \to \infty} H_{n+1}(x) = \infty$ and $H_{n+1}(\beta_1) < 0$, it follows that $H_{n+1}(x)$ has a zero $\beta_1^* > \beta_1$. If n is even, then $H_{n+1}(\beta_n) > 0$. Since $n + 1$ is odd, $H_{n+1}(x)$ is a polynomial of odd degree, and so $\lim_{x \to -\infty} H_{n+1}(x) = -\infty$. It follows that $H_{n+1}(x)$ has a zero $\beta_{n+1}^* < \beta_n$. Similarly, if n is odd, $H_{n+1}(\beta_n) < 0$ and the even polynomial $H_{n+1}(x)$ has a zero $\beta_{n+1}^* < \beta_n$. Thus, $H_{n+1}(x)$ has $n + 1$ distinct real zeros. This completes the proof.

Lemma 3.2 Let $n \geq 1$ and $f(x)$ be a polynomial of degree at most $n - 1$. Then

$$\int_{-\infty}^{\infty} e^{-x^2} H_n(x) f(x) dx = 0.$$

Proof. This is by induction on n.) If $n = 1$, then $H_n(x) = x$ and $f(x)$ is constant, say, $f(x) = a_0$, so

$$\int_{-\infty}^{\infty} e^{-x^2} H_n(x) f(x) dx = a_0 \int_{-\infty}^{\infty} e^{-x^2} x \, dx = 0.$$

Now assume that the lemma is true for n, and let $f(x)$ be a polynomial of degree at most n. Then $f'(x)$ is a polynomial of degree at most $n - 1$. Integrating by parts, we obtain

$$\int_{-\infty}^{\infty} e^{-x^2} H_{n+1}(x) f(x) dx = \left(\frac{-1}{2}\right)^{n+1} \int_{-\infty}^{\infty} \frac{d^{n+1}}{dx^{n+1}} \left(e^{-x^2}\right) f(x) dx$$

$$= \left(\frac{-1}{2}\right)^{n+1} \int_{-\infty}^{\infty} \frac{d^n}{dx^n} \left(e^{-x^2}\right) f'(x) dx$$

$$= \left(\frac{-1}{2}\right) \int_{-\infty}^{\infty} e^{-x^2} H_n(x) f'(x) dx$$

$$= 0.$$

This completes the proof.

Lemma 3.3 For $n \geq 0$,

$$c_n = \frac{1}{\sqrt{\pi}} \int_{-\infty}^{\infty} e^{-x^2} x^n dx = \begin{cases} \frac{n!}{2^n (n/2)!} & \text{if } n \text{ is even} \\ 0 & \text{if } n \text{ is odd.} \end{cases} \tag{3.3}$$

Proof. This is by induction on n. For $n = 0$, we have

$$\int_{-\infty}^{\infty} e^{-x^2} dx = \sqrt{\pi}$$

and so $c_0 = 1$. For $n = 1$, the function $e^{-x^2}x$ is odd, and so

$$\int_{-\infty}^{\infty} e^{-x^2} x\, dx = 0$$

and $c_1 = 0$. Now let $n \geq 2$, and assume that the lemma holds for $n - 2$. Integrating by parts, we obtain

$$c_n = \frac{1}{\sqrt{\pi}} \int_{-\infty}^{\infty} e^{-x^2} x^n dx$$

$$= \left(\frac{n-1}{2}\right) \frac{1}{\sqrt{\pi}} \int_{-\infty}^{\infty} e^{-x^2} x^{n-2} dx$$

$$= \left(\frac{n-1}{2}\right) c_{n-2}.$$

If n is odd, then $c_{n-2} = 0$ and so $c_n = 0$. If n is even,

$$c_n = \left(\frac{n-1}{2}\right) c_{n-2}$$

$$= \left(\frac{n-1}{2}\right) \frac{(n-2)!}{2^{n-2}((n-2)/2)!}$$

$$= \frac{n!}{2^n (n/2)!}.$$

This completes the proof.

Lemma 3.4 *Let $n \geq 1$, let $\beta_1, \ldots, \beta_n$ be n distinct real numbers, and let $c_0, c_1, \ldots, c_{n-1}$ be the numbers defined by (3.3). The system of linear equations*

$$\sum_{j=1}^{n} \beta_j^k x_j = c_k \qquad \text{for } k = 0, 1, \ldots, n-1 \qquad (3.4)$$

has a unique solution $\rho_1, \ldots, \rho_n$. If $r(x)$ is a polynomial of degree at most $n - 1$, then

$$\sum_{j=1}^{n} r(\beta_j)\rho_j = \frac{1}{\sqrt{\pi}} \int_{-\infty}^{\infty} e^{-x^2} r(x) dx.$$

Proof. The existence and uniqueness of the solution $\rho_1, \ldots, \rho_n$ follows immediately from the fact that the determinant of the system of linear equations

$$
\begin{array}{ccccccc}
x_1 & + & x_2 & + \cdots + & x_n & = & c_0 \\
\beta_1 x_1 & + & \beta_2 x_2 & + \cdots + & \beta_n x_n & = & c_1 \\
\beta_1^2 x_1 & + & \beta_2^2 x_2 & + \cdots + & \beta_n^2 x_n & = & c_2 \\
& & & \vdots & & & \\
\beta_1^{n-1} x_1 & + & \beta_2^{n-1} x_2 & + \cdots + & \beta_n^{n-1} x_n & = & c_{n-1}
\end{array}
$$

is the Vandermonde determinant

$$
\begin{vmatrix}
1 & 1 & \cdots & 1 \\
\beta_1 & \beta_2 & \cdots & \beta_n \\
\beta_1^2 & \beta_2^2 & \cdots & \beta_n^2 \\
\vdots & \vdots & & \vdots \\
\beta_1^{n-1} & \beta_2^{n-1} & \cdots & \beta_n^{n-1}
\end{vmatrix}
= \prod_{1 \le i < j \le n} (\beta_j - \beta_i) \neq 0.
$$

Let $r(x) = \sum_{k=0}^{n-1} a_k x^k$. Then

$$
\sum_{j=1}^{n} r(\beta_j)\rho_j = \sum_{j=1}^{n} \sum_{k=0}^{n-1} a_k \beta_j^k \rho_j
$$

$$
= \sum_{k=0}^{n-1} a_k \sum_{j=1}^{n} \beta_j^k \rho_j
$$

$$
= \sum_{k=0}^{n-1} a_k c_k
$$

$$
= \frac{1}{\sqrt{\pi}} \sum_{k=0}^{n-1} a_k \int_{-\infty}^{\infty} e^{-x^2} x^k dx
$$

$$
= \frac{1}{\sqrt{\pi}} \int_{-\infty}^{\infty} e^{-x^2} r(x) dx.
$$

This completes the proof.

Lemma 3.5 *Let* $n \ge 1$, *let* $\beta_1, \ldots, \beta_n$ *be the* n *distinct real roots of the Hermite polynomial* $H_n(x)$, *and let* $\rho_1, \ldots, \rho_n$ *be the solution of the system of linear equations (3.4). Let* $f(x)$ *be a polynomial of degree at most* $2n - 1$. *Then*

$$
\sum_{j=1}^{n} f(\beta_j)\rho_j = \frac{1}{\sqrt{\pi}} \int_{-\infty}^{\infty} e^{-x^2} f(x) dx.
$$

Proof. By the division algorithm for polynomials, there exist polynomials $q(x)$ and $r(x)$ of degree at most $n - 1$ such that

$$
f(x) = H_n(x)q(x) + r(x).
$$

Since $H_n(\beta_j) = 0$ for $j = 1, \ldots, n$, we have

$$
f(\beta_j) = H_n(\beta_j)q(\beta_j) + r(\beta_j) = r(\beta_j),
$$

and so, by Lemma 3.4 and Lemma 3.2,

$$
\sum_{j=1}^{n} f(\beta_j)\rho_j = \sum_{j=1}^{n} r(\beta_j)\rho_j
$$

$$= \frac{1}{\sqrt{\pi}} \int_{-\infty}^{\infty} e^{-x^2} r(x)dx$$

$$= \frac{1}{\sqrt{\pi}} \int_{-\infty}^{\infty} e^{-x^2} H_n(x)q(x)dx + \frac{1}{\sqrt{\pi}} \int_{-\infty}^{\infty} e^{-x^2} r(x)dx$$

$$= \frac{1}{\sqrt{\pi}} \int_{-\infty}^{\infty} e^{-x^2} f(x)dx.$$

This completes the proof.

Lemma 3.6 *Let $n \geq 1$, let $\beta_1, \ldots, \beta_n$ be the n distinct real roots of the Hermite polynomial $H_n(x)$, and let $\rho_1, \ldots, \rho_n$ be the solution of the linear system (3.4). Then*

$$\rho_i > 0 \qquad for \ i = 1, \ldots, n.$$

Proof. Since

$$H_n(x) = \prod_{j=1}^{n} (x - \beta_j),$$

it follows that, for $i = 1, \ldots, n$,

$$f_i(x) = \left(\frac{H_n(x)}{x - \beta_i} \right)^2 = \prod_{\substack{j=1 \\ j \neq i}}^{n} (x - \beta_j)^2$$

is a monic polynomial of degree $2n - 2$ such that $f_i(x) \geq 0$ for all x. Therefore,

$$\frac{1}{\sqrt{\pi}} \int_{-\infty}^{\infty} e^{-x^2} f_i(x)dx > 0.$$

Since $f_i(\beta_i) > 0$ and $f_i(\beta_j) = 0$ for $j \neq i$, we have, by Lemma 3.5,

$$f_i(\beta_i)\rho_i = \sum_{j=1}^{n} f_i(\beta_j)\rho_j$$

$$= \frac{1}{\sqrt{\pi}} \int_{-\infty}^{\infty} e^{-x^2} f_i(x)dx$$

$$> 0.$$

This completes the proof.

Lemma 3.7 *Let $n \geq 1$, and let $c_0, c_1, \ldots, c_{n-1}$ be the rational numbers defined by (3.3). There exist pairwise distinct rational numbers $\beta_1^*, \ldots, \beta_n^*$ and positive rational numbers $\rho_1^*, \ldots, \rho_n^*$ such that*

$$\sum_{j=1}^{n} (\beta_j^*)^k \rho_j^* = c_k \qquad for \ k = 0, 1, \ldots, n - 1.$$

Proof. By Lemma 3.4, for any set of n pairwise distinct real numbers $\beta_1, \ldots, \beta_n$, the system of n linear equations in n unknowns

$$\sum_{j=1}^{n} \beta_j^k x_j = c_k \qquad \text{for } k = 0, 1, \ldots, n-1$$

has a unique solution $(\rho_1, \ldots, \rho_n)$. Let $\mathcal{R}$ be the open subset of $\mathbf{R}^n$ consisting of all points $(\beta_1, \ldots, \beta_n)$ such that $\beta_i \neq \beta_j$ for $i \neq j$, and let $\Phi : \mathcal{R} \to \mathbf{R}^n$ be the function that sends $(\beta_1, \ldots, \beta_n)$ to $(\rho_1, \ldots, \rho_n)$. By Cramer's rule for solving linear equations, we can express each ρ_j as a rational function of $\beta_1, \ldots, \beta_n$, and so the function

$$\Phi(\beta_1, \ldots, \beta_n) = (\rho_1, \ldots, \rho_n)$$

is continuous. Let $\mathbf{R}_+^n$ be the open subset of $\mathbf{R}^n$ consisting of all points $(x_1, \ldots, x_n)$ such that $x_i > 0$ for $i = 1, \ldots, n$. By Lemma 3.6, if $\beta_1, \ldots, \beta_n$ are the n zeros of $H_n(x)$, then $(\beta_1, \ldots, \beta_n) \in \mathcal{R}$ and

$$\Phi(\beta_1, \ldots, \beta_n) = (\rho_1, \ldots, \rho_n) \in \mathbf{R}_+^n.$$

Since $\mathbf{R}_+^n$ is an open subset of $\mathbf{R}^n$, it follows that $\Phi^{-1}(\mathbf{R}_+^n)$ is an open neighborhood of $(\beta_1, \ldots, \beta_n)$ in $\mathcal{R}$. Since the points with rational coordinates are dense in $\mathcal{R}$, it follows that this neighborhood contains a rational point $(\beta_1^*, \ldots, \beta_n^*)$. Let

$$(\rho_1^*, \ldots, \rho_n^*) = \Phi(\beta_1^*, \ldots, \beta_n^*) \in \mathbf{R}_+^n.$$

Since each number ρ_i^* can be expressed as a rational function with rational coefficients of the rational numbers $\beta_1^*, \ldots, \beta_n^*$, it follows that each of the positive numbers ρ_i^* is rational. This completes the proof.

Lemma 3.8 *Let $n \geq 1$, let $c_0, c_1, \ldots, c_{n-1}$ be the numbers defined by (3.3), let $\beta_1, \ldots, \beta_n$ be n distinct real numbers, and let $\rho_1, \ldots, \rho_n$ be the solution of the linear system (3.4). For every positive integer r and for $m = 1, 2, \ldots, n-1$,*

$$c_m \left(x_1^2 + \cdots + x_r^2 \right)^{m/2} . = \sum_{j_1=1}^{n} \cdots \sum_{j_r=1}^{n} \rho_{j_1} \cdots \rho_{j_r} \left(\beta_{j_1} x_1 + \cdots + \beta_{j_r} x_r \right)^m$$

is a polynomial identity.

Proof. The proof is an exercise in algebraic manipulation and the multinomial theorem. We have

$$\sum_{j_1=1}^{n} \cdots \sum_{j_r=1}^{n} \rho_{j_1} \cdots \rho_{j_r} \left(\beta_{j_1} x_1 + \cdots + \beta_{j_r} x_r \right)^m$$

$$= \sum_{j_1=1}^{n} \cdots \sum_{j_r=1}^{n} \rho_{j_1} \cdots \rho_{j_r} \sum_{\substack{\mu_1 + \cdots + \mu_r = m \\ \mu_i \geq 0}} \frac{m!}{\mu_1! \cdots \mu_r!} (\beta_{j_1} x_1)^{\mu_1} \cdots (\beta_{j_r} x_r)^{\mu_r}$$

$$= m! \sum_{j_1=1}^{n} \cdots \sum_{j_r=1}^{n} \sum_{\substack{\mu_1+\cdots+\mu_r=m \\ \mu_i \geq 0}} \frac{x_1^{\mu_1}}{\mu_1!} \left(\beta_{j_1}^{\mu_1} \rho_{j_1}\right) \cdots \frac{x_r^{\mu_r}}{\mu_r!} \left(\beta_{j_r}^{\mu_r} \rho_{j_r}\right)$$

$$= m! \sum_{\substack{\mu_1+\cdots+\mu_r=m \\ \mu_i \geq 0}} \sum_{j_1=1}^{n} \cdots \sum_{j_r=1}^{n} \prod_{i=1}^{r} \frac{x_i^{\mu_i}}{\mu_i!} \left(\beta_{j_i}^{\mu_i} \rho_{j_i}\right)$$

$$= m! \sum_{\substack{\mu_1+\cdots+\mu_r=m \\ \mu_i \geq 0}} \prod_{i=1}^{r} \left(\frac{x_i^{\mu_i}}{\mu_i!} \sum_{j=1}^{n} \beta_j^{\mu_i} \rho_j\right)$$

$$= m! \sum_{\substack{\mu_1+\cdots+\mu_r=m \\ \mu_i \geq 0}} \prod_{i=1}^{r} \frac{c_{\mu_i} x_i^{\mu_i}}{\mu_i!}.$$

By Lemma 3.3, $c_m = 0$ if m is odd. If m is odd and $\mu_1 + \cdots + \mu_r = m$, then μ_i must be odd for some i, and so

$$\sum_{j_1=1}^{n} \cdots \sum_{j_r=1}^{n} \rho_{j_1} \cdots \rho_{j_r} \left(\beta_{j_1} x_1 + \cdots + \beta_{j_r} x_r\right)^m = 0.$$

This proves the lemma for odd m. If m is even, then we need only consider partitions of m into even parts $\mu_i = 2\nu_i$. Inserting the expressions for the numbers c_n from (3.3), we obtain

$$\sum_{j_1=1}^{n} \cdots \sum_{j_r=1}^{n} \rho_{j_1} \cdots \rho_{j_r} \left(\beta_{j_1} x_1 + \cdots + \beta_{j_r} x_r\right)^m$$

$$= m! \sum_{\substack{2\nu_1+\cdots+2\nu_r=m \\ \nu_i \geq 0}} \prod_{i=1}^{r} \frac{c_{2\nu_i} x_i^{2\nu_i}}{(2\nu_i)!}$$

$$= m! \sum_{\substack{\nu_1+\cdots+\nu_r=m/2 \\ \nu_i \geq 0}} \prod_{i=1}^{r} \frac{(2\nu_i)!}{2^{2\nu_i} \nu_i!} \frac{x_i^{2\nu_i}}{(2\nu_i)!}$$

$$= \frac{m!}{2^m} \sum_{\substack{\nu_1+\cdots+\nu_r=m/2 \\ \nu_i \geq 0}} \prod_{i=1}^{r} \frac{x_i^{2\nu_i}}{\nu_i!}$$

$$= \frac{m!}{2^m (m/2)!} (m/2)! \sum_{\substack{\nu_1+\cdots+\nu_r=m/2 \\ \nu_i \geq 0}} \prod_{i=1}^{r} \frac{(x_i^2)^{\nu_i}}{\nu_i!}$$

$$= c_m \sum_{\substack{\nu_1+\cdots+\nu_r=m/2 \\ \nu_i \geq 0}} \frac{(m/2)!}{\nu_1! \cdots \nu_r!} \left(x_1^2\right)^{\nu_1} \cdots \left(x_r^2\right)^{\nu_r}$$

$$= c_m \left(x_1^2 + \cdots x_r^2\right)^{m/2}.$$

This proves the polynomial identity.

Theorem 3.4 (Hilbert's identity) *For every $k \geq 1$ and $r \geq 1$ there exist an integer M and positive rational numbers a_i and integers $b_{i,j}$ for $i = 1, \ldots, M$ and $j = 1, \ldots, r$ such that*

$$(x_1^2 + \cdots + x_r^2)^k = \sum_{i=1}^{M} a_i \left(b_{i,1}x_1 + \cdots + b_{i,r}x_r\right)^{2k}. \qquad (3.5)$$

Proof. Choose $n > 2k$, and let $\beta_1^*, \ldots, \beta_n^*, \rho_1^*, \ldots, \rho_n^*$ be the rational numbers constructed in Lemma 3.7. Then $\beta_1^*, \ldots, \beta_n^*$ are pairwise distinct and $\rho_1^*, \ldots, \rho_n^*$ are positive. We use these numbers in Lemma 3.8 with $m = 2k$ and obtain the polynomial identity

$$c_{2k} \left(x_1^2 + \cdots + x_r^2\right)^k = \sum_{j_1=1}^{n} \cdots \sum_{j_r=1}^{n} \rho_{j_1}^* \cdots \rho_{j_r}^* \left(\beta_{j_1}^* x_1 + \cdots + \beta_{j_r}^* x_r\right)^{2k}.$$

Let q be a common denominator of the n fractions $\beta_1^*, \ldots, \beta_n^*$. Then $q\beta_j^*$ is an integer for all j, and

$$\left(x_1^2 + \cdots + x_r^2\right)^k = \sum_{j_1=1}^{n} \cdots \sum_{j_r=1}^{n} \frac{\rho_{j_1}^* \cdots \rho_{j_r}^*}{c_{2k}q^{2k}} \left(q\beta_{j_1}^* x_1 + \cdots + q\beta_{j_r}^* x_r\right)^{2k}$$

is a polynomial identity of Hilbert type. This completes the proof.

Lemma 3.9 *Let $k \geq 1$. If there exist positive rational numbers $a_1, \ldots, a_M$ such that every sufficiently large integer n can be written in the form*

$$n = \sum_{i=1}^{M} a_i y_i^k, \qquad (3.6)$$

where $x_1, \ldots, x_M$ are nonnegative integers, then Waring's problem is true for exponent k.

Proof. Choose n_0 such that every integer $n \geq n_0$ can be represented in the form (3.6). Let q be the least common denominator of the fractions $a_1, \ldots, a_M$. Then $qa_i \in \mathbf{Z}$ for $i = 1, \ldots, M$, and qn is a sum of $\sum_{i=1}^{M} qa_i$ nonnegative kth powers for every $n \geq n_0$. Since every integer $N \geq qn_0$ can be written in the form $N = qn + r$, where $n \geq n_0$ and $0 \leq r \leq q - 1$, it follows that N can be written as the sum of $\sum_{i=1}^{M} qa_i + q - 1$ nonnegative kth powers. Clearly, every nonnegative integer $N < qn_0$ can be written as the sum of a bounded number of kth powers, and so Waring's problem holds for k. This completes the proof.

The following notation is due to Stridsberg: Let $\sum_{i=1}^{M} a_i x_i^k$ be a fixed diagonal form of degree k with positive rational coefficients $a_1, \ldots, a_M$. We write $n = \sum(k)$ if there exist nonnegative integers $x_1, \ldots, x_M$ such that

$$n = \sum_{i=1}^{M} a_i x_i^k. \qquad (3.7)$$

We let $\sum(k)$ denote any integer of the form (3.7). Then $\sum(k) + \sum(k) = \sum(k)$ and $\sum(2k) = \sum(k)$. Lemma 3.9 can be restated as follows: If $n = \sum(k)$ for every sufficiently large nonnegative integer n, then Waring's problem is true for exponent k.

Theorem 3.5 *If Waring's problem holds for k, then Waring's problem holds for $2k$.*

Proof. We use Hilbert's identity (3.5) for k with $r = 4$:

$$(x_1^2 + \cdots + x_4^2)^k = \sum_{i=1}^{M} a_i \left(b_{i,1}x_1 + \cdots + b_{i,4}x_4\right)^{2k}.$$

Let y be a nonnegative integer. By Lagrange's theorem, there exist nonnegative integers x_1, x_2, x_3, x_4 such that

$$y = x_1^2 + x_2^2 + x_3^2 + x_4^2,$$

and so

$$y^k = \sum_{i=1}^{M} a_i z_i^{2k}, \tag{3.8}$$

where

$$z_i = b_{i,1}x_1 + \cdots + b_{i,4}x_4$$

is a nonnegative integer. This means that

$$y^k = \sum(2k)$$

for every nonnegative integer y. If Waring's problem is true for k, then every nonnegative integer is the sum of a bounded number of kth powers, and so every nonnegative integer is the sum of a bounded number of numbers of the form $\sum(2k)$. By Lemma 3.9, Waring's problem holds for exponent $2k$. This completes the proof.

3.3   A proof by induction

We shall use Hilbert's identity to obtain Waring's problem for all exponents $k \geq 2$. The proof is by induction on k. The starting point is Lagrange's theorem that every nonnegative integer is the sum of four squares. This is the case where $k = 2$. We shall prove that if $k > 2$ and Waring's problem is true for every exponent less than k, then it is also true for k.

Lemma 3.10 *Let $k \geq 2$ and $0 \leq \ell \leq k$. There exist positive integers $B_{0,\ell}$, $B_{1,\ell}$, ..., $B_{\ell-1,\ell}$ depending only on k and ℓ such that*

$$x^{2\ell} T^{k-\ell} + \sum_{i=0}^{\ell-1} B_{i,\ell} x^{2i} T^{k-i} = \sum(2k)$$

for all integers x and T satisfying

$$x^2 \le T.$$

Proof. We begin with Hilbert's identity for exponent $k + \ell$ with $r = 5$:

$$(x_1^2 + \cdots + x_5^2)^{k+\ell} = \sum_{i=1}^{M_\ell} a_i \left(b_{i,1}x_1 + \cdots + b_{i,5}x_5\right)^{2k+2\ell},$$

where the integers M_ℓ and $b_{i,j}$ and the positive rational numbers a_i depend only on k and ℓ. Let U be a nonnegative integer. By Lagrange's theorem, we can write

$$U = x_1^2 + x_2^2 + x_3^2 + x_4^2$$

for nonnegative integers x_1, x_2, x_3, x_4. Let $x_5 = x$. We obtain the polynomial identity

$$(x^2 + U)^{k+\ell} = \sum_{i=1}^{M_\ell} a_i (b_i x + c_i)^{2k+2\ell}, \tag{3.9}$$

where the numbers M_ℓ, a_i, and $b_i = b_{i,5}$ depend only on k and ℓ, and the integers $c_i = b_{i,1}x_1 + \cdots + b_{i,4}x_4$ depend on k, ℓ, and U. Note that $2\ell \le k + \ell$ since $\ell \le k$. Differentiating the polynomial on the left side of (3.9) 2ℓ times, we obtain (see Exercise 6)

$$\frac{d^{2\ell}}{dx^{2\ell}} \left((x^2 + U)^{k+\ell}\right) = \sum_{i=0}^{\ell} A_{i,\ell} x^{2i} (x^2 + U)^{k-i},$$

where the $A_{i,\ell}$ are positive integers that depend only on k and ℓ. Differentiating the polynomial on the right side of (3.9) 2ℓ times, we obtain

$$\frac{d^{2\ell}}{dx^{2\ell}} \left(\sum_{i=1}^{M_\ell} a_i (b_i x + c_i)^{2k+2\ell}\right)$$

$$= \sum_{i=1}^{M_\ell} (2k+1)(2k+2)\cdots(2k+2\ell)b_i^{2\ell}a_i(b_i x + c_i)^{2k}$$

$$= \sum_{i=1}^{M_\ell} a_i'(b_i x + c_i)^{2k}$$

$$= \sum_{i=1}^{M_\ell} a_i' y_i^{2k},$$

where $y_i = |b_i x + c_i|$ is a nonnegative integer and

$$a_i' = (2k+1)(2k+2)\cdots(2k+2\ell)b_i^{2\ell}a_i$$

is a nonnegative rational number depending only on k and ℓ. It follows that, if x and U are integers and $U \ge 0$, then there exist nonnegative integers $y_1, \ldots, y_{M_\ell}$ such that

$$\sum_{i=0}^{\ell} A_{i,\ell} x^{2i} (x^2 + U)^{k-i} = \sum_{i=1}^{M_\ell} a_i' y_i^{2k}.$$

Let x and T be nonnegative integers such that $x^2 \leq T$. Since $A_{\ell,\ell}$ is a positive integer, it follows that $x^2 \leq A_{\ell,\ell}T$, and so

$$U = A_{\ell,\ell}T - x^2$$

is a nonnegative integer. With this choice of U, we have

$$\sum_{i=0}^{\ell} A_{i,\ell}x^{2i}(x^2+U)^{k-i} = \sum_{i=0}^{\ell} A_{i,\ell}x^{2i}(A_{\ell,\ell}T)^{k-i}$$

$$= \sum_{i=0}^{\ell} A_{i,\ell}A_{\ell,\ell}^{k-i}x^{2i}T^{k-i}$$

$$= A_{\ell,\ell}^{k-\ell+1} \sum_{i=0}^{\ell} A_{i,\ell}A_{\ell,\ell}^{\ell-i-1}x^{2i}T^{k-i}$$

$$= A_{\ell,\ell}^{k-\ell+1} \sum_{i=0}^{\ell} B_{i,\ell}x^{2i}T^{k-i},$$

where $B_{\ell,\ell} = 1$ and

$$B_{i,\ell} = A_{i,\ell}A_{\ell,\ell}^{\ell-i-1}$$

is a positive integer for $i = 0, \ldots, \ell - 1$. Let

$$a_i' = \frac{a_i'}{A_{\ell,\ell}^{k-\ell+1}}.$$

Then

$$x^{2\ell}T^{k-\ell} + \sum_{i=0}^{\ell-1} B_{\ell,\ell}x^{2i}T^{k-i} = \sum_{i=1}^{M_\ell} a_i'y_i^{2k} = \sum(2k).$$

This completes the proof.

Theorem 3.6 (Hilbert–Waring) *The set of nonnegative kth powers is a basis of finite order for every positive integer k.*

Proof. This is by induction on k. The case $k = 1$ is clear, and the case $k = 2$ is Theorem 1.1 (Lagrange's theorem). Let $k \geq 3$, and suppose that the set of ℓth powers is a basis of finite order for every $\ell < k$. By Theorem 3.5, the set of (2ℓ)-th powers is a basis of finite order for $\ell = 1, 2, \ldots, k - 1$. Therefore, there exists an integer r such that, for every nonnegative integer n and for $\ell = 1, \ldots, k - 1$, the equation

$$n = x_1^{2\ell} + \cdots + x_r^{2\ell}$$

is solvable in nonnegative integers $x_{1,\ell}, \ldots, x_{r,\ell}$. (For example, we could let $r = \max\{g(2\ell) : \ell = 1, 2, \ldots, k - 1\}$.)

Let $T \geq 2$. Choose integers $C_1, \ldots, C_{k-1}$ such that

$$0 \leq C_\ell < T \quad \text{for } \ell = 1, \ldots, k - 1.$$

There exist nonnegative integers $x_{j,\ell}$ for $j = 1, \ldots, r$ and $\ell = 1, \ldots, k - 1$ such that

$$x_1^{2\ell} + \cdots + x_r^{2\ell} = C_{k-\ell}. \tag{3.10}$$

Then

$$x_{j,\ell}^2 \leq \sum_{j=1}^{r} x_{j,\ell}^{2i} \leq C_{k-\ell} < T$$

for $j = 1, \ldots, r$, $\ell = 1, \ldots, k - 1$, and $i = 1, \ldots, \ell$. By Lemma 3.10, there exist positive integers $B_{i,\ell}$ depending only on k and ℓ such that

$$x_{j,\ell}^{2\ell} T^{k-\ell} + \sum_{i=0}^{\ell-1} B_{i,\ell} x_{j,\ell}^{2i} T^{k-i} = \sum(2k) = \sum(k). \tag{3.11}$$

Summing (3.11) for $j = 1, \ldots, r$ and using (3.10), we obtain

$$C_{k-\ell} T^{k-\ell} + \sum_{i=0}^{\ell-1} B_{i,\ell} T^{k-i} \sum_{j=1}^{r} x_{j,\ell}^{2i}$$

$$= C_{k-\ell} T^{k-\ell} + T^{k-\ell+1} \sum_{i=0}^{\ell-1} B_{i,\ell} T^{\ell-1-i} \sum_{j=1}^{r} x_{j,\ell}^{2i}$$

$$= C_{k-\ell} T^{k-\ell} + D_{k-\ell+1} T^{k-\ell+1}$$

$$= \sum(k),$$

where

$$D_{k-\ell+1} = \sum_{i=0}^{\ell-1} B_{i,\ell} T^{\ell-1-i} \sum_{j=1}^{r} x_{j,\ell}^{2i}$$

for $\ell = 1, \ldots, k - 1$. The integer $D_{k-\ell+1}$ is completely determined by k, ℓ, T, and $C_{k-\ell}$ and is independent of C_{k-i} for $i \neq \ell$. Let

$$B^* = \max\{B_{i,\ell} : \ell = 1, \ldots, k - 1 \text{ and } i = 0, 1, \ldots, \ell - 1\}.$$

Then

$$0 \leq C_{k-\ell} T^{k-\ell} + D_{k-\ell+1} T^{k-\ell+1}$$

$$= C_{k-\ell} T^{k-\ell} + \sum_{i=0}^{\ell-1} B_{i,\ell} T^{k-i} \sum_{j=1}^{r} x_{j,\ell}^{2i}$$

$$< B^* \left(T^{k-\ell+1} + rT^k + \sum_{i=1}^{\ell-1} T^{k-i+1} \right)$$

$$= B^* \left(rT^k + T^{k-\ell+1} \sum_{i=0}^{\ell-1} T^i \right)$$

$$< B^* \left(rT^k + \frac{T^{k+1}}{T - 1} \right)$$

$$\leq (r + 2) B^* T^k,$$

since $T/(T - 1) \leq 2$ for $T \geq 2$. Let

$$C_k = D_1 = 0.$$

Then

$$\sum_{\ell=1}^{k-1} \left(C_{k-\ell} T^{k-\ell} + D_{k-\ell+1} T^{k-\ell+1} \right) = \sum_{\ell=1}^{k} (C_\ell + D_\ell) T^\ell = \sum(k)$$

and

$$0 \leq \sum_{\ell=1}^{k} (C_\ell + D_\ell) T^\ell < (k - 1)(r + 2) B^* T^k = E^* T^k,$$

where the integer

$$E^* = (k - 1)(r + 2) B^*$$

is determined by k and is independent of T. If we choose

$$T \geq E^*,$$

then

$$0 \leq \sum_{\ell=1}^{k} (C_\ell + D_\ell) T^\ell < E^* T^k < T^{k+1},$$

and so the expansion of $\sum_{\ell=1}^{k} (C_\ell + D_\ell) T^\ell$ to base T is of the form

$$\sum_{\ell=1}^{k} (C_\ell + D_\ell) T^\ell = E_1 T + \cdots + E_{k-1} T^{k-1} + E_k T^k, \tag{3.12}$$

where

$$0 \leq E_i < T \quad \text{for } i = 1, \ldots, k - 1$$

and

$$0 \leq E_k < E^*.$$

In this way, every choice of a $(k - 1)$-tuple $(C_1, \ldots, C_{k-1})$ of integers in $\{0, 1, \ldots, T - 1\}$ determines another $(k - 1)$-tuple $(E_1, \ldots, E_{k-1})$ of integers in $\{0, 1, \ldots, T - 1\}$. We shall prove that this map of $(k - 1)$-tuples is bijective.

It suffices to prove it is surjective. Let $(E_1, \ldots, E_{k-1})$ be a $(k - 1)$-tuple of integers in $\{0, 1, \ldots, T - 1\}$. There is a simple algorithm that generates integers $C_1, C_2, \ldots, C_{k-1} \in \{0, 1, \ldots, T - 1\}$ such that (3.12) is satisfied for some nonnegative integer $E_k < E^*$. Let $C_1 = E_1$ and $I_2 = 0$. Since $D_1 = 0$, we have

$$(C_1 + D_1)T = E_1 T + I_2 T^2.$$

The integer C_1 determines the integer D_2. Choose $C_2 \in \{0, 1, \ldots, T - 1\}$ such that

$$C_2 + D_2 + I_2 \equiv E_2 \pmod{T}.$$

Then
$$C_2 + D_2 + I_2 = E_2 + I_3 T$$

for some integer I_3, and

$$\sum_{\ell=1}^{2}(C_\ell + D_\ell)T^\ell = \sum_{\ell=1}^{2} E_\ell T^\ell + I_3 T^3.$$

The integer C_2 determines D_3. Choose $C_3 \in \{0, 1, \ldots, T-1\}$ such that

$$C_3 + D_3 + I_3 \equiv E_3 \pmod{T}.$$

Then
$$C_3 + D_3 + I_3 = E_3 + I_4 T$$

for some integer I_4, and

$$\sum_{\ell=1}^{3}(C_\ell + D_\ell)T^\ell = \sum_{\ell=1}^{3} E_\ell T^\ell + I_4 T^4.$$

Let $2 \leq j \leq k - 1$, and suppose that we have constructed integers I_j and

$$C_1, \ldots, C_{j-1} \in \{0, 1, \ldots, T-1\}$$

such that

$$\sum_{\ell=1}^{j-1}(C_\ell + D_\ell)T^\ell = \sum_{\ell=1}^{j-1} E_\ell T^\ell + I_j T^j.$$

There exists a unique integer $C_j \in \{0, 1, \ldots, T-1\}$ such that

$$C_j + D_j + I_j \equiv E_j \pmod{T}.$$

Then
$$C_j + D_j + I_j = E_j + I_{j+1} T$$

for some integer I_{j+1}, and

$$\sum_{\ell=1}^{j}(C_\ell + D_\ell)T^\ell = \sum_{\ell=1}^{j} E_\ell T^\ell + I_{j+1} T^{j+1}.$$

It follows by induction that this procedure generates a unique sequence of integers $C_1, C_2, \ldots, C_{k-1} \in \{0, 1, \ldots, T-1\}$ such that

$$\sum_{\ell=1}^{k-1}(C_\ell + D_\ell)T^\ell = \sum_{\ell=1}^{k-1} E_\ell T^\ell + I_k T^k.$$

Since $C_k = 0$ and C_{k-1} determines D_k, we have

$$0 \leq \sum_{\ell=1}^{k}(C_\ell + D_\ell)T^\ell = \sum_{\ell=1}^{k-1} E_\ell T^\ell + (D_k + I_k)T^k = \sum_{\ell=1}^{k} E_\ell T^\ell < E^* T^k,$$

where $D_k + I_k = E_k$. Since

$$0 \leq \sum_{\ell=1}^{k-1} E_\ell T^\ell < T^k,$$

it follows that

$$0 \leq E_k < E^*$$

and

$$\sum_{\ell=1}^{k-1} E_\ell T^\ell + E^* T^k < (1 + E^*) T^k \leq 2E^* T^k. \tag{3.13}$$

Recall that

$$\sum_{\ell=1}^{k} E_\ell T^\ell = \sum_{\ell=1}^{k} (C_\ell + D_\ell) T^\ell = \sum(k).$$

Since E^* depends only on k and not on T, it follows that

$$(E^* - E_k) T^k = \sum(k),$$

and so

$$\sum_{\ell=1}^{k-1} E_\ell T^\ell + E^* T^k = \sum(k) \tag{3.14}$$

for every $(k-1)$-tuple $(E_1, \ldots, E_{k-1})$ of integers $E_\ell \in \{0, 1, \ldots, T-1\}$. Choose the integer $T_0 > 5E^*$ so that

$$4(T+1)^k \leq 5T^k \qquad \text{for all } T \geq T_0.$$

We shall prove that if $T \geq T_0$ and if $(F_0, F_1, \ldots, F_{k-1})$ is any k-tuple of integers in $\{0, 1, \ldots, T-1\}$, then

$$F_0 + F_1 T + \cdots + F_{k-1} T^{k-1} + 4E^* T^k = \sum(k).$$

We use the following trick. Let $E'_0 \in \{0, 1, \ldots, T-1\}$. Applying (3.13) with $T+1$ in place of T, we obtain

$$E'_0 (T+1) + E^* (T+1)^k < (T+1)^2 + E^* (T+1)^k$$
$$\leq (1 + E^*)(T+1)^k$$
$$\leq 2E^* (T+1)^k. \tag{3.15}$$

Applying (3.14) with $T+1$ in place of T, we obtain

$$E'_0 (T+1) + E^* (T+1)^k = \sum(k). \tag{3.16}$$

Adding equations (3.14) and (3.16), we see that for every choice of k integers

$$E'_0, E_1, \ldots, E_{k-1} \in \{0, 1, \ldots, T-1\},$$

we have

$$F^* = \left(E_1 T + \cdots + E_{k-1} T^{k-1} + E^* T^k\right) + \left(E_0'(T+1) + E^*(T+1)^k\right)$$

$$= (E_0' + E^*) + (E_1 + E_0' + kE^*)T + \sum_{\ell=2}^{k-1} \left(E_\ell + \binom{k}{\ell} E^*\right) T^\ell + 2E^* T^k$$

$$= \sum(k).$$

Moreover, it follows from (3.13) and (3.15) that

$$0 \leq F^* < 4E^*(T+1)^k \leq 5E^* T^k < T^{k+1}$$

since $4(T+1)^k \leq 5T^k$ and $T \geq T_0 > 5E^*$. Given any k integers

$$F_0, F_1, \ldots, F_{k-1} \in \{0, 1, \ldots, T-1\},$$

we can again apply our algorithm (see Exercise 7) to obtain integers F_k and

$$E_0', E_1, E_2, \ldots, E_{k-1} \in \{0, 1, \ldots, T-1\}$$

such that

$$F_0 + F_1 T + \cdots + F_{k-1} T^{k-1} + F_k T^k$$

$$= E_1 T + \cdots + E_{k-1} T^{k-1} + E^* T^k + E_0'(T+1) + E^*(T+1)^k$$

$$= \sum(k),$$

where F_k is an integer that satisfies

$$0 \leq F_k < 5E^*.$$

After the addition of $(5E^* - F_k)T^k = \sum(k)$, we obtain

$$F_0 + F_1 T + \cdots + F_{k-1} T^{k-1} + 5E^* T^k = \sum(k)$$

for all $T \geq T_0$ and for all choices of $F_0, F_1, \ldots, F_{k-1} \in \{0, 1, \ldots, T-1\}$. This proves that $n = \sum(k)$ if $T \geq T_0$ and

$$5E^* T^k \leq n < (5E^* + 1)T^k.$$

There exists an integer $T_1 \geq T_0$ such that

$$5E^*(T+1)^k < (5E^* + 1)T^k \qquad \text{for all } T \geq T_1.$$

Then $n = \sum(k)$ if $T \geq T_1$ and

$$5E^* T^k \leq n < 5E^*(T+1)^k. \tag{3.17}$$

Since every integer $n \geq 5E^* T_1^k$ satisfies inequality (3.17) for some $T \geq T_1$, we have

$$n = \sum(k) \qquad \text{for all } n \geq 5E^* T_1^k.$$

It follows from Lemma 3.9 that Waring's problem holds for exponent k. This completes the proof of the Hilbert–Waring theorem.

3.4   Notes

The polynomial identities in Theorems 3.1, 3.2, and 3.3 are due to Liouville [79, pages 112–115], Fleck [40], and Hurwitz [65], respectively. Hurwitz's observations [65] on polynomial identities appeared in 1908.

Hilbert [56] published his proof of Waring's problem in 1909 in a paper dedicated to the memory of Minkowski. The original proof was quickly simplified by several authors. The proof of Hilbert's identity given in this book is due to Hausdorff [52], and the inductive argument that allows us to go from exponent k to exponent $k+1$ is due to Stridsberg [120]. Oppenheim [94] contains an excellent account of the Hausdorff–Stridsberg proof of Hilbert's theorem. Schmidt [105] introduced a convexity argument to prove Hilbert's identity. This is the argument that Ellison [28] uses in his excellent survey paper on Waring's problem. Dress [25] gives a different proof of the Hilbert–Waring theorem that involves a clever application of the easier Waring's problem to avoid induction on the exponent k. Rieger [102] used Hilbert's method to obtain explicit estimates for $g(k)$.

3.5   Exercises

1. (Euler) Let $[x]$ denote the integer part of x, and let

$$q = \left[\left(\frac{3}{2} \right)^k \right].$$

   Prove that
   $$g(k) \geq 2^k + q - 2.$$
   Hint: Consider the number $N = q2^k - 1$.

2. Verify the polynomial identity in Theorem 3.2, and obtain an explicit upper bound for $g(6)$.

3. Verify the polynomial identity in Theorem 3.3, and obtain an explicit upper bound for $g(8)$.

4. (Schur) Verify the polynomial identity
   $$22,680(x_1^2 + x_2^2 + x_3^2 + x_4^2)^5$$
   $$= 9 \sum (2x_i)^{10} + 180 \sum (x_i \pm x_j)^{10} + \sum (2x_i \pm x_j \pm x_k)^{10}$$
   $$+9 \sum (x_1 \pm x_2 \pm x_3 \pm x_4)^{10}.$$

5. Show that every integer of the form $22,680a^5$ is the sum of 2316 nonnegative integral 10th powers.

6. Let k, ℓ, and U be integers such that $0 \leq \ell \leq k$. Let

$$f(x) = (x^2 + U)^{k+\ell}.$$

Show that there exist positive integers $A_0, A_1, \ldots, A_\ell$ depending only on k and ℓ such that

$$\frac{d^{2\ell} f}{dx^{2\ell}} = \sum_{i=0}^{\ell} A_i x^{2i} (x^2 + U)^{k-i}.$$

7. Let $k \geq 1$, $T \geq 2$, and D_i, E_i be integers for $i = 0, 1, \ldots, k-1$. Prove that there exist unique integers $C_0, \ldots, C_{k-1}$ and I_k such that

$$0 \leq C_i < T \quad \text{for } i = 0, 1, \ldots, k-1$$

and

$$\sum_{\ell=0}^{k-1} (C_\ell + D_\ell) T^\ell = \sum_{\ell=0}^{k-1} E_\ell T^\ell + I_k T^k.$$

8. This is an exercise in notation: Prove that $\sum(2k) = \sum(k)$ but $\sum(k) \neq \sum(2k)$.

4

Weyl's inequality

The analytic method of Hardy and Littlewood (sometimes called the 'circle method') was developed for the treatment of *additive problems* in the theory of numbers. These are problems which concern the representation of a large number as a sum of numbers of some specified type. The number of summands may be either fixed or unrestricted; in the latter case we speak of *partition problems*. The most famous additive problem is Waring's Problem, where the specified numbers are kth powers .... The most important single tool for the investigation of Waring's Problem, and indeed many other problems in the analytic theory of numbers, is Weyl's inequality.

H. Davenport [18]

4.1 Tools

The purpose of this chapter is to develop some analytical tools that will be needed to prove the Hardy–Littlewood asymptotic formula for Waring's problem and other results in additive number theory. The most important of these tools are two inequalities for exponential sums, Weyl's inequality and Hua's lemma. We shall also introduce partial summation, infinite products, and Euler products.

We begin with the following simple result about approximating real numbers by rationals with small denominators. Recall that $[x]$ denotes the integer part of the real number x and that $\{x\}$ denotes the fractional part of x.

Theorem 4.1 (Dirichlet) *Let α and Q be real numbers, $Q \geq 1$. There exist integers a and q such that*

$$1 \leq q \leq Q, \qquad (a, q) = 1,$$

and

$$\left| \alpha - \frac{a}{q} \right| < \frac{1}{qQ}.$$

Proof. Let $N = [Q]$. Suppose that $\{q\alpha\} \in [0, 1/(N+1))$ for some positive integer $q \leq N$. If $a = [q\alpha]$, then

$$0 \leq \{q\alpha\} = q\alpha - [q\alpha] = q\alpha - a < \frac{1}{N+1},$$

and so

$$\left| \alpha - \frac{a}{q} \right| < \frac{1}{q(N+1)} < \frac{1}{qQ} \leq \frac{1}{q^2}.$$

Similarly, if $\{q\alpha\} \in [N/(N+1), 1)$ for some positive integer $q \leq N$ and if $a = [q\alpha] + 1$, then

$$\frac{N}{N+1} \leq \{q\alpha\} = q\alpha - a + 1 < 1$$

implies that

$$|q\alpha - a| \leq \frac{1}{N+1}$$

and so

$$\left| \alpha - \frac{a}{q} \right| \leq \frac{1}{q(N+1)} < \frac{1}{qQ} \leq \frac{1}{q^2}.$$

If

$$(q\alpha) \in \left[\frac{1}{N+1}, \frac{N}{N+1} \right)$$

for all $q = 1, \ldots, N$, then each of the N real numbers $\{q\alpha\}$ lies in one of the $N - 1$ intervals

$$\left[\frac{i}{N+1}, \frac{i+1}{N+1} \right) \qquad \text{for } i = 1, \ldots N - 1.$$

By Dirichlet's box principle, there exist integers $i \in [1, N-1]$ and $q_1, q_2 \in [1, N]$ such that

$$1 \leq q_1 < q_2 \leq N$$

and

$$\{q_1\alpha\}, \{q_2\alpha\} \in \left[\frac{i}{N+1}, \frac{i+1}{N+1} \right).$$

Let

$$q = q_2 - q_1 \in [1, N-1]$$

and

$$a = [q_2\alpha] - [q_1\alpha].$$

Then

$$|q\alpha - a| = |(q_2\alpha - [q_2\alpha]) - (q_1\alpha - [q_1\alpha])| = |\{q_2\alpha\} - \{q_1\alpha\}| < \frac{1}{N+1} < \frac{1}{Q}.$$

This completes the proof.

4.2 Difference operators

The *forward difference operator* Δ_d is the linear operator defined on functions f by the formula

$$\Delta_d(f)(x) = f(x+d) - f(x).$$

For $\ell \geq 2$, we define the *iterated difference operator* $\Delta_{d_\ell, d_{\ell-1}, \ldots, d_1}$ by

$$\Delta_{d_\ell, d_{\ell-1}, \ldots, d_1} = \Delta_{d_\ell} \circ \Delta_{d_{\ell-1}, \ldots, d_1} = \Delta_{d_\ell} \circ \Delta_{d_{\ell-1}} \circ \cdots \circ \Delta_{d_1}.$$

For example,

$$\begin{aligned}
\Delta_{d_2, d_1}(f)(x) &= \Delta_{d_2}\left(\Delta_{d_1}(f)\right)(x) \\
&= \left(\Delta_{d_1}(f)\right)(x+d_2) - \left(\Delta_{d_1}(f)\right)(x) \\
&= f(x+d_2+d_1) - f(x+d_2) - f(x+d_1) + f(x)
\end{aligned}$$

and

$$\begin{aligned}
\Delta_{d_3, d_2, d_1}(f)(x) &= f(x+d_3+d_2+d_1) - f(x+d_3+d_2) \\
&\quad - f(x+d_3+d_1) - f(x+d_2+d_1) \\
&\quad + f(x+d_3) + f(x+d_2) + f(x+d_1) - f(x).
\end{aligned}$$

We let $\Delta^{(\ell)}$ be the iterated difference operator $\Delta_{1,\ldots,1}$ with $d_i = 1$ for $i = 1, \ldots, \ell$. Then

$$\Delta^{(2)}(f)(x) = f(x+2) - 2f(x+1) + f(x)$$

and

$$\Delta^{(3)}(f)(x) = f(x+3) - 3f(x+2) + 3f(x+1) - f(x).$$

Lemma 4.1 *Let $\ell \geq 1$. Then*

$$\Delta^{(\ell)}(f)(x) = \sum_{j=0}^{\ell}(-1)^{\ell-j}\binom{\ell}{j}f(x+j).$$

Proof. This is by induction on ℓ. If the lemma holds for ℓ, then

$$\Delta^{(\ell+1)}(f)(x)$$
$$= \Delta\left(\Delta^{(\ell)}(f)\right)(x)$$

$$= \Delta \left(\sum_{j=0}^{\ell} (-1)^{\ell-j} \binom{\ell}{j} f(x+j) \right)$$

$$= \sum_{j=0}^{\ell} (-1)^{\ell-j} \binom{\ell}{j} \Delta(f)(x+j)$$

$$= \sum_{j=0}^{\ell} (-1)^{\ell-j} \binom{\ell}{j} f(x+j+1) + \sum_{j=0}^{\ell} (-1)^{\ell+1-j} \binom{\ell}{j} f(x+j)$$

$$= \sum_{j=1}^{\ell+1} (-1)^{\ell+1-j} \binom{\ell}{j-1} f(x+j) + \sum_{j=0}^{\ell} (-1)^{\ell+1-j} \binom{\ell}{j} f(x+j)$$

$$= f(x+\ell+1) + \sum_{j=1}^{\ell} (-1)^{\ell+1-j} \left(\binom{\ell}{j-1} + \binom{\ell}{j} \right) f(x+j) + (-1)^{\ell+1} f(x).$$

This completes the proof.

We shall compute the polynomial obtained by applying an iterated difference operator to the power function $f(x) = x^k$.

Lemma 4.2 *Let $k \geq 1$ and $1 \leq \ell \leq k$. Let $\Delta_{d_\ell,...,d_1}$ be an iterated difference operator. Then*

$$\Delta_{d_\ell,...,d_1}(x^k) = \sum_{\substack{j_1+\cdots+j_\ell+j=k \\ j\geq 0, j_1,...,j_\ell\geq 1}} \frac{k!}{j! j_1! \cdots j_\ell!} d_1^{j_1} \cdots d_\ell^{j_\ell} x^j \qquad (4.1)$$

$$= d_1 \cdots d_\ell \, p_{k-\ell}(x),$$

where $p_{k-\ell}(x)$ is a polynomial of degree $k - \ell$ and leading coefficient $k(k - 1)\cdots(k - \ell + 1)$. If $d_1, \ldots, d_l$ are integers, then $p_{k-\ell}(x)$ is a polynomial with integer coefficients.

Proof. This is by induction on ℓ. For $\ell = 1$, we have

$$\Delta_{d_1}(x^k) = (x + d_1)^k - x^k$$

$$= \sum_{j=0}^{k-1} \binom{k}{j} d_1^{k-j} x^j$$

$$= \sum_{\substack{j_1+j=k \\ j\geq 0, j_1\geq 1}} \frac{k!}{j! j_1!} d_1^{j_1} x^j.$$

Let $1 \leq \ell \leq k - 1$, and assume that formula (4.1) holds for ℓ. Then

$$\Delta_{d_{\ell+1},d_\ell,...,d_1}(x^k)$$

$$= \Delta_{d_{\ell+1}} \left(\Delta_{d_\ell,...,d_1}(x^k) \right)$$

$$= \sum_{\substack{j_1+\cdots+j_\ell+m=k \\ m\geq 0, j_1,...,j_\ell\geq 1}} \frac{k!}{m! j_1! \cdots j_\ell!} d_1^{j_1} \cdots d_\ell^{j_\ell} \Delta_{d_{\ell+1}}(x^m)$$

$$= \sum_{\substack{j_1+\cdots+j_\ell+m=k \\ m,j_1,\ldots,j_\ell \geq 1}} \frac{k!}{m!j_1!\cdots j_\ell!} d_1^{j_1} \cdots d_\ell^{j_\ell} \sum_{\substack{j_{\ell+1}+j=m \\ j \geq 0, j_{\ell+1} \geq 1}} \frac{m!}{j!j_{\ell+1}!} d_{\ell+1}^{j_{\ell+1}} x^j$$

$$= \sum_{\substack{j_1+\cdots+j_\ell+m=k \\ m,j_1,\ldots,j_\ell \geq 1}} \sum_{\substack{j_{\ell+1}+j=m \\ j \geq 0, j_{\ell+1} \geq 1}} \frac{k!}{j!j_1!\cdots j_\ell!j_{\ell+1}!} d_1^{j_1} \cdots d_\ell^{j_\ell} d_{\ell+1}^{j_{\ell+1}} x^j$$

$$= \sum_{\substack{j_1+\cdots+j_\ell+j_{\ell+1}+j=k \\ j \geq 0, j_1,\ldots,j_\ell,j_{\ell+1} \geq 1}} \frac{k!}{j!j_1!\cdots j_\ell!j_{\ell+1}!} d_1^{j_1} \cdots d_\ell^{j_\ell} d_{\ell+1}^{j_{\ell+1}} x^j.$$

Since the multinomial coefficients $k!/j!j_1!\cdots j_\ell!$ are integers, it follows that if $d_1, \ldots, d_\ell$ are integers, then the polynomial $p_{k-\ell}(x)$ has integer coefficients. This completes the proof.

Lemma 4.3 *Let $k \geq 2$. Then*

$$\Delta_{d_{k-1},\ldots,d_1}(x^k) = d_1 \ldots d_{k-1} k! \left(x + \frac{d_1 + \cdots + d_{k-1}}{2} \right).$$

Proof. This follows immediately from Lemma 4.2.

Lemma 4.4 *Let $\ell \geq 1$ and $\Delta_{d_\ell,d_{\ell-1},\ldots,d_1}$ be an iterated difference operator. Let $f(x) = \alpha x^k + \cdots$ be a polynomial of degree k. Then*

$$\Delta_{d_\ell,\ldots,d_1}(f)(x) = d_1 \cdots d_\ell \left(k(k-1)\cdots(k-\ell+1)\alpha x^{k-\ell} + \cdots \right)$$

if $1 \leq \ell \leq k$ and

$$\Delta_{d_\ell,d_{\ell-1},\ldots,d_1}(f)(x) = 0$$

if $\ell > k$. In particular, if $\ell = k-1$ and $d_1 \cdots d_{k-1} \neq 0$, then

$$\Delta_{d_{k-1},\ldots,d_1}(f)(x) = d_1 \cdots d_{k-1} k! \alpha x + \beta$$

is a polynomial of degree one.

Proof. Let $f(x) = \sum_{j=1}^k \alpha_j x^j$, where $\alpha_k = \alpha$. Since the difference operator Δ is linear, it follows that

$$\Delta_{d_\ell,\ldots,d_1}(f)(x) = \sum_{j=0}^k \alpha_j \Delta_{d_\ell,\ldots,d_1}(x^j)$$

$$= d_1 \cdots d_\ell \left(\frac{k!}{(k-\ell)!} \alpha x^{k-\ell} + \cdots \right).$$

This completes the proof.

Lemma 4.5 *Let $1 \leq \ell \leq k$. If*

$$-P \leq d_1, \ldots, d_\ell, x \leq P,$$

then

$$\Delta_{d_\ell,\ldots,d_1}(x^k) \ll P^k,$$

where the implied constant depends only on k.

Proof. It follows from Lemma 4.2 that

$$\left|\Delta_{d_\ell,\dots,d_1}(x^k)\right| \le \sum_{\substack{j_1+\dots+j_\ell+j=k \\ j\ge0,j_1,\dots,j_\ell\ge1}} \frac{k!}{j!j_1!\cdots j_\ell!} P^{j_1+\dots+j_\ell+j}$$

$$\le \sum_{\substack{j_1+\dots+j_\ell+j=k \\ j,j_1,\dots,j_\ell\ge0}} \frac{k!}{j!j_1!\cdots j_\ell!} P^k$$

$$= (\ell+1)^k P^k$$

$$\le (k+1)^k P^k$$

$$\ll P^k.$$

This completes the proof.

4.3   Easier Waring's problem

Here is a simple application of difference operators.

Waring's problem states that every nonnegative integer can be written as the sum of a bounded number of nonnegative kth powers. We can ask the following similar question: Is it true that every integer can be written as the sum or difference of a bounded number of kth powers? If the answer is "yes," then for every k there exists a smallest integer $v(k)$ such that the equation

$$n = \pm x_1^k \pm x_2^k \cdots \pm x_{v(k)}^k \tag{4.2}$$

has a solution in integers for every integer n. This is called the *easier Waring's problem*, and it is, indeed, much easier to prove the existence of $v(k)$ than to prove the existence of $g(k)$. It is still an unsolved problem, however, to determine the exact value of $v(k)$ for any $k \ge 3$.

Theorem 4.2 (Easier Waring's problem) *Let $k \ge 2$. Then $v(k)$ exists, and*

$$v(k) \le 2^{k-1} + \frac{k!}{2}.$$

Proof. Applying the $(k-1)$-st forward difference operator to the polynomial $f(x) = x^k$, we obtain from Lemma 4.1 and Lemma 4.3 that

$$\Delta^{(k-1)}(x^k) = k!x + m = \sum_{\ell=0}^{k-1}(-1)^{k-1-\ell}\binom{k-1}{\ell}(x+\ell)^k,$$

where $m = (k-1)!\binom{k}{2}$. In this way, every integer of the form $k!x+m$ can be written as the sum or difference of at most

$$\sum_{\ell=0}^{k-1}\binom{k-1}{\ell} = 2^{k-1}$$

kth powers of integers. For any integer n, we can choose integers q and r such that

$$n - m = k!q + r,$$

where

$$-\frac{k!}{2} < r \le \frac{k!}{2}.$$

Since r is the sum or difference of exactly $|r|$ kth powers 1^k, it follows that n can be written as the sum of at most $2^{k-1} + k!/2$ integers of the form $\pm x^k$. This completes the proof.

4.4   Fractional parts

Let $[\alpha]$ denote the integer part of the real number α and let $\{\alpha\}$ denote the fractional part of α. Then $[\alpha] \in \mathbf{Z}$, $\{\alpha\} \in [0, 1)$, and

$$\alpha = [\alpha] + \{\alpha\}.$$

The distance from the real number α to the nearest integer is denoted

$$\|\alpha\| = \min\left(|n - \alpha| : n \in \mathbf{Z}\right) = \inf(\{\alpha\}, 1 - \{\alpha\}).$$

Then $\|\alpha\| \in [0, 1/2]$, and

$$\alpha = n \pm \|\alpha\|$$

for some integer n. It follows that

$$|\sin \pi\alpha| = \sin \pi\}\alpha\|$$

for all real numbers α. The triangle inequality

$$\|\alpha + \beta\| \le \|\alpha\| + \|\beta\| \tag{4.3}$$

holds for all real numbers α and β (see Exercise 2).

The following two very simple lemmas are at the core of Weyl's inequality for exponential sums, and Weyl's inequality, in turn, is at the core of our application of the circle method to Waring's problem. Recall that $\exp(t) = e^t$ and $e(t) = \exp(2\pi it) = e^{2\pi it}$.

Lemma 4.6 *If* $0 < \alpha < 1/2$, *then*

$$2\alpha < \sin \pi\alpha < \pi\alpha.$$

Proof. Let $s(\alpha) = \sin \pi\alpha - 2\alpha$. Then $s(0) = s(1/2) = 0$. If $s(\alpha) = 0$ for some $\alpha \in (0, 1/2)$, then $s'(\alpha) = \pi \cos \pi\alpha - 2$ would have at least two zeros in $(0, 1/2)$, which is impossible because $s'(\alpha)$ decreases monotonically from $\pi - 2$ to -2 in this interval. Since $s(1/4) = (\sqrt{2} - 1)/2 > 0$, it follows that $s(\alpha) > 0$ for all $\alpha \in (0, \pi/2)$. This gives the lower bound. The proof of the upper bound is similar.

Lemma 4.7 *For every real number α and all integers $N_1 < N_2$,*

$$\sum_{n=N_1+1}^{N_2} e(\alpha n) \ll \min(N_2 - N_1, \|\alpha\|^{-1}).$$

Proof. Since $|e(\alpha n)| = 1$ for all integers n, we have

$$\left| \sum_{n=N_1+1}^{N_2} e(\alpha n) \right| \leq \sum_{n=N_1+1}^{N_2} 1 = N_2 - N_1.$$

If $\alpha \notin \mathbf{Z}$, then $\|\alpha\| > 0$ and $e(\alpha) \neq 1$. Since the sum is also a geometric progression, we have

$$\left| \sum_{n=N_1+1}^{N_2} e(\alpha n) \right| = \left| e(\alpha(N_1 + 1)) \sum_{n=0}^{N_2-N_1-1} e(\alpha)^n \right|$$

$$= \left| \frac{e(\alpha(N_2 - N_1)) - 1}{e(\alpha) - 1} \right|$$

$$\leq \frac{2}{|e(\alpha) - 1|}$$

$$= \frac{2}{|e(\alpha/2) - e(-\alpha/2)|}$$

$$= \frac{2}{|2i \sin \pi\alpha|}$$

$$= \frac{1}{|\sin \pi\alpha|}$$

$$= \frac{1}{\sin(\pi \|\alpha\|)}$$

$$\leq \frac{1}{2\|\alpha\|}.$$

This completes the proof.

Lemma 4.8 *Let α be a real number, and let q and a be integers such that $q \geq 1$ and $(a, q) = 1$. If*

$$\left| \alpha - \frac{a}{q} \right| \leq \frac{1}{q^2},$$

then

$$\sum_{1 \leq r \leq q/2} \frac{1}{\|\alpha r\|} \ll q \log q.$$

Proof. The lemma holds for $q = 1$,

$$\sum_{1 \leq r \leq q/2} \frac{1}{\|\alpha r\|} = 0.$$

Therefore, we can assume that $q \geq 2$. For each integer r, there exist integers $s(r) \in [0, q/2]$ and $m(r)$ such that

$$\frac{s(r)}{q} = \left\| \frac{ar}{q} \right\| = \pm \left(\frac{ar}{q} - m(r) \right).$$

Since $(a, q) = 1$, it follows that $s(r) = 0$ if and only if $r \equiv 0 \pmod q$, and so $s(r) \in [1, q/2]$ if $r \in [1, q/2]$. Let

$$\alpha - \frac{a}{q} = \frac{\theta}{q^2},$$

where $-1 \leq \theta \leq 1$. Then

$$\alpha r = \frac{ar}{q} + \frac{\theta r}{q^2} = \frac{ar}{q} + \frac{\theta'}{2q},$$

where

$$|\theta'| = \left| \frac{2\theta r}{q} \right| \leq |\theta| \leq 1.$$

It follows from (4.3) that

$$
\begin{aligned}
\| \alpha r \| &= \left\| \frac{ar}{q} + \frac{\theta'}{2q} \right\| \\
&= \left| m(r) \pm \frac{s(r)}{q} + \frac{\theta'}{2q} \right\| \\
&= \left\| \frac{s(r)}{q} \pm \frac{\theta'}{2q} \right\| \\
&\geq \left\| \frac{s(r)}{q} \right\| - \left\| \frac{\theta'}{2q} \right\| \\
&\geq \frac{s(r)}{q} - \frac{1}{2q} \\
&\geq \frac{1}{2q}.
\end{aligned}
$$

Let $1 \leq r_1 \leq r_2 \leq q/2$. We shall show that $s(r_1) = s(r_2)$ if and only if $r_1 = r_2$. If

$$\left\| \frac{ar_1}{q} \right\| = \left\| \frac{ar_2}{q} \right\|,$$

then

$$\pm \left(\frac{ar_1}{q} - m(r_1) \right) = \pm \left(\frac{ar_2}{q} - m(r_2) \right)$$

and so

$$ar_1 \equiv \pm ar_2 \pmod q.$$

Since $(a, q) = 1$ and $1 \leq r_1 \leq r_2 \leq q/2$, we have

$$r_1 \equiv \pm r_2 \pmod{q}$$

and so

$$r_1 = r_2.$$

It follows that

$$\left\{ \left\| \frac{ar}{q} \right\| : 1 \leq r \leq \frac{q}{2} \right\} = \left\{ \frac{s(r)}{q} : 1 \leq r \leq \frac{q}{2} \right\} = \left\{ \frac{s}{q} : 1 \leq s \leq \frac{q}{2} \right\}.$$

Therefore,

$$\sum_{1 \leq r \leq q/2} \frac{1}{\|ar\|} \leq \sum_{1 \leq r \leq q/2} \frac{1}{\frac{s(r)}{q} - \frac{1}{2q}}$$

$$= \sum_{1 \leq s \leq q/2} \frac{1}{\frac{s}{q} - \frac{1}{2q}}$$

$$= 2q \sum_{1 \leq s \leq q/2} \frac{1}{2s - 1}$$

$$\leq 2q \sum_{1 \leq s \leq q/2} \frac{1}{s}$$

$$\ll q \log q.$$

This completes the proof.

Lemma 4.9 *Let α be a real number. If*

$$\left| \alpha - \frac{a}{q} \right| \leq \frac{1}{q^2},$$

where $q \geq 1$ and $(a, q) = 1$, then for any nonnegative real number V and nonnegative integer h, we have

$$\sum_{r=1}^{q} \min \left(V, \frac{1}{\|\alpha(hq + r)\|} \right) \ll V + q \log q.$$

Proof. Let

$$\alpha = \frac{a}{q} + \frac{\theta}{q^2},$$

where

$$-1 \leq \theta \leq 1.$$

Then

$$\alpha(hq + r) = ah + \frac{ar}{q} + \frac{\theta h}{q} + \frac{\theta r}{q^2}$$

$$= ah + \frac{ar}{q} + \frac{[\theta h] + \{\theta h\}}{q} + \frac{\theta r}{q^2}$$

$$= ah + \frac{ar + [\theta h] + \delta(r)}{q},$$

where
$$-1 \leq \delta(r) = \{\theta h\} + \frac{\theta r}{q} < 2.$$

For each $r = 1, \ldots, q$ there is a unique integer r' such that

$$\{\alpha(hq + r)\} = \frac{ar + [\theta h] + \delta(r)}{q} - r'.$$

Let

$$0 \leq t \leq 1 - \frac{1}{q}.$$

If

$$t \leq \{\alpha(hq + r)\} \leq t + \frac{1}{q},$$

then

$$qt \leq ar - qr' + [\theta h] + \delta(r) \leq qt + 1.$$

This implies that

$$ar - qr' \leq qt - [\theta h] + 1 - \delta(r) \leq qt - [\theta h] + 2$$

and

$$ar - qr' \geq qt - [\theta h] - \delta(r) > qt - [\theta h] - 2.$$

Thus, $ar - qr'$ lies in the half-open interval J of length 4, where

$$J = (qt - [\theta h] - 2, qt - [\theta h] + 2].$$

This interval contains exactly four distinct integers. If $1 \leq r_1 \leq r_2 \leq q$ and

$$ar_1 - qr_1' = ar_2 - qr_2',$$

then

$$ar_1 \equiv ar_2 \pmod{q}.$$

Since $(a, q) = 1$, we have

$$r_1 \equiv r_2 \pmod{q}$$

and so

$$r_1 = r_2.$$

It follows that for any $t \in [0, (q - 1)/q]$, there are at most four integers $r \in [1, q]$ such that

$$\{\alpha(hq + r)\} \in [t, t + (1/q)].$$

We observe that

$$\|\alpha(hq + r)\| \in [t, t + (1/q)]$$

if and only if either

$$\{\alpha(hq + r)\} \in [t, t + (1/q)]$$

or

$$1 - \{\alpha(hq + r)\} \in [t, t + (1/q)].$$

The latter inclusion is equivalent to

$$\{\alpha(hq + r)\} \in [t', t' + (1/q)],$$

where

$$0 \leq t' = 1 - \frac{1}{q} - t \leq 1 - \frac{1}{q}.$$

It follows that for any $t \in [0, (q - 1)/q]$, there are at most eight integers $r \in [1, q]$ for which

$$\|\alpha(hq + r)\| \in [t, t + (1/q)].$$

In particular, if we let $J(s) = [s/q, (s + 1)/q]$ for $s = 0, 1, \ldots$, then

$$\|\alpha(hq + r)\| \in J(s)$$

for at most eight $r \in [1, q]$.

We apply this fact to estimate the sum

$$\sum_{1 \leq r \leq q} \min\left(V, \frac{1}{\|\alpha(hq + r)\|}\right).$$

If $\|\alpha(hq + r)\| \in J(0) = [0, 1/q]$, then we use the inequality

$$\min\left(V, \frac{1}{\|\alpha(hq + r)\|}\right) \leq V.$$

If $\|\alpha(hq + r)\| \in J(s)$ for some $s \geq 1$, then we use the inequality

$$\min\left(V, \frac{1}{\|\alpha(hq + r)\|}\right) \leq \frac{1}{\|\alpha(hq + r)\|} \leq \frac{q}{s}.$$

Since $\|\alpha(hq + r)\| \in J(s)$ for some $s < q/2$, it follows that

$$\sum_{1 \leq r \leq q} \min\left(V, \frac{1}{\|\alpha(hq + r)\|}\right) \leq 8V + 8 \sum_{1 \leq s < q/2} \frac{q}{s}$$

$$\ll V + q \log q.$$

This completes the proof.

Lemma 4.10 *Let α be a real number. If*

$$\left|\alpha - \frac{a}{q}\right| \leq \frac{1}{q^2},$$

where $q \geq 1$ and $(a, q) = 1$, then for any real number $U \geq 1$ and positive integer n we have

$$\sum_{1 \leq k \leq U} \min\left(\frac{n}{k}, \frac{1}{\|\alpha k\|}\right) \ll \left(\frac{n}{q} + U + q\right) \log 2qU.$$

Proof. We can write k in the form

$$k = hq + r,$$

where

$$1 \le r \le q$$

and

$$0 \le h < \frac{U}{q}.$$

Then

$$S = \sum_{1 \le k \le U} \min\left(\frac{n}{k}, \frac{1}{\|\alpha k\|}\right)$$

$$\le \sum_{0 \le h < U/q} \sum_{1 \le r \le q} \min\left(\frac{n}{hq + r}, \frac{1}{\|\alpha(hq + r)\|}\right).$$

If $h = 0$ and $1 \le r \le q/2$, then Lemma 4.8 gives

$$\sum_{1 \le r \le q/2} \min\left(\frac{n}{r}, \frac{1}{\|\alpha r\|}\right) \le \sum_{1 \le r \le q/2} \frac{1}{\|\alpha r\|} \ll q \log q.$$

For the remaining terms, we have

$$\frac{1}{hq + r} < \frac{2}{(h + 1)q},$$

since either $h \ge 1$ and

$$hq + r > hq \ge \frac{(h + 1)q}{2}$$

or $h = 0, q/2 < r \le q$, and

$$hq + r = r > \frac{q}{2} = \frac{(h + 1)q}{2}.$$

Therefore,

$$S \ll q \log q + \sum_{0 \le h < U/q} \sum_{1 \le r \le q} \min\left(\frac{n}{(h + 1)q}, \frac{1}{\|\alpha(hq + r)\|}\right). \qquad (4.4)$$

Note that

$$\frac{U}{q} + 1 \le U + q \le 2\max(q, U) \le 2qU.$$

Estimating the inner sum by Lemma 4.9 with $V = n/(h + 1)q$, we obtain

$$S \ll q \log q + \sum_{0 \le h < U/q} \sum_{1 \le r \le q} \min\left(\frac{n}{(h + 1)q}, \frac{1}{\|\alpha(hq + r)\|}\right)$$

$$\ll q\log q + \sum_{0\le h<U/q}\left(\frac{n}{(h+1)q}+q\log q\right)$$

$$\ll q\log q + \frac{n}{q}\sum_{0\le h<U/q}\frac{1}{h+1}+\left(\frac{U}{q}+1\right)q\log q$$

$$\ll q\log q + \frac{n}{q}\log\left(\frac{U}{q}+1\right)+U\log q+q\log q$$

$$\ll \left(\frac{n}{q}+U+q\right)\log 2qU.$$

This completes the proof.

Lemma 4.11 *Let α be a real number. If*

$$\left|\alpha-\frac{a}{q}\right|\le\frac{1}{q^2},$$

where $q\ge 1$ and $(a,q)=1$, then for any real numbers U and n we have

$$\sum_{1\le k\le U}\min\left(n,\frac{1}{\|\alpha k\|}\right)\ll\left(q+U+n+\frac{Un}{q}\right)\max\{1,\log q\}.$$

Proof. This is almost exactly the same as the proof of Lemma 4.10. We have

$$S = \sum_{1\le k\le U}\min\left(n,\frac{1}{\|\alpha k\|}\right)$$

$$\le \sum_{0\le h<U/q}\sum_{1\le r\le q}\min\left(n,\frac{1}{\|\alpha(hq+r)\|}\right)$$

$$\le q\log q + \sum_{0\le h<U/q}\left(n+\sum_{1\le s<q/2}\frac{q}{s}\right)$$

$$\ll q\log q + \sum_{0\le h<U/q}(n+q\log q)$$

$$\ll q\log q + \left(\frac{U}{q}+1\right)(n+q\log q)$$

$$\ll q\log q + U\log q + n + \frac{Un}{q}$$

$$\ll \left(q+U+n+\frac{Un}{q}\right)\max\{1,\log q\}.$$

This completes the proof.

4.5   Weyl's inequality and Hua's lemma

In this section, we denote by $[M, N]$ the interval of integers m such that $M \leq m \leq N$. For any real number t, the complex conjugate of $e(t) = e^{2\pi i t}$ is $\overline{e(t)} = e(-t)$.

Lemma 4.12   *Let N_1, N_2, and N be integers such that $N_1 < N_2$ and $0 \leq N_2 - N_1 \leq N$. Let $f(n)$ be a real-valued arithmetic function, and let*

$$S(f) = \sum_{n=N_1+1}^{N_2} e(f(n)).$$

Then

$$|S(f)|^2 = \sum_{|d|<N} S_d(f),$$

where

$$S_d(f) = \sum_{n \in I(d)} e(\Delta_d(f)(n))$$

and $I(d)$ is an interval of consecutive integers contained in $[N_1 + 1, N_2]$.

Proof. For any integer d, let

$$I(d) = [N_1 + 1 - d, N_2 - d] \cap [N_1 + 1, N_2].$$

Squaring the absolute value of the exponential sum, we get

$$
\begin{aligned}
|S(f)|^2 &= S(f)\overline{S(f)} \\
&= \sum_{m=N_1+1}^{N_2} e(f(m)) \sum_{n=N_1+1}^{N_2} \overline{e(f(n))} \\
&= \sum_{n=N_1+1}^{N_2} \sum_{m=N_1+1}^{N_2} e(f(m) - f(n)) \\
&= \sum_{n=N_1+1}^{N_2} \sum_{d=N_1+1-n}^{N_2-n} e(f(n+d) - f(n)) \\
&= \sum_{n=N_1+1}^{N_2} \sum_{d=N_1+1-n}^{N_2-n} e(\Delta_d(f)(n)) \\
&= \sum_{d=-(N_2-N_1-1)}^{N_2-N_1-1} \sum_{n \in I(d)} e(\Delta_d(f)(n)) \\
&= \sum_{|d|<N} \sum_{n \in I(d)} e(\Delta_d(f)(n)) \\
&= \sum_{|d|<N} S_d(f).
\end{aligned}
$$

This completes the proof.

Lemma 4.13 *Let N_1, N_2, N, and ℓ be integers such that $\ell \geq 1$, $N_1 < N_2$, and $0 \leq N_2 - N_1 \leq N$. Let $f(n)$ be a real-valued arithmetic function, and let*

$$S(f) = \sum_{n=N_1+1}^{N_2} e(f(n)).$$

Then

$$|S(f)|^{2^\ell} \leq (2N)^{2^\ell - \ell - 1} \sum_{|d_1| < N} \cdots \sum_{|d_\ell| < N} S_{d_\ell,\ldots,d_1}(f),$$

where

$$S_{d_\ell,\ldots,d_1}(f) = \sum_{n \in I(d_\ell,\ldots,d_1)} e\left(\Delta_{d_\ell,\ldots,d_1}(f)(n)\right) \qquad (4.5)$$

and $I(d_\ell, \ldots, d_1)$ is an interval of consecutive integers contained in $[N_1 + 1, N_2]$.

Proof. This is by induction on ℓ. The case $\ell = 1$ is Lemma 4.12. Now assume that the result is true for $\ell \geq 1$. Using the Cauchy–Schwarz inequality, we obtain

$$|S(f)|^{2^{\ell+1}} = \left(|S(f)|^{2^\ell}\right)^2$$

$$\leq \left((2N)^{2^\ell - \ell - 1} \sum_{|d_1| < N} \cdots \sum_{|d_\ell| < N} |S_{d_\ell,\ldots,d_1}(f)|\right)^2$$

$$= (2N)^{2^{\ell+1} - 2\ell - 2} \left(\sum_{|d_1| < N} \cdots \sum_{|d_\ell| < N} |S_{d_\ell,\ldots,d_1}(f)|\right)^2$$

$$\leq (2N)^{2^{\ell+1} - 2\ell - 2}(2N)^\ell \sum_{|d_1| < N} \cdots \sum_{|d_\ell| < N} |S_{d_\ell,\ldots,d_1}(f)|^2,$$

where $S_{d_\ell,\ldots,d_1}(f)$ is an exponential sum of the form (4.5). By Lemma 4.12, for each $d_1, \ldots, d_\ell$, there is an interval

$$I(d_{\ell+1}, d_\ell, \ldots, d_1) \subseteq I(d_\ell, \ldots, d_1) \subseteq [N_1 + 1, N_2]$$

such that

$$|S_{d_\ell,\ldots,d_1}(f)|^2 = \left|\sum_{n \in I(d_1,\ldots,d_\ell)} e\left(\Delta_{d_\ell,\ldots,d_1}(f)(n)\right)\right|^2$$

$$= \sum_{|d_{\ell+1}| < N} \sum_{n \in I(d_{\ell+1}, d_\ell, \ldots, d_1)} e\left(\Delta_{d_{\ell+1}, d_\ell, \ldots, d_1}(f)(n)\right)$$

$$= \sum_{|d_{\ell+1}| < N} S_{d_{\ell+1}, d_\ell, \ldots, d_1}(f),$$

and so

$$|S(f)|^{2^{\ell+1}} \leq (2N)^{2^{\ell+1} - (\ell+1) - 1} \sum_{|d_1| < N} \cdots \sum_{|d_\ell| < N} \sum_{|d_{\ell+1}| < N} S_{d_{\ell+1}, d_\ell, \ldots, d_1}(f).$$

This completes the proof.

Lemma 4.14 *Let $k \geq 1$, $K = 2^{k-1}$, and $\varepsilon > 0$. Let $f(x) = \alpha x^k + \cdots$ be a polynomial of degree k with real coefficients. If*

$$S(f) = \sum_{n=1}^{N} e(f(n)),$$

then

$$|S(f)|^K \ll N^{K-1} + N^{K-k+\varepsilon} \sum_{m=1}^{k!N^{k-1}} \min\left(N, \|m\alpha\|^{-1}\right),$$

where the implied constant depends on k and ε.

Proof. Applying Lemma 4.13 with $\ell = k - 1$, we obtain

$$|S(f)|^K \leq (2N)^{K-k} \sum_{|d_1|<N} \cdots \sum_{|d_{k-1}|<N} |S_{d_{k-1},\dots,d_1}(f)|,$$

where

$$S_{d_{k-1},\dots,d_1}(f) = \sum_{n \in I(d_{k-1},\dots,d_1)} e\left(\Delta_{d_{k-1},\dots,d_1}(f)(n)\right)$$

and $I(d_{k-1}, \dots, d_1)$ is an interval of integers contained in $[1, N]$. Since $|e(t)| = 1$ for all real t, we have the upper bound

$$|S_{d_{k-1},\dots,d_1}(f)| \leq \sum_{n \in I(d_{k-1},\dots,d_1)} |e\left(\Delta_{d_{k-1},\dots,d_1}(f)(n)\right)| \leq N.$$

By Lemma 4.4, for any nonzero integers $d_1, \dots, d_{k-1}$, the difference operator $\Delta_{d_{k-1},\dots,d_1}$ applied to the polynomial $f(x)$ of degree k produces the linear polynomial

$$\Delta_{d_1,\dots,d_{k-1}}(f)(x) = d_{k-1} \cdots d_1 k! \alpha x + \beta = \lambda x + \beta,$$

where

$$\lambda = d_{k-1} \cdots d_1 k! \alpha$$

and $\beta \in \mathbf{R}$. Let $I(d_{k-1}, \dots, d_1) = [N_1 + 1, N_2]$. By Lemma 4.7,

$$
\begin{aligned}
|S_{d_{k-1},\dots,d_1}(f)| &= \left| \sum_{n \in I(d_{k-1},\dots,d_1)} e\left(\Delta_{d_{k-1},d_{k-2},\dots,d_1}(f)(n)\right) \right| \\
&= \left| \sum_{n=N_1+1}^{N_2} e(\lambda n + \beta) \right| \\
&= \left| \sum_{n=N_1+1}^{N_2} e(\lambda n) \right| \\
&\ll \frac{1}{\|\lambda\|} \\
&= \frac{1}{\|d_{k-1} \cdots d_1 k! \alpha\|}.
\end{aligned}
$$

It follows that

$$|S_{d_{k-1},\ldots,d_1}(f)| \leq \min(N, \|d_1 \cdots d_{k-1}k!\alpha\|^{-1}).$$

Therefore,

$$|S(f)|^K \leq (2N)^{K-k} \sum_{|d_1|<N} \cdots \sum_{|d_{k-1}|<N} |S_{d_{k-1},\ldots,d_1}(f)|$$

$$\leq (2N)^{K-k} \sum_{|d_1|<N} \cdots \sum_{|d_{k-1}|<N} \min(N, \|d_1 \cdots d_{k-1}k!\alpha\|^{-1}).$$

Since there are fewer than $(k-1)(2N)^{k-2}$ choices of $d_1, \ldots, d_{k-1}$ such that $d_1 \cdots d_{k-1} = 0$, and each such choice contributes N to the sum, it follows that

$$|S(f)|^K \leq (2N)^{K-k}(k-1)(2N)^{k-2}N$$
$$+ (2N)^{K-k} \sum_{1\leq|d_1|<N} \cdots \sum_{1\leq|d_{k-1}|<N} \min(N, \|d_1 \cdots d_{k-1}k!\alpha\|^{-1})$$

$$\leq k(2N)^{K-1}$$
$$+ 2^{k-1}N^{K-k} \sum_{1\leq d_1<N} \cdots \sum_{1\leq d_{k-1}<N} \min(N, \|d_1 \cdots d_{k-1}k!\alpha\|^{-1})$$

$$\ll N^{K-1} + N^{K-k} \sum_{d_1=1}^{N} \cdots \sum_{d_{k-1}=1}^{N} \min\left(N, \|d_1 \cdots d_{k-1}k!\alpha\|^{-1}\right),$$

where the implied constant depends only on k. Since

$$1 \leq d_1 \cdots d_{k-1}k! \leq k!N^{k-1}$$

and the divisor function $\tau(m)$ satisfies $\tau(m) \ll_\varepsilon m^\varepsilon$ for every $\varepsilon > 0$, it follows that the number of representations of an integer m in the form $d_1 \cdots d_{k-1}k!$ is $\ll m^\varepsilon \ll N^\varepsilon$. Therefore,

$$|S(f)|^K \ll N^{K-1} + N^{K-k} \sum_{d_1=1}^{N} \cdots \sum_{d_{k-1}=1}^{N} \min\left(N, \|d_{k-1} \cdots d_1 k!\alpha\|^{-1}\right)$$

$$\ll N^{K-1} + N^{K-k+\varepsilon} \sum_{m=1}^{k!N^{k-1}} \min\left(N, \|m\alpha\|^{-1}\right),$$

where the implied constant depends on k and ε. This completes the proof.

Theorem 4.3 (Weyl's inequality) *Let* $f(x) = \alpha x^k + \cdots$ *be a polynomial of degree* $k \geq 2$ *with real coefficients, and suppose that* α *has the rational approximation* a/q *such that*

$$\left|\alpha - \frac{a}{q}\right| \leq \frac{1}{q^2},$$

where $q \geq 1$ and $(a, q) = 1$. Let

$$S(f) = \sum_{n=1}^{N} e(f(n)).$$

Let $K = 2^{k-1}$ and $\varepsilon > 0$. Then

$$S(f) \ll N^{1+\varepsilon} \left(N^{-1} + q^{-1} + N^{-k}q\right)^{1/K},$$

where the implied constant depends on k and ε.

Proof. Since $|S(f)| \leq N$, the result is immediate if $q \geq N^k$. Thus, we can assume that

$$1 \leq q < N^k,$$

and so

$$\log q \ll \log N \ll N^{\varepsilon}.$$

By Lemma 4.14, we have

$$|S(f)|^K \ll N^{K-1} + N^{K-k+\varepsilon} \sum_{m=1}^{k!N^{k-1}} \min\left(N, \|m\alpha\|^{-1}\right).$$

By Lemma 4.11, we have

$$
\sum_{m=1}^{k!N^{k-1}} \min\left(N, \|m\alpha\|^{-1}\right) \ll \left(q + k!N^{k-1} + N + \frac{k!N^k}{q}\right) \max\{1, \log q\}
$$
$$
\ll \left(q + N^{k-1} + \frac{N^k}{q}\right) \log N
$$
$$
\ll N^k \left(qN^{-k} + N^{-1} + q^{-1}\right) N^{\varepsilon}.
$$

Therefore,

$$
|S(f)|^K \ll N^{K-1} + N^{K+\varepsilon} \left(qN^{-k} + N^{-1} + q^{-1}\right)
$$
$$
\ll N^{K+\varepsilon} \left(qN^{-k} + N^{-1} + q^{-1}\right).
$$

This completes the proof.

Theorem 4.4 *Let $k \geq 2$, and let a/q be a rational number with $q \geq 1$ and $(a, q) = 1$. Then*

$$S(q, a) = \sum_{x=1}^{q} e(ax^k/q) \ll q^{1-1/K+\varepsilon}.$$

Proof. Apply Weyl's inequality with $f(x) = ax^k/q$ and $N = q$. We obtain

$$S(q, a) \ll q^{1+\varepsilon}(q^{-1} + q^{-k+1})^{1/K} \ll q^{1-1/K+\varepsilon}.$$

This completes the proof.

Theorem 4.5 *Let $k \geq 2$. There exists $\delta > 0$ with the following property: If $N \geq 2$ and a/q is a rational number such that $(a, q) = 1$ and*

$$N^{1/2} \leq q \leq N^{k-1/2},$$

then

$$\sum_{n=1}^{N} e(an^k/q) \ll N^{1-\delta}.$$

Proof. Applying Weyl's inequality with $f(x) = ax^k/q$, we obtain

$$
\begin{aligned}
S(f) &\ll N^{1+\varepsilon} \left(N^{-1} + q^{-1} + N^{-k}q \right)^{1/K} \\
&\leq N^{1+\varepsilon} \left(N^{-1} + N^{-1/2} + N^{-1/2} \right)^{1/K} \\
&\leq N^{1-1/2K+\varepsilon} \\
&\leq N^{1-\delta}
\end{aligned}
$$

for any $\delta < 1/2K$. This completes the proof.

Theorem 4.6 (Hua's lemma) *For $k \geq 2$, let*

$$T(\alpha) = \sum_{n=1}^{N} e(\alpha n^k).$$

Then

$$\int_0^1 |T(\alpha)|^{2^k} \, d\alpha \ll N^{2^k - k + \varepsilon}.$$

Proof. We shall prove by induction on j that

$$\int_0^1 |T(\alpha)|^{2^j} \, d\alpha \ll N^{2^j - j + \varepsilon}$$

for $j = 1, \ldots, k$. The case $j = 1$ is clear since

$$\int_0^1 |T(\alpha)|^2 \, d\alpha = \sum_{m=1}^{N} \sum_{n=1}^{N} \int_0^1 e(\alpha(m^k - n^k)) d\alpha = N.$$

Let $1 \leq j \leq k - 1$, and assume that the result holds for j. Let $f(x) = \alpha x^k$. By Lemma 4.2,

$$\Delta_{d_j,\ldots,d_1}(f)(x) = \alpha d_j \cdots d_1 p_{k-j}(x),$$

where $p_{k-j}(x)$ is a polynomial of degree $k - j$ with integer coefficients. Applying Lemma 4.13 with $N_1 = 0$, $N_2 = N$, and $S(f) = T(\alpha)$, we obtain

$$
\begin{aligned}
|T(\alpha)|^{2^j} &\leq (2N)^{2^j-j-1} \sum_{|d_1|<N} \cdots \sum_{|d_j|<N} \sum_{n\in I(d_j,\ldots,d_1)} e\left(\Delta_{d_j,\ldots,d_1}(f)(n)\right) \\
&= (2N)^{2^j-j-1} \sum_{|d_1|<N} \cdots \sum_{|d_j|<N} \sum_{n\in I(d_j,\ldots,d_1)} e\left(\alpha d_j \cdots d_1 p_{k-j}(n)\right),
\end{aligned}
$$

where $I(d_j, \ldots, d_1)$ is an interval of consecutive integers contained in $[1, N]$. It follows that

$$|T(\alpha)|^{2^j} \le N^{2^j - j - 1} \sum_d r(d)e(\alpha d),\tag{4.6}$$

where $r(d)$ is the number of factorizations of d in the form

$$d = d_j \cdots d_1 p_{k-j}(n)$$

with $|d_i| \le N$ and $n \in I(d_j, \ldots, d_1)$. Since $d \ll N^k$ by Lemma 4.5, we have

$$r(d) \ll |d|^\varepsilon \ll N^\varepsilon$$

for $d \ne 0$. Since $p_{k-j}(x)$ is a polynomial of degree $k - j \ge 1$, there are at most $k - j$ integers x such that $p_{k-j} = 0$, and so

$$r(0) \ll N^j.$$

Similarly, since

$$
\begin{aligned}
|T(\alpha)|^{2^j} &= T(\alpha)^{2^{j-1}} T(-\alpha)^{2^{j-1}} \\
&= \left(\sum_{x=1}^N e(-\alpha x^k) \right)^{k-1} \left(\sum_{y=1}^N e(\alpha y^k) \right)^{k-1} \\
&= \sum_{x_1=1}^N \cdots \sum_{x_{j-1}=1}^N \sum_{y_1=1}^N \cdots \sum_{y_{j-1}=1}^N e\left(\alpha \left(\sum_{i=1}^{j-1} x_i^k - \sum_{i=1}^{j-1} y_i^k \right) \right) \\
&= \sum_d s(d)e(-\alpha d),
\end{aligned}
$$

where $s(d)$ is the number of representations of d in the form

$$d = \sum_{i=1}^{j-1} y_i^k - \sum_{i=1}^{j-1} x_i^k,$$

with $1 \le x_i, y_i \le N$ for $i = 1, \ldots, j - 1$. Then

$$\sum_d s(d) = |T(0)|^{2^j} = N^{2^j}$$

and, by the induction hypothesis,

$$s(0) = \int_0^1 |T(\alpha)|^{2^j} \, d\alpha \ll N^{2^j - j + \varepsilon}.$$

It follows from (4.6) that

$$\int_0^1 |T(\alpha)|^{2^{j+1}} \, d\alpha = \int_0^1 |T(\alpha)|^{2^j} \, |T(\alpha)|^{2^j} \, d\alpha$$

$$\leq N^{2^j-j-1} \int_0^1 \sum_{d'} r(d')e(\alpha d') \sum_d s(d)e(-\alpha d)d\alpha$$

$$= N^{2^j-j-1} \sum_d r(d)s(d)$$

$$= N^{2^j-j-1} r(0)s(0) + N^{2^j-j-1} \sum_{d\neq0} r(d)s(d)$$

$$\ll N^{2^j-j-1} N^j N^{2^j-j+\varepsilon} + N^{2^j-j-1} N^\varepsilon \sum_{d\neq0} s(d)$$

$$\ll N^{2^{j+1}-(j+1)+\varepsilon} + N^{2^j-j-1} N^\varepsilon N^{2^j}$$

$$\ll N^{2^{j+1}-(j+1)+\varepsilon}.$$

This completes the proof.

4.6 Notes

The material in this chapter is well-known. For the original proofs of Weyl's inequality and Hua's lemma, see Weyl [141] and Hua [62], respectively. Davenport [18], Schmidt [106], and Vaughan [125] are standard and excellent introductions to the circle method in additive number theory.

The easier Waring's problem was introduced by Wright [150].

4.7 Exercises

1. Prove that

$$\|x\| = \|-x\| = \|n+x\|$$

for all $x \in \mathbf{R}$ and $n \in \mathbf{Z}$. Let (x) denote the fractional part of x. Graph $f(x) = (x) + \|x\|$ for $0 \leq x \leq 1$.

2. Prove that

$$\|\alpha + \beta\| \leq \|\alpha\| + \|\beta\|$$

for all $\alpha, \beta \in \mathbf{R}$.

3. Let $\ell \geq 1$, and let Δ_ℓ denote the iterated difference operator $\Delta_{1,1,\dots,1}$. Prove that

$$\Delta_\ell(f)(x) = \sum_{j=0}^\ell (-1)^{\ell-j} \binom{\ell}{j} f(x+j).$$

4. Let $\Delta_{d_\ell,\dots,d_1}$ be an iterated difference operator. Find a general formula to express $\Delta_{d_\ell,\dots,d_1}(f)(x)$.

5. Let $\ell \geq 2$, let σ be a permutation of $\{1, 2, \ldots, \ell\}$, and let $\Delta_{d_\ell, \ldots, d_1}$ be an iterated difference operator. Prove that

$$\Delta_{d_{\sigma(\ell)}, \ldots, d_{\sigma(1)}} = \Delta_{d_\ell, \ldots, d_1}.$$

5

The Hardy–Littlewood asymptotic formula

... using essentially the same techniques as Hardy and Littlewood's but in a different way and introducing certain additional considerations, we shall derive the same result with incomparable brevity and simplicity.

I. M. Vinogradov [131]

5.1 The circle method

For any positive integers k and s, let $r_{k,s}(N)$ denote the number of representations of N as the sum of s positive kth powers, that is, the number of s-tuples $(x_1, \ldots, x_s)$ of positive integers such that

$$N = x_1^k + \cdots + x_s^k.$$

Waring's problem is to prove that every nonnegative integer is the sum of a bounded number of kth powers. Since $1 = 1^k$ is a kth power, this is equivalent to showing that

$$r_{k,s}(N) > 0$$

for some s and for all sufficiently large integers N. Hilbert gave the first proof of Waring's problem in 1909. Ten years later, Hardy and Littlewood succeeded in finding a beautiful asymptotic formula for $r_{k,s}(N)$. They proved that for $s \geq s_0(k)$, there exists $\delta = \delta(s, k) > 0$ such that

$$r_{k,s}(N) = \mathfrak{S}(N)\Gamma\left(1 + \frac{1}{k}\right)^{s} \Gamma\left(\frac{s}{k}\right)^{-1} N^{(s/k)-1} + O(N^{(s/k)-1-\delta}), \qquad (5.1)$$

where $\Gamma(x)$ is the Gamma function and $\mathfrak{S}(N)$ is the "singular series," an arithmetic function that is uniformly bounded above and below by positive constants depending only on k and s. We shall prove that the asymptotic formula (5.1) holds for $s_0(k) = 2^k + 1$.

Hardy and Littlewood used the "circle method" to obtain their result. The idea at the heart of the circle method is simple. Let A be any set of nonnegative integers. The generating function for A is

$$f(z) = \sum_{a \in A} z^{a}.$$

We can consider $f(z)$ either as a formal power series in z or as the Taylor series of an analytic function that converges in the open unit disc $|z| < 1$. In both cases,

$$f(z)^{s} = \sum_{N=0}^{\infty} r_{A,s}(N) z^{N},$$

where $r_{A,s}(N)$ is the number of representations of N as the sum of s elements of A, that is, the number of solutions of the equation

$$N = a_1 + a_2 + \cdots + a_s$$

with

$$a_1, a_2, \cdots, a_s \in A.$$

By Cauchy's theorem, we can recover $r_{A,s}(N)$ by integration:

$$r_{A,s}(N) = \frac{1}{2\pi i} \int_{|z|=\rho} \frac{f(z)^{s}}{z^{N+1}} dz$$

for any $\rho \in (0, 1)$.

This is the original form of the "circle method" introduced by Hardy, Littlewood, and Ramanujan in 1918–20. They evaluated the integral by dividing the circle of integration into two disjoint sets, the "major arcs" and the "minor arcs." In the classical applications to Waring's problem, the integral over the minor arcs is negligible, and the integral over the major arcs provides the main term in the estimate for $r_{A,s}(N)$.

Vinogradov greatly simplified and improved the circle method. He observed that in order to study $r_{A,s}(N)$, it is possible to replace the power series $f(z)$ with the polynomial

$$p(z) = \sum_{\substack{a \in A \\ a \leq N}} z^{a}.$$

Then

$$p(z)^s = \sum_{m=0}^{sN} r_{A,s}^{(N)}(m)z^m,$$

where $r_{A,s}^{(N)}(m)$ is the number of representations of m as the sum of s elements of A not exceeding N. In particular, since the elements of A are nonnegative, we have $r_{A,s}^{(N)}(m) = r_{A,s}(m)$ for $m \le N$ and $r_{A,s}^{(N)}(m) = 0$ for $m > sN$. If we let

$$z = e(\alpha) = e^{2\pi i \alpha},$$

then we obtain the trigonometric polynomial

$$F(\alpha) = p(e(\alpha)) = \sum_{\substack{a \in A \\ a \le N}} e(a\alpha)$$

and

$$F(\alpha)^s = \sum_{m=0}^{sN} r_{A,s}^{(N)}(m)e(m\alpha).$$

From the basic orthogonality relation for the functions $e(n\alpha)$,

$$\int_0^1 e(m\alpha)e(-n\alpha)d\alpha = \begin{cases} 1 & \text{if } m = n \\ 0 & \text{if } m \ne n, \end{cases}$$

we obtain

$$r_{A,s}(N) = \int_0^1 F(\alpha)^s e(-N\alpha)d\alpha.$$

In applications, of course, the hard part is to estimate the integral.

To apply the circle method to Waring's problem, let $k \ge 2$ and A be the set of positive kth powers. Let $r_{k,s}(N)$ denote the number of representations of N as the sum of s positive kth powers. Let

$$P = [N^{1/k}].$$

Then

$$F(\alpha) = \sum_{\substack{a \in A \\ a \le N}} e(\alpha a) = \sum_{n=1}^{P} e(\alpha n^k)$$

and

$$r_{k,s}(N) = \int_0^1 F(\alpha)^s e(-\alpha N)d\alpha.$$

5.2   Waring's problem for $k = 1$

For $k = 1$, there is an explicit formula for $r_{1,s}(N)$.

Theorem 5.1 *Let $s \geq 1$. Then*

$$r_{1,s}(N) = \binom{N-1}{s-1} = \frac{N^{s-1}}{(s-1)!} + O\left(N^{s-2}\right)$$

for all positive integers N.

Proof. Let $N \geq s$. We observe that

$$N = a_1 + \cdots + a_s$$

is a decomposition of N into s positive parts if and only if

$$N - s = (a_1 - 1) + \cdots + (a_s - 1)$$

is a decomposition of N into s nonnegative parts. Therefore,

$$r_{1,s}(N) = R_{1,s}(N - s),$$

where $R_{1,s}(N)$ denotes the number of representations of N as the sum of s non-negative integers.

We shall give two proofs of the theorem. The first is combinatorial. We begin by computing $R_{1,s}(N)$ for every nonnegative integer N. Let $N = a_1 + \cdots + a_s$ be a partition into nonnegative integers. Imagine a row of $N + s - 1$ boxes. We color the first a_1 boxes red, the next box blue, the next a_2 boxes red, the next box blue, and so on. There will be exactly $s - 1$ blue boxes. Conversely, if we choose $s - 1$ of the $N + s - 1$ boxes and color them blue, and if we color the remaining N boxes red, then we have a partition of N into s nonnegative parts as follows. Let a_1 be the number of red boxes before the first blue box, a_2 the number of red boxes between the first and second blue boxes, and, in general, for $j = 2, \ldots, s - 1$, let a_j be the number of red boxes that are between the $(j - 1)$-st and jth blue boxes. Let a_s be the number of red boxes that come after the last blue box. This establishes a one-to-one correspondence between the subsets of size $s - 1$ of the $N + s - 1$ boxes and the representations of N as the sum of s nonnegative integers. Therefore, the number of decompositions of N into s nonnegative parts is the binomial coefficient $\binom{N+s-1}{s-1}$. It follows that

$$r_{1,s}(N) = R_{1,s}(N - s) = \binom{N-1}{s-1}.$$

This gives the first proof of the theorem.

There is also a simple analytic proof. The series

$$f(z) = \sum_{N=0}^{\infty} z^N = \frac{1}{1-z}$$

converges for $|z| < 1$, and

$$f(z)^s = \sum_{N=0}^{\infty} R_{1,s}(N) z^N.$$

We also have

$$
\begin{aligned}
f(z)^s &= \frac{1}{(1-z)^s} \\
&= \frac{1}{(s-1)!} \frac{d^{s-1}}{dz^{s-1}} \left(\frac{1}{1-z} \right) \\
&= \frac{1}{(s-1)!} \frac{d^{s-1}}{dz^{s-1}} \left(\sum_{N=0}^{\infty} z^N \right) \\
&= \sum_{N=s-1}^{\infty} \frac{N(N-1)\cdots(N-s+2)}{(s-1)!} z^{N-s+1} \\
&= \sum_{N=s-1}^{\infty} \binom{N}{s-1} z^{N-s+1} \\
&= \sum_{N=0}^{\infty} \binom{N+s-1}{s-1} z^N.
\end{aligned}
$$

Therefore,

$$R_{1,s}(N) = \binom{N+s-1}{s-1}.$$

This completes the proof.

5.3   The Hardy–Littlewood decomposition

For $k \geq 2$ there is no easy way to compute–or even to estimate–$r_{k,s}(N)$ for large N. It was a great achievement of Hardy and Littlewood to obtain an asymptotic formula for $r_{k,s}(N)$ for all $k \geq 2$ and $s \geq s_0(k)$. In this chapter, we shall prove the Hardy–Littlewood asymptotic formula for $s \geq 2^k + 1$. For $N \geq 2^k$, let

$$P = \left[N^{1/k} \right] \tag{5.2}$$

and

$$F(\alpha) = \sum_{m=1}^{P} e(\alpha m^k). \tag{5.3}$$

The trigonometric polynomial $F(\alpha)$ is the generating function for representing N as the sum of kth powers. The basis of the circle method is the simple formula

$$r_{k,s}(N) = \int_0^1 F(\alpha)^s e(-N\alpha) d\alpha. \tag{5.4}$$

We cannot compute this integral explicitly in terms of elementary functions. By carefully estimating the integral, however, we shall derive the Hardy–Littlewood asymptotic formula.

The first step is to decompose the unit interval $[0, 1]$ into two disjoint sets, called the *major arcs* $\mathfrak{M}$ and the *minor arcs* m, and to evaluate the integral separately over both sets. The major arcs will consist of all real numbers $\alpha \in [0, 1]$ that can, in a certain sense, be "well approximated" by rational numbers, and the minor arcs consist of the numbers $\alpha \in [0, 1]$ that cannot be well approximated. Although most of the mass of the unit interval lies in the minor arcs, it will follow from Weyl's inequality and Hua's lemma that the integral of $f(\alpha)^s e(-N\alpha)$ over the minor arcs is negligible. The integral over the major arcs will factor into the product of two terms: the "singular integral" $J(N)$ and the "singular series" $\mathfrak{S}(N)$. The singular integral will be evaluated in terms of the Gamma function, and the singular series will be estimated by elementary number theory.

The major and minor arcs are constructed as follows. Let $N \geq 2^k$. Then $P = [N^{1/k}] \geq 2$. Choose

$$0 < \nu < 1/5.$$

For

$$1 \leq q \leq P^\nu,$$

$$0 \leq a \leq q,$$

and

$$(a, q) = 1,$$

we let

$$\mathfrak{M}(q, a) = \left\{ \alpha \in [0, 1] : \left| \alpha - \frac{a}{q} \right| \leq \frac{1}{P^{k-\nu}} \right\}$$

and

$$\mathfrak{M} = \bigcup_{1 \leq q \leq P^\nu} \bigcup_{\substack{a=0 \\ (a,q)=1}}^{q} \mathfrak{M}(q, a).$$

The interval $\mathfrak{M}(q, a)$ is called a *major arc*, and $\mathfrak{M}$ is the set of all major arcs. We see that

$$\mathfrak{M}(1, 0) = \left[0, \frac{1}{P^{k-\nu}} \right],$$

$$\mathfrak{M}(1, 1) = \left[1 - \frac{1}{P^{k-\nu}}, 1 \right],$$

and

$$\mathfrak{M}(q, a) = \left[\frac{a}{q} - \frac{1}{P^{k-\nu}}, \frac{a}{q} + \frac{1}{P^{k-\nu}} \right]$$

for $q \geq 2$. The major arcs consist of all real numbers $\alpha \in [0, 1]$ that are well approximated by rationals in the sense that they are close, within distance $P^{\nu-k}$, to a rational number with denominator no greater than P^ν.

If $\alpha \in \mathfrak{M}(q, a) \cap \mathfrak{M}(q', a')$ and $a/q \neq a'/q'$, then $|aq' - a'q| \geq 1$ and

$$\frac{1}{P^{2v}} \leq \frac{1}{qq'}$$

$$\leq \left| \frac{a}{q} - \frac{a'}{q'} \right|$$

$$\leq \left| \alpha - \frac{a}{q} \right| + \left| \alpha - \frac{a'}{q'} \right|$$

$$\leq \frac{2}{P^{k-v}},$$

which is impossible for $P \geq 2$ and $k \geq 2$. Therefore, the major arcs $\mathfrak{M}(q, a)$ are pairwise disjoint.

The measure of the set $\mathfrak{M}(1, 0) \cup \mathfrak{M}(1, 1)$ is $2P^{v-k}$, and, for every $q \geq 2$ and $(a, q) = 1$, the measure of the major arc $\mathfrak{M}(q, a)$ is $2P^{v-k}$. For every $q \geq 2$ there are exactly $\varphi(q)$ positive integers a such that $1 \leq a \leq q$ and $(q, a) = 1$. It follows that the measure of the set $\mathfrak{M}$ of major arcs is

$$\mu(\mathfrak{M}) = \frac{2}{P^{k-v}} \sum_{1 \leq q \leq P^v} \varphi(q) \leq \frac{2}{P^{k-v}} \sum_{1 \leq q \leq P^v} q$$

$$\leq \frac{2}{P^{k-v}} \frac{P^v(P^v + 1)}{2} \leq \frac{2}{P^{k-3v}}, \tag{5.5}$$

which goes to zero as P goes to infinity.

The set

$$\mathfrak{m} = [0, 1] \setminus \mathfrak{M}$$

is called the set of *minor arcs*. This set is a finite union of open intervals and consists of all $\alpha \in [0, 1]$ that are not well approximated by rationals. The measure of the set of minor arcs is

$$\mu(\mathfrak{m}) = 1 - \mu(\mathfrak{M}) > 1 - \frac{2}{P^{k-3v}}.$$

Even though the measure of the set $\mathfrak{m}$ is large in the sense that it tends to 1 as P tends to infinity, we shall prove in the next section that the integral over the minor arcs contributes only a negligible amount to $r_{k,s}(N)$.

5.4 The minor arcs

We shall now show that the integral over the minor arcs is small.

Theorem 5.2 *Let $k \geq 2$ and $s \geq 2^k + 1$. There exists $\delta_1 > 0$ such that*

$$\int_{\mathfrak{m}} F(\alpha)^s e(-N\alpha) d\alpha = O\left(P^{s-k-\delta_1}\right),$$

where the implied constant depends only on k and s.

Proof. By Dirichlet's theorem (Theorem 4.1) with $Q = P^{k-\nu}$, to every real number α there corresponds a fraction a/q such that

$$1 \le q \le P^{k-\nu}, \qquad (a, q) = 1,$$

and

$$\left| \alpha - \frac{a}{q} \right| \le \frac{1}{q P^{k-\nu}} \le \min\left(\frac{1}{P^{k-\nu}}, \frac{1}{q^2} \right).$$

If $\alpha \in \mathfrak{m}$, then $\alpha \notin \mathfrak{M}(1, 0) \cup \mathfrak{M}(1, 1)$, so

$$\frac{1}{P^{k-\nu}} < \alpha < 1 - \frac{1}{P^{k-\nu}}$$

and $1 \le a \le q - 1$. If $q \le P^\nu$, then

$$\left| \alpha - \frac{a}{q} \right| \le \frac{1}{P^{k-\nu}}$$

implies that

$$\alpha \in \mathfrak{M}(q, a) \subseteq \mathfrak{M} = [0, 1] \setminus \mathfrak{m},$$

which is absurd. Therefore,

$$P^\nu < q \le P^{k-\nu}.$$

Let

$$K = 2^{k-1}. \tag{5.6}$$

It follows from Weyl's inequality (Theorem 4.3) with $f(x) = \alpha x^k$ that

$$F(\alpha) \ll P^{1+\varepsilon} \left(P^{-1} + q^{-1} + P^{-k} q \right)^{1/K}$$

$$\ll P^{1+\varepsilon} \left(P^{-1} + P^{-\nu} + P^{-k} P^{k-\nu} \right)^{1/K}$$

$$\ll P^{1+\varepsilon-\nu/K}.$$

Applying Hua's lemma (Theorem 4.6), we obtain

$$\left| \int_{\mathfrak{m}} F(\alpha)^s e(-n\alpha) d\alpha \right| = \left| \int_{\mathfrak{m}} F(\alpha)^{s-2^k} F(\alpha)^{2^k} e(-n\alpha) d\alpha \right|$$

$$\le \int_{\mathfrak{m}} |F(\alpha)|^{s-2^k} |F(\alpha)|^{2^k} d\alpha$$

$$\le \max_{\alpha \in \mathfrak{m}} |F(\alpha)|^{s-2^k} \int_0^1 |F(\alpha)|^{2^k} d\alpha$$

$$\ll \left(P^{1+\varepsilon-\nu/K} \right)^{s-2^k} P^{2^k-k+\varepsilon}$$

$$= P^{s-k-\delta_1},$$

where

$$\delta_1 = \frac{\nu(s - 2^k)}{K} - (s - 2^k + 1)\varepsilon > 0$$

if $\varepsilon > 0$ is chosen sufficiently small. This completes the proof.

5.5 The major arcs

We introduce the auxiliary functions

$$v(\beta) = \sum_{m=1}^{N} \frac{1}{k} m^{1/k-1} e(\beta m)$$

and

$$S(q, a) = \sum_{r=1}^{q} e(ar^k/q).$$

We shall prove that if α lies in the major arc $\mathfrak{M}(q, a)$, then $F(\alpha)$ is the product of $S(q, a)/q$ and $v(\alpha - a/q)$, plus a small error term. We begin by estimating these functions.

Clearly, $|S(q, a)| \leq q$. By Weyl's inequality (Theorem 4.4), we have

$$S(q, a) \ll q^{1-1/K+\varepsilon}$$

and

$$\frac{S(q, a)}{q} \ll q^{-1/K+\varepsilon}, \tag{5.7}$$

where the implied constant depends only on ε.

Lemma 5.1 If $|\beta| \leq 1/2$, then

$$v(\beta) \ll \min(P, |\beta|^{-1/k}).$$

Proof. The function

$$f(x) = \frac{1}{k} x^{1/k-1}$$

is positive, continuous, and decreasing for $x \geq 1$. By Lemma A.2, it follows that

$$|v(\beta)| \leq \sum_{m=1}^{N} \frac{1}{k} m^{1/k-1}$$

$$\leq \int_{1}^{N} k^{-1} x^{1/k-1} dx + f(1)$$

$$< N^{1/k}$$

$$\ll P.$$

If $|\beta| \leq 1/N$, then $P \leq N^{1/k} \leq |\beta|^{-1/k}$ and $v(\beta) \ll \min(P, |\beta|^{-1/k})$. Suppose that $1/N < |\beta| \leq 1/2$. Then $|\beta|^{-1/k} \ll P$. Let $M = \left[|\beta|^{-1}\right]$. Then

$$M \leq \frac{1}{\beta} < M + 1 \leq N.$$

Let $U(t) = \sum_{m \le t} e(\beta m)$. By Lemma 4.7, we have $U(t) \ll \|\beta\|^{-1} = |\beta|^{-1}$. By partial summation (Theorem A.4),

$$\sum_{m=M+1}^{N} \frac{1}{k} m^{1/k-1} e(\beta m) = f(N)U(N) - f(M)U(M) - \int_{M}^{N} U(t) f'(t) dt$$

$$\ll \frac{M^{1/k-1}}{|\beta|}$$

$$\le |\beta|^{-1/k}$$

$$\ll \min(P, |\beta|^{-1/k}).$$

Therefore,

$$v(\beta) = \sum_{m=1}^{M} \frac{1}{k} m^{1/k-1} e(\beta m) + \sum_{m=M+1}^{N} \frac{1}{k} m^{1/k-1} e(\beta m)$$

$$\ll \min(P, |\beta|^{-1/k}).$$

This completes the proof.

Lemma 5.2 *Let q and a be integers such that $1 \le q \le P^v$, $0 \le a \le q$, and $(a, q) = 1$. If $\alpha \in \mathfrak{M}(q, a)$, then*

$$F(\alpha) = \left(\frac{S(q, a)}{q}\right) v\left(\alpha - \frac{a}{q}\right) + O(P^{2v}).$$

Proof. Let $\beta = \alpha - a/q$. Then $|\beta| \le P^{v-k}$ and

$$F(\alpha) - \frac{S(q, a)}{q} v(\beta)$$

$$= \sum_{m=1}^{P} e(\alpha m^k) - \frac{S(q, a)}{q} \sum_{m=1}^{N} \frac{1}{k} m^{1/k-1} e(\beta m)$$

$$= \sum_{m=1}^{P} e\left(\frac{am^k}{q}\right) e(\beta m^k) - \frac{S(q, a)}{q} \sum_{m=1}^{N} \frac{1}{k} m^{1/k-1} e(\beta m)$$

$$= \sum_{m=1}^{N} u(m) e(\beta m),$$

where

$$u(m) = \begin{cases} e(am/q) - (S(q, a)/q) k^{-1} m^{1/k-1} & \text{if } m \text{ is a } k\text{th power} \\ -(S(q, a)/q) k^{-1} m^{1/k-1} & \text{otherwise.} \end{cases}$$

We shall estimate the last sum. Let $y \ge 1$. Since $|S(q, a)| \le q$, we have

$$\sum_{1 \le m \le y} e(am^k/q) = \sum_{r=1}^{q} e(ar^k/q) \sum_{\substack{1 \le m \le y \\ m \equiv r \pmod q}} 1$$

$$= S(q, a) \left(\frac{y}{q} + O(1) \right)$$

$$= y \left(\frac{S(q, a)}{q} \right) + O(q).$$

Let $t \geq 1$. Since $v(\beta) \ll P$, we have

$$U(t) = \sum_{1 \leq m \leq t} u(m)$$

$$= \sum_{1 \leq m \leq t^{1/k}} e(am^k/q) - \frac{S(q, a)}{q} \sum_{1 \leq m \leq t} \frac{1}{k} m^{1/k-1}$$

$$= t^{1/k} \left(\frac{S(q, a)}{q} \right) + O(q) - \left(\frac{S(q, a)}{q} \right) (t^{1/k} + O(1))$$

$$= O(q).$$

By partial summation,

$$\sum_{m=1}^{N} u(m) e(\beta m) = e(\beta N) U(N) - 2\pi i \beta \int_{1}^{N} e(\beta t) U(t) dt$$

$$= O(q) - 2\pi i \beta \int_{1}^{N} e(\beta t) O(q) dt$$

$$\ll q + |\beta| N q$$

$$\ll (1 + |\beta| N) q$$

$$\ll (1 + P^{v-k} P^k) P^v$$

$$\ll P^{2v}.$$

This completes the proof.

Theorem 5.3 *Let*

$$\mathfrak{S}(N, Q) = \sum_{1 \leq q \leq Q} \sum_{\substack{a=1 \\ (a,q)=1}}^{q} \left(\frac{S(q, a)}{q} \right)^s e(-Na/q)$$

and

$$J^*(N) = \int_{-P^{v-k}}^{P^{v-k}} v(\beta)^s e(-N\beta) d\beta.$$

Let $\mathfrak{M}$ *denote the set of major arcs. Then*

$$\int_{\mathfrak{M}} F(\alpha)^s e(-N\alpha) d\alpha = \mathfrak{S}(N, P^v) J^*(N) + O\left(P^{s-k-\delta_2} \right),$$

where $\delta_2 = (1 - 5v)/k > 0.$

Proof. Let $\alpha \in \mathfrak{M}(q, a)$ and

$$\beta = \alpha - \frac{a}{q}.$$

Let

$$V = V(\alpha, q, a) = \frac{S(q, a)}{q} v\left(\alpha - \frac{a}{q}\right) = \frac{S(q, a)}{q} v(\beta).$$

Since $|S(q, a)| \leq q$, we have $|V| \ll |v(\beta)| \ll P$ by Lemma 5.1. Let $F = F(\alpha)$. Then $|F| \leq P$. Since $F - V = O(P^{2\nu})$ by Lemma 5.2, it follows that

$$F^s - V^s = (F - V)\left(F^{s-1} + F^{s-2}V + \cdots + V^{s-1}\right)$$
$$\ll P^{2\nu} P^{s-1}$$
$$= P^{s-1+2\nu}.$$

Since $\mu(\mathfrak{M}) \ll P^{3\nu - k}$ by (5.5), it follows that

$$\int_{\mathfrak{M}} |F^s - V^s| \, d\alpha \ll P^{3\nu - k} P^{s-1+2\nu} = P^{s-k-\delta_2},$$

where $\delta_2 = 1 - 5\nu > 0$. Therefore,

$$\int_{\mathfrak{M}} F(\alpha)^s e(-N\alpha) d\alpha$$
$$= \int_{\mathfrak{M}} V(\alpha, q, a)^s e(-N\alpha) d\alpha + O\left(P^{s-k-\delta_2}\right)$$
$$= \sum_{1 \leq q \leq P^\nu} \sum_{\substack{a=0 \\ (a,q)=1}}^{q} \int_{\mathfrak{M}(q,a)} V(\alpha, q, a)^s e(-N\alpha) d\alpha + O\left(P^{s-k-\delta_2}\right).$$

For $q \geq 2$, we have

$$\int_{\mathfrak{M}(q,a)} V(\alpha, q, a)^s e(-N\alpha) d\alpha$$
$$= \int_{a/q - P^{\nu-k}}^{a/q + P^{\nu-k}} V(\alpha, q, a)^s e(-N\alpha) d\alpha$$
$$= \int_{-P^{\nu-k}}^{P^{\nu-k}} V(\beta + a/q, q, a)^s e(-N(\beta + a/q)) d\beta$$
$$= \left(\frac{S(q, a)}{q}\right)^s e(-Na/q) \int_{-P^{\nu-k}}^{P^{\nu-k}} v(\beta)^s e(-N\beta) d\beta$$
$$= \left(\frac{S(q, a)}{q}\right)^s e(-Na/q) J^*(N).$$

For $q = 1$ we have $V(\alpha, 1, 0) = v(\alpha)$ and $V(\alpha, 1, 1) = v(\alpha - 1)$. Therefore,

$$\int_{\mathfrak{M}(1,0)} V(\alpha, q, a)^s e(-N\alpha) d\alpha + \int_{\mathfrak{M}(1,1)} V(\alpha, q, a)^s e(-N\alpha) d\alpha$$

$$= \int_0^{P^{v-k}} v(\alpha)^s e(-N\alpha) d\alpha + \int_{1-P^{v-k}}^1 v(\alpha - 1)^s e(-N\alpha) d\alpha$$

$$= \int_0^{P^{v-k}} v(\beta)^s e(-N\beta) d\beta + \int_{-P^{v-k}}^0 v(\beta)^s e(-N\beta) d\beta$$

$$= J^*(N).$$

Therefore,

$$\int_{\mathfrak{M}} F(\alpha)^s e(-N\alpha) d\alpha$$

$$= \sum_{1 \leq q \leq P^v} \sum_{\substack{a=1 \\ (a,q)=1}}^q \left(\frac{S(q,a)}{q} \right)^s e(-Na/q) J^*(N) + O\left(P^{s-k-\delta_2} \right)$$

$$= \mathfrak{S}(N, P^v) J^*(N) + O\left(P^{s-k-\delta_2} \right).$$

This completes the proof.

5.6 The singular integral

Next we consider the integral

$$J(N) = \int_{-1/2}^{1/2} v(\beta)^s e(-\beta N) d\beta. \tag{5.8}$$

This is called the *singular integral* for Waring's problem.

Theorem 5.4 *There exists $\delta_3 > 0$ such that*

$$J(N) \ll P^{s-k}$$

and

$$J^*(N) = J(N) + O\left(P^{s-k-\delta_3} \right).$$

Proof. By Lemma 5.1,

$$J(N) \ll \int_0^{1/2} \min(P, |\beta|^{-1/k})^s d\beta$$

$$= \int_0^{1/N} \min(P, |\beta|^{-1/k})^s d\beta + \int_{1/N}^{1/2} \min(P, |\beta|^{-1/k})^s d\beta$$

$$= \int_0^{1/N} P^s d\beta + \int_{1/N}^{1/2} \beta^{-s/k} d\beta$$

$$\ll P^{s-k}$$

and

$$J(N) - J^*(N) = \int_{P^{\nu-k} \le |\beta| \le 1/2} v(\beta)^s e(-N\beta) d\beta$$

$$\ll \int_{P^{\nu-k}}^{1/2} |v(\beta)|^s d\beta$$

$$\ll \int_{P^{\nu-k}}^{1/2} \beta^{-s/k} d\beta$$

$$\ll P^{(k-\nu)(s/k-1)}$$

$$= P^{s-k-\delta_3},$$

where $\delta_3 = \nu(s/k - 1) > 0$. This completes the proof.

Lemma 5.3 *Let α and β be real numbers such that $0 < \beta < 1$ and $\alpha \ge \beta$. Then*

$$\sum_{m=1}^{N-1} m^{\beta-1}(N-m)^{\alpha-1} = N^{\alpha+\beta-1}\frac{\Gamma(\alpha)\Gamma(\beta)}{\Gamma(\alpha+\beta)} + O\left(N^{\alpha-1}\right),$$

where the implied constant depends only on β.

Proof. The function

$$g(x) = x^{\beta-1}(N-x)^{\alpha-1}$$

is positive and continuous on $(0, N)$, integrable on $[0, N]$, and

$$\int_0^N g(x)dx = \int_0^N x^{\beta-1}(N-x)^{\alpha-1}dx$$

$$= N^{\alpha+\beta-1}\int_0^1 t^{\beta-1}(1-t)^{\alpha-1}dt$$

$$= N^{\alpha+\beta-1}B(\alpha, \beta)$$

$$= N^{\alpha+\beta-1}\frac{\Gamma(\alpha)\Gamma(\beta)}{\Gamma(\alpha+\beta)},$$

where $B(\alpha, \beta)$ is the Beta function and $\Gamma(\alpha)$ is the Gamma function.
    If $\alpha \ge 1$, then

$$f'(x) = g(x)\left(\frac{\beta-1}{x} - \frac{\alpha-1}{N-x}\right) < 0$$

and so $g(x)$ is decreasing on $(0, N)$ and

$$\int_1^N g(x)dx < \sum_{m=1}^{N-1} g(x) < \int_0^{N-1} g(x)dx.$$

Therefore,

$$0 < \int_0^N g(x)dx - \sum_{m=1}^{N-1} g(m)$$

$$< \int_0^1 g(x)dx$$

$$= \int_0^1 x^{\beta-1}(N-x)^{\alpha-1}dx$$

$$\leq N^{\alpha-1} \int_0^1 x^{\beta-1}dx$$

$$= \frac{N^{\alpha-1}}{\beta}.$$

If $0 < \beta \leq \alpha < 1$, then $0 < \alpha + \beta < 2$ and $g(x)$ has a local minimum at

$$c = \frac{(1-\beta)N}{2-\alpha-\beta} \in [N/2, N).$$

Since $g(x)$ is strictly decreasing for $x \in (0, c)$, it follows that

$$\sum_{m=1}^{[c]} g(m) < \int_0^c g(x)dx$$

and

$$\sum_{m=1}^{[c]} g(m) \geq \int_1^{[c]} g(x)dx + g([c])$$

$$> \int_1^c g(x)dx$$

$$> \int_0^c g(x)dx - \frac{N^{\alpha-1}}{\beta}.$$

Similarly, since $g(x)$ is increasing for $x \in (c, N)$, it follows that

$$\sum_{m=[c]+1}^{N-1} g(m) < \int_c^N g(x)dx$$

and

$$\sum_{m=[c]+1}^{N-1} g(m) \geq \int_{[c]+1}^{N-1} g(x)dx + g([c]+1)$$

$$> \int_c^{N-1} g(x)dx$$

$$> \int_c^N g(x)dx - \frac{N^{\beta-1}}{\alpha}.$$

Therefore,

$$0 < \int_0^N g(x)dx - \sum_{m=1}^{N-1} g(m) < \frac{N^{\alpha-1}}{\beta} + \frac{N^{\beta-1}}{\alpha} \leq \frac{2N^{\alpha-1}}{\beta}.$$

This completes the proof.

Theorem 5.5 *If $s \geq 2$, then*

$$J(N) = \Gamma\left(1 + \frac{1}{k}\right)^s \Gamma\left(\frac{s}{k}\right)^{-1} N^{s/k-1} + O\left(N^{(s-1)/k-1}\right).$$

Proof. Let

$$J_s(N) = \int_{-1/2}^{1/2} v(\beta)^s e(-N\beta) d\beta$$

for $s \geq 1$. We shall compute this integral by induction on s. Since

$$v(\beta) = \sum_{m=1}^{N} \frac{1}{k} m^{1/k-1} e(\beta m),$$

it follows that

$$v(\beta)^s = k^{-s} \sum_{m_1=1}^{N} \cdots \sum_{m_s=1}^{N} (m_1 \cdots m_s)^{1/k-1} e((m_1 + \cdots + m_s)\beta)$$

and so

$$J_s(N) = k^{-s} \sum_{m_1=1}^{N} \cdots \sum_{m_s=1}^{N} (m_1 \cdots m_s)^{1/k-1} \int_{-1/2}^{1/2} e((m_1 + \cdots + m_s - N)\beta) d\beta$$

$$= k^{-s} \sum_{\substack{m_1+\cdots+m_s=N \\ 1 \leq m_i \leq N}} (m_1 \cdots m_s)^{1/k-1}.$$

In particular, for $s = 2$, we apply Lemma 5.3 with $\alpha = \beta = 1/k$ and obtain

$$J_2(N) = k^{-2} \sum_{m=1}^{N-1} m^{1/k-1}(N-m)^{1/k-1}$$

$$= \frac{(1/k)^2 \Gamma(1/k)^2}{\Gamma(2/k)} N^{2/k-1} + O(N^{1/k-1})$$

$$= \frac{\Gamma(1+1/k)^2}{\Gamma(2/k)} N^{2/k-1} + O(N^{1/k-1}).$$

This proves the result in the case where $s = 2$.

If $s \geq 2$ and the theorem holds for s, then

$$J_{s+1}(N) = \int_{-1/2}^{1/2} v(\beta)^{s+1} e(-N\beta) d\beta$$

$$= \int_{-1/2}^{1/2} v(\beta)v(\beta)^s e(-N\beta) d\beta$$

$$= \int_{-1/2}^{1/2} \sum_{m=1}^{N} \frac{1}{k} m^{1/k-1} e(\beta m) v(\beta)^s e(-N\beta) d\beta$$

$$= \sum_{m=1}^{N} \frac{1}{k} m^{1/k-1} \int_{-1/2}^{1/2} v(\beta)^s e(-(N-m)\beta) d\beta$$

$$= \sum_{m=1}^{N} \frac{1}{k} m^{1/k-1} J_s(N-m)$$

$$= \frac{\Gamma(1+1/k)^s}{\Gamma(s/k)} \sum_{m=1}^{N-1} \frac{1}{k} m^{1/k-1} (N-m)^{s/k-1}$$

$$+ O\left(\sum_{m=1}^{N-1} \frac{1}{k} m^{1/k-1} (N-m)^{(s-1)/k-1} \right).$$

Applying Lemma 5.3 to the main term (with $\alpha = s/k$ and $\beta = 1/k$) and the error term (with $\alpha = (s-1)/k$ and $\beta = 1/k$), we obtain

$$\sum_{m=1}^{N-1} \frac{1}{k} m^{1/k-1} (N-m)^{s/k-1} = \frac{(1/k)\Gamma(1/k)\Gamma(s/k)}{\Gamma((s+1)/k)} N^{(s+1)/k-1} + O\left(N^{s/k-1}\right)$$

and

$$\sum_{m=1}^{N-1} \frac{1}{k} m^{1/k-1} (N-m)^{(s-1)/k-1} = O\left(N^{s/k-1}\right).$$

This gives

$$J_{s+1}(N) = \frac{(1/k)\Gamma(1/k)\Gamma(s/k)}{\Gamma((s+1)/k)} \frac{\Gamma(1+1/k)^s}{\Gamma(s/k)} N^{(s+1)/k-1} + O\left(N^{s/k-1}\right)$$

$$= \frac{\Gamma(1+1/k)^{s+1}}{\Gamma((s+1)/k)} N^{(s+1)/k-1} + O\left(N^{s/k-1}\right).$$

This completes the induction.

5.7 The singular series

In Theorem 5.3, we introduced the function

$$\mathfrak{S}(N, Q) = \sum_{1 \leq q \leq Q} A_N(q),$$

where

$$A_N(q) = \sum_{\substack{a=1 \\ (a,q)=1}}^{q} \left(\frac{S(q,a)}{q} \right)^s e\left(\frac{-Na}{q} \right).$$

We define the *singular series* for Waring's problem as the arithmetic function

$$\mathfrak{S}(N) = \sum_{q=1}^{\infty} A_N(q).$$

Let

$$0 < \varepsilon < \frac{1}{sK}.$$

Since $s \geq 2^k + 1 = 2K + 1$, we have

$$\frac{s}{K} - 1 - s\varepsilon \geq 1 + \frac{1}{K} - s\varepsilon = 1 + \delta_4,$$

where

$$\delta_4 = \frac{1}{K} - s\varepsilon > 0.$$

By (5.7),

$$A_N(q) \ll \frac{q}{q^{s/K - s\varepsilon}} \leq \frac{1}{q^{1+\delta_4}}, \qquad (5.9)$$

and so the singular series $\sum_q A_N(q)$ converges absolutely and uniformly with respect to N. In particular, there exists a constant $c_2 = c_2(k, s)$ such that

$$|\mathfrak{S}(N)| < c_2 \qquad (5.10)$$

for all positive integers N. Moreover,

$$\mathfrak{S}(N) - \mathfrak{S}(N, P^\nu) = \sum_{q > P^\nu} A_N(q)$$

$$\ll \sum_{q > P^\nu} \frac{1}{q^{1+\delta_4}}$$

$$\ll P^{-\nu\delta_4}.$$

We shall show that $\mathfrak{S}(N)$ is a positive real number for all N and that there exists a positive constant c_1 depending only on k and s such that

$$0 < c_1 < \mathfrak{S}(N) < c_2$$

for all positive integers N. The proof is a nice exercise in elementary number theory. We begin by showing that $A_N(q)$ is a multiplicative function of q.

Lemma 5.4 *Let* $(q, r) = 1$. *Then*

$$S(qr, ar + bq) = S(q, a)S(r, b).$$

Proof. Since $(q, r) = 1$, the sets $\{xr : 1 \leq x \leq q\}$ and $\{yq : 1 \leq y \leq r\}$ are complete residue systems modulo q and r, respectively. Because every congruence class modulo qr can be written uniquely in the form $xr + yq$, where $1 \leq x \leq q$ and $1 \leq y \leq r$, it follows that

$$S(qr, ar + bq) = \sum_{m=1}^{qr} e\left(\frac{(ar + bq)m^k}{qr}\right)$$

$$= \sum_{x=1}^{q} \sum_{y=1}^{r} e\left(\frac{(ar+bq)(xr+yq)^k}{qr}\right)$$

$$= \sum_{x=1}^{q} \sum_{y=1}^{r} e\left(\left(\frac{(ar+bq)}{qr}\right) \sum_{\ell=0}^{k} \binom{k}{\ell}(xr)^\ell(yq)^{k-\ell}\right)$$

$$= \sum_{x=1}^{q} \sum_{y=1}^{r} e\left(\left(\frac{(ar+bq)}{qr}\right)((xr)^k+(yq)^k)\right)$$

$$= \sum_{x=1}^{q} \sum_{y=1}^{r} e\left(\frac{a(xr)^k}{q}\right) e\left(\frac{b(yq)^k}{r}\right)$$

$$= \sum_{x=1}^{q} e\left(\frac{ax^k}{q}\right) \sum_{y=1}^{r} e\left(\frac{by^k}{r}\right)$$

$$= S(q,a)S(r,b).$$

This completes the proof.

Lemma 5.5 *If* $(q,r) = 1$, *then*

$$A_N(qr) = A_N(q)A_N(r),$$

that is, the function $A_N(q)$ *is multiplicative.*

Proof. If c and qr are relatively prime, then c is congruent modulo qr to a number of the form $ar + bq$, where $(a,q) = (b,r) = 1$. It follows from Lemma 5.4 that

$$A_N(qr) = \sum_{\substack{c=1 \\ (c,qr)=1}}^{qr} \left(\frac{S(qr,c)}{qr}\right)^s e\left(-\frac{cN}{qr}\right)$$

$$= \sum_{\substack{a=1 \\ (a,q)=1}}^{q} \sum_{\substack{b=1 \\ (b,q)=1}}^{r} \left(\frac{S(qr, ar+bq)}{qr}\right)^s e\left(-\frac{(ar+bq)N}{qr}\right)$$

$$= \sum_{\substack{a=1 \\ (a,q)=1}}^{q} \sum_{\substack{b=1 \\ (b,q)=1}}^{r} \left(\frac{S(q,a)}{q}\right)^s \left(\frac{S(r,b)}{r}\right)^s e\left(-\frac{aN}{q}\right) e\left(-\frac{bN}{r}\right)$$

$$= \sum_{\substack{a=1 \\ (a,q)=1}}^{q} \left(\frac{S(q,a)}{q}\right)^s e\left(-\frac{aN}{q}\right) \sum_{\substack{b=1 \\ (b,q)=1}}^{r} \left(\frac{S(r,b)}{r}\right)^s e\left(-\frac{bN}{r}\right)$$

$$= A_N(q)A_N(r).$$

This completes the proof.

For any positive integer q, we let $M_N(q)$ denote the number of solutions of the congruence

$$x_1^k + \cdots + x_s^k \equiv N \pmod{q}$$

in integers x_i such that $1 \le x_i \le q$ for $i = 1, \ldots, q$.

Lemma 5.6 *Let $s \geq 2^k + 1$. For every prime p, the series*

$$\chi_N(p) = 1 + \sum_{h=1}^{\infty} A_N(p^h) \qquad (5.11)$$

converges, and

$$\chi_N(p) = \lim_{h \to \infty} \frac{M_N(p^h)}{p^{h(s-1)}}. \qquad (5.12)$$

Proof. The convergence of the series (5.11) follows immediately from inequality (5.9). If $(a, q) = d$, then

$$S(q, a) = \sum_{x=1}^{q} e\left(\frac{ax^k}{q}\right) = \sum_{x=1}^{q} e\left(\frac{(a/d)x^k}{q/d}\right)$$

$$= d \sum_{x=1}^{q/d} e\left(\frac{(a/d)x^k}{q/d}\right) = dS(q/d, a/d).$$

Since

$$\frac{1}{q} \sum_{a=1}^{q} e\left(\frac{am}{q}\right) = \begin{cases} 1 & \text{if } m \equiv 0 \pmod{q} \\ 0 & \text{if } m \not\equiv 0 \pmod{q}, \end{cases}$$

it follows that for any integers $x_1, \ldots, x_s$

$$\frac{1}{q} \sum_{a=1}^{q} e\left(\frac{a(x_1^k + \cdots + x_s^k - N)}{q}\right) = \begin{cases} 1 & \text{if } x_1^k + \cdots + x_s^k \equiv N \pmod{q} \\ 0 & \text{if } x_1^k + \cdots + x_s^k \not\equiv N \pmod{q} \end{cases}$$

and so

$$M_N(q) = \sum_{x_1=1}^{q} \cdots \sum_{x_s=1}^{q} \frac{1}{q} \sum_{a=1}^{q} e\left(\frac{a(x_1^k + \cdots + x_s^k - N)}{q}\right)$$

$$= \frac{1}{q} \sum_{a=1}^{q} \sum_{x_1=1}^{q} \cdots \sum_{x_s=1}^{q} e\left(\frac{a(x_1^k + \cdots + x_s^k - N)}{q}\right)$$

$$= \frac{1}{q} \sum_{a=1}^{q} \sum_{x_1=1}^{q} e\left(\frac{ax_1^k}{q}\right) \cdots \sum_{x_s=1}^{q} e\left(\frac{ax_s^k}{q}\right) e\left(\frac{-aN}{q}\right)$$

$$= \frac{1}{q} \sum_{a=1}^{q} S(q, a)^s e\left(\frac{-aN}{q}\right)$$

$$= \frac{1}{q} \sum_{d|q} \sum_{\substack{a=1 \\ (a,q)=d}}^{q} S(q, a)^s e\left(\frac{-aN}{q}\right)$$

$$= \frac{1}{q} \sum_{d|q} \sum_{\substack{a=1 \\ (a,q)=d}}^{q} d^s S(q/d, a/d)^s e\left(\frac{-(a/d)N}{q/d}\right)$$

$$= \frac{1}{q} \sum_{d|q} \sum_{\substack{a=1 \\ (a,q)=d}}^{q} q^s \left(\frac{S(q/d, a/d)}{q/d} \right)^s e\left(\frac{-(a/d)N}{q/d} \right)$$

$$= q^{s-1} \sum_{d|q} A_N(q/d).$$

Therefore,

$$\sum_{d|q} A_N(q/d) = q^{1-s} M_N(q)$$

for all $q \geq 1$. In particular, for $q = p^h$ we have

$$1 + \sum_{j=1}^{h} A_N(p^j) = \sum_{d|p^h} A_N(p^h/d) = p^{h(1-s)} M_N(p^h)$$

and so

$$\chi_N(p) = \lim_{h \to \infty} \left(1 + \sum_{j=1}^{h} A_N(p^j) \right)$$

$$= \lim_{h \to \infty} p^{h(1-s)} M_N(p^h).$$

This completes the proof.

Lemma 5.7 *If $s \geq 2^k + 1$, then*

$$\mathfrak{S}(N) = \prod_p \chi_N(p). \tag{5.13}$$

Moreover, there exists a constant c_2 depending only on k and s such that

$$0 < \mathfrak{S}(N) < c_2$$

for all N, and there exists a prime p_0 depending only on k and s such that

$$1/2 \leq \prod_{p > p_0} \chi_N(p) \leq 3/2 \tag{5.14}$$

for all $N \geq 1$.

Proof. We proved that if $s > 2^k + 1$, then

$$A_N(q) \ll \frac{1}{q^{1+\delta_4}},$$

where δ_4 depends only on k and s, and so the series $\sum_q A_N(q)$ converges absolutely. Since the function $A_N(q)$ is multiplicative, Theorem A.28 immediately implies the convergence of the Euler product (5.13). In particular, $\chi_N(p) \neq 0$ for all N and

p. Since $\chi_N(p)$ is nonnegative by (5.12), it follows that $\chi_N(p)$ is a positive real number for all N and p, and so the singular series $\mathfrak{S}(N)$ is positive. Again, by (5.9),

$$0 < \mathfrak{S}(N) \le \sum_{q=1}^{\infty} \frac{1}{q^{1+\delta_4}} = c_2 < \infty$$

and

$$|\chi_N(p) - 1| \le \sum_{h=1}^{\infty} |A_N(p^h)| \ll \sum_{h=1}^{\infty} \frac{1}{p^{h(1+\delta_4)}} \ll \frac{1}{p^{1+\delta_4}}.$$

Therefore, there exists a constant c depending only on k and s such that

$$1 - \frac{c}{p^{1+\delta_4}} \le \chi_n(p) \le 1 + \frac{c}{p^{1+\delta_4}}$$

for all N and p. Inequality (5.14) follows from the convergence of the infinite products $\prod_p (1 \pm cp^{-1-\delta_4})$. This completes the proof.

We want to show that $\mathfrak{S}(N)$ is bounded away from 0 uniformly for all N. By inequality (5.14), it suffices to show, for every prime p, that $\chi_N(p)$ is uniformly bounded away from 0.

Let p be a prime, and let

$$k = p^\tau k_0,$$

where $\tau \ge 0$ and $(p, k_0) = 1$. We define

$$\gamma = \begin{cases} \tau + 1 & \text{if } p > 2 \\ \tau + 2 & \text{if } p = 2. \end{cases}$$

Lemma 5.8 *Let m be an integer not divisible by p. If the congruence $x^k \equiv m$ (mod p^γ) is solvable, then the congruence $y^k \equiv m$ (mod p^h) is solvable for every $h \ge \gamma$.*

Proof. There are two cases. In the first case, p is an odd prime. For $h \ge \gamma = \tau + 1$, we have

$$(k, \varphi(p^h)) = (k_0 p^\tau, (p-1)p^{h-1}) = (k_0, p-1)p^\tau = (k, \varphi(p^\gamma)).$$

The congruence classes modulo p^h that are relatively prime to p form a cyclic group of order $\varphi(p^h) = (p-1)p^{h-1}$. Let g be a generator of this cyclic group, that is, a primitive root modulo p^h. Then g is also a primitive root modulo p^γ. Let $x^k \equiv m$ (mod p^γ). Then $(x, p) = 1$, and we can choose integers r and u such that

$$x \equiv g^u \pmod{p^h}$$

and

$$m \equiv g^r \pmod{p^h}.$$

Then

$$ku \equiv r \pmod{\varphi(p^\gamma)},$$

and so

$$r \equiv 0 \quad (\text{mod } (k, \varphi(p^\gamma)))$$

and

$$r \equiv 0 \quad (\text{mod } (k, \varphi(p^h))).$$

Therefore, there exists an integer v such that

$$kv \equiv r \quad (\text{mod } \varphi(p^h)).$$

Let $y = g^v$. Then $y^k \equiv m \pmod{p^h}$.

In the second case, $p = 2$ and so m and x are odd. If $\tau = 0$, then k is odd. As y runs through the set of odd congruence classes modulo 2^h, so does y^k, and the congruence $y^k \equiv m \pmod{2^h}$ is solvable for all $h \geq 1$. If $\tau \geq 1$, then k is even and $m \equiv x^k \equiv 1 \pmod 4$. Also, $x^k = (-x)^k$, and so we can assume that $x \equiv 1 \pmod 4$. The congruence classes modulo 2^h that are congruent to 1 modulo 4 form a cyclic subgroup of order 2^{h-2}, and 5 is a generator of this subgroup. Choose integers r and u such that

$$m \equiv 5^r \quad (\text{mod } 2^h)$$

and

$$x \equiv 5^u \quad (\text{mod } 2^h).$$

Then $x^k \equiv m \pmod{2^\gamma}$ is equivalent to

$$ku \equiv r \quad (\text{mod } 2^{\gamma-2}),$$

and so r is divisible by $(k, 2^\tau) = 2^\tau = (k, 2^{h-2})$. It follows that there exists an integer v such that

$$kv \equiv r \quad (\text{mod } 2^{h-2}).$$

Let $y = 5^v$. Then $y^k \equiv m \pmod{2^h}$. This completes the proof.

Lemma 5.9 *Let p be prime. If there exist integers $a_1, \ldots, a_s$, not all divisible by p, such that*

$$a_1^k + \cdots + a_s^k \equiv N \quad (\text{mod } p^\gamma),$$

then

$$\chi_N(p) \geq \frac{1}{p^{\gamma(1-s)}} > 0.$$

Proof. Suppose that $a_1 \not\equiv 0 \pmod p$. Let $h > \gamma$. For each $i = 2, \ldots, s$ there exist $p^{h-\gamma}$ pairwise incongruent integers x_i such that

$$x_i \equiv a_i \quad (\text{mod } p^h).$$

Since the congruence

$$x_1^k \equiv N - x_2^k - \cdots - x_s^k \quad (\text{mod } p^\gamma)$$

is solvable with $x_1 = a_1 \not\equiv 0 \pmod{p}$, it follows from Lemma 5.8 that the congruence

$$x_1^k \equiv N - x_2^k - \cdots - x_s^k \pmod{p^h}.$$

This implies that

$$M_N(p^h) \geq p^{(h-\gamma)(s-1)},$$

and so

$$\chi_N(p) = \lim_{h \to \infty} \frac{M_N(p^h)}{p^{h(s-1)}} \geq \frac{1}{p^{\gamma(s-1)}} > 0.$$

This completes the proof.

Lemma 5.10 *If $s \geq 2k$ for k odd or $s \geq 4k$ for k even, then*

$$\chi_N(p) \geq p^{\gamma(1-s)} > 0.$$

Proof. By Lemma 5.9, it suffices to prove that the congruence

$$a_1^k + \cdots + a_s^k \equiv N \pmod{p^\gamma} \tag{5.15}$$

is solvable in integers a_i not all divisible by p. If N is not divisible by p and the congruence is solvable, then at least one of the integers a_i is prime to p. If N is divisible by p, then it suffices to show that the congruence

$$a_1^k + \cdots + a_{s-1}^k + 1^k \equiv N \pmod{p^\gamma}$$

has a solution in integers. This is equivalent to solving the congruence

$$a_1^k + \cdots + a_{s-1}^k \equiv N - 1 \pmod{p^\gamma}.$$

In this case, $(N - 1, p) = 1$. Therefore, it suffices to prove that, for $(N, p) = 1$, the congruence (5.15) is solvable in integers for $s \geq 2k - 1$ if p is odd and for $s \geq 4k - 1$ if p is even.

Let p be an odd prime and g be a primitive root modulo p^γ. The order of g is $\varphi(p^\gamma) = (p - 1)p^{\gamma-1} = (p - 1)p^\tau$. Let $(m, p) = 1$. The integer m is a kth power residue modulo p^γ if and only if there exists an integer x such that

$$x^k \equiv m \pmod{p^\gamma}.$$

Let $m \equiv g^r \pmod{p^\gamma}$. Then m is a kth power residue if and only if there exists an integer v such that $x \equiv g^v \pmod{p^\gamma}$ and

$$kv \equiv r \pmod{(p - 1)p^\tau}.$$

Since $k = k_0 p^\tau$ with $(k_0, p) = 1$, it follows that this congruence is solvable if and only if

$$r \equiv 0 \pmod{(k_0, p - 1)p^\tau},$$

and so there are

$$\frac{\varphi(p^\gamma)}{(k_0, p-1)p^\tau} = \frac{p-1}{(k_0, p-1)}$$

distinct kth power residues modulo p^γ. Let $s(N)$ denote the smallest integer s for which the congruence (5.15) is solvable, and let $C(j)$ denote the set of all congruence classes N modulo p^γ such that $(N, p) = 1$ and $s(N) = j$. In particular, $C(1)$ consists precisely of the kth power residues modulo p^γ. If $(m, p) = 1$ and $N' = m^k N$, then $s(N') = s(N)$. It follows that the sets $C(j)$ are closed under multiplication by kth power residues, and so, if $C(j)$ is nonempty, then $|C(j)| \geq (p-1)/(k_0, p-1)$. Let n be the largest integer such that the set $C(n)$ is nonempty. Let $j < n$ and let N be the smallest integer such that $(N, p) = 1$ and $s(N) > j$. Since p is an odd prime, it follows that $N - i$ is prime to p for $i = 1$ or 2, and $s(N - i) \leq j$. Since $N = (N - 1) + 1^k$ and $N = (N - 2) + 1^k + 1^k$, it follows that

$$j + 1 \leq s(N) \leq s(N - i) + 2 \leq j + 2$$

and so $s(N - i) = j$ or $j - 1$. This implies that no two consecutive sets $C(j)$ are nonempty for $j = 1, \ldots, n$, and so the number of nonempty sets $C(j)$ is at least $(n + 1)/2$. Since the sets $C(j)$ are pairwise disjoint, it follows that

$$(p - 1)p^\tau = \varphi(p^\gamma) = \sum_{\substack{j=1 \\ C(j) \neq \emptyset}}^{n} |C(j)| \geq \frac{n+1}{2} \frac{p-1}{(k_0, p-1)},$$

and so

$$n \leq 2(k_0, p-1)p^\tau - 1 \leq 2k - 1.$$

Therefore, $s(N) \leq 2k - 1$ if p is an odd prime and N is prime to p.

Let $p = 2$. If k is odd, then every odd integer is a kth power residue modulo 2^γ, so $s(N) = 1$ for all odd integers N. If k is even, then $k = 2^\tau k_0$ with $\tau \geq 1$, and $\gamma = \tau + 2$. We can assume that $1 \leq N \leq 2^\gamma - 1$. If

$$s = 2^\gamma - 1 = 4 \cdot 2^\tau - 1 \leq 4k - 1,$$

then congruence (5.15) can always be solved by choosing $a_i = 1$ for $i = 1, \ldots, N$ and $a_i = 0$ for $i = N + 1, \ldots, s$. Therefore, $s(N) \leq 4k - 1$ for all odd N. This completes the proof.

Theorem 5.6 *There exist positive constants* $c_1 = c_1(k, s)$ *and* $c_2 = c_2(k, s)$ *such that*

$$c_1 < \mathfrak{S}(N) < c_2.$$

Moreover, for all sufficiently large integers N,

$$\mathfrak{S}(N, P^\nu) = \mathfrak{S}(N) + O\left(P^{-\nu\delta_4}\right).$$

Proof. The only part of the theorem that we have not yet proved is the lower bound for $\mathfrak{S}(N)$. However, we showed that there exists a prime $p_0 = p_0(k, s)$ such that

$$1/2 \leq \prod_{p > p_0} \chi_N(p) \leq 3/2$$

for all $N \geq 1$. Since

$$\chi_N(p) \geq p^{\gamma(1-s)} > 0$$

for all primes p and all N, it follows that

$$\mathfrak{S}(N) = \prod_p \chi_N(p) > \frac{1}{2} \prod_{p \leq p_0} \chi_N(p) \geq \frac{1}{2} \prod_{p \leq p_0} p^{\gamma(1-s)} = c_1 > 0.$$

This completes the proof.

5.8   Conclusion

We are now ready to prove the Hardy–Littlewood asymptotic formula.

Theorem 5.7 (Hardy–Littlewood) *Let $k \geq 2$ and $s \geq 2^k + 1$. Let $r_{k,s}(N)$ denote the number of representations of N as the sum of s kth powers of positive integers. There exists $\delta = \delta(k, s) > 0$ such that*

$$r_{k,s}(N) = \mathfrak{S}(N)\Gamma\left(1 + \frac{1}{k}\right)^s \Gamma\left(\frac{s}{k}\right)^{-1} N^{(s/k)-1} + O(N^{(s/k)-1-\delta}),$$

where the implied constant depends only on k and s, and $\mathfrak{S}(N)$ is an arithmetic function such that

$$c_1 < \mathfrak{S}(N) < c_2$$

for all N, where c_1 and c_2 are positive constants that depend only on k and s.

Proof. Let $\delta_0 = \min(1, \delta_1, \delta_2, \delta_3, v\delta_4)$. By Theorems 5.2–5.6, we have

$$r_{k,s}(N) = \int_0^1 F(\alpha)^s e(-\alpha N) d\alpha$$

$$= \int_{\mathfrak{M}} F(\alpha)^s e(-\alpha N) d\alpha + \int_{\mathfrak{m}} F(\alpha)^s e(-\alpha N) d\alpha$$

$$= \mathfrak{S}(N, P^v) J^*(N) + O\left(P^{s-k-\delta_2}\right) + O\left(P^{s-k-\delta_1}\right)$$

$$= \left(\mathfrak{S}(N) + O\left(P^{-v\delta_4}\right)\right)\left(J(N) + O\left(P^{s-k-\delta_3}\right)\right) + O\left(P^{s-k-\delta_2}\right)$$

$$\quad + O\left(P^{s-k-\delta_1}\right)$$

$$= \mathfrak{S}(N)J(N) + O\left(P^{s-k-\delta_0}\right)$$

$$= \mathfrak{S}(N)\Gamma\left(1 + \frac{1}{k}\right)^s \Gamma\left(\frac{s}{k}\right)^{-1} N^{s/k-1} + O\left(N^{(s-1)/k-1}\right)$$

$$+O\left(N^{s/k-1-\delta_0/k}\right)$$
$$= \mathfrak{S}(N)\Gamma\left(1+\frac{1}{k}\right)^s\Gamma\left(\frac{s}{k}\right)^{-1}N^{s/k-1}+O\left(N^{s/k-1-\delta}\right),$$

where $\delta = \delta_0/k$. This completes the proof.

5.9   Notes

The circle method was invented by Hardy and Ramanujan [50] to obtain the asymptotic formula for the partition function $p(N)$, which counts the number of unordered representations of a positive integer N as the sum of any number of positive integers. The circle method was also applied to study the number of representations of an integer as a sum of squares. See, for example, Hardy [45], and the particularly important work of Kloosterman [71, 72, 73].

In a classic series of papers, "Some problems of 'Partitio Numerorum',", Hardy and Littlewood [47, 48] applied the circle method to Waring's problem. Vinogradov [131, 134, 135] subsequently simplified and strengthened their method. This chapter gives the classical proof of the Hardy–Littlewood formula for $s \geq s_0(k) = 2^k + 1$. There is a vast literature on applications of the circle method to Waring's problem as well as to other problems in additive number theory. The books of Davenport [18], Hua [64], Vaughan [125], and Vinogradov [135] are excellent references.

There have been great technological improvements in the circle method in recent years, particularly by the Anglo-Michigan school (for example, Vaughan and Wooley [126, 127, 128, 129, 130, 147, 148]). In particular, Wooley [146] proved that

$$G(k) < k(\log k + \log\log k + O(1)).$$

Another interesting recent result concerns the range of validity of the Hardy–Littlewood asymptotic formula. Let $\tilde{G}(k)$ denote the smallest integer s_0 such that the Hardy-Littlewood asymptotic formula (5.1) holds for all $s \geq s_0$. Ford [41] proved that

$$\tilde{G}(k) \leq k^2(\log k + \log\log k + O(1)).$$

For other recent developments in the circle method, see Heath-Brown [54, 55], Hooley [59, 60, 61], and Schmidt [107].

5.10   Exercises

1. Show that for $k = 1$ the Hardy–Littlewood asymptotic formula is consistent with Theorem 5.1.

2. Let $k \geq 2$. Show that the number of positive integers not exceeding x that can be written as the sum of k nonnegative kth powers is $x/k! + O\left(x^{(k-1)/k}\right)$. Show that

$$G(k) \geq k + 1.$$

Hint: If $n \leq x$ is a sum of k kth powers, then

$$n = a_1^k + a_2^k + \cdots + a_k^k,$$

where

$$0 \leq a_1 \leq a_2 \leq \cdots \leq a_k \leq x^{1/k},$$

and the number of such expressions is given by a binomial coefficient.

3. Let $f(x)$ be a polynomial of degree $k \geq 2$ with integral coefficients, and let

$$S_f(q, a) = \sum_{r=1}^{q} e(af(r)/q).$$

Prove that if $(q, r) = 1$, then

$$S_f(qr, ar + bq) = S_f(q, a) S_f(r, b).$$

4. Let $R_{k,s}(N)$ denote the number of representations of an integer N as the sum of s nonnegative kth powers. State and prove an asymptotic formula for $R_{k,s}(N)$.

Part II

The Goldbach conjecture

6
Elementary estimates for primes

Brun's method is perhaps our most powerful elementary tool in number theory.

P. Erdős [34]

6.1 Euclid's theorem

Before beginning to study sums of primes, we need some elementary results about the distribution of prime numbers.

Let $s = \sigma + it$ be a complex number with real part σ and imaginary part t. To every sequence of complex numbers $a_1, a_2, \ldots$ is associated the *Dirichlet series*

$$F(s) = \sum_{n=1}^{\infty} \frac{a_n}{n^s}.$$

If the series $F(s)$ converges absolutely for some complex number $s_0 = \sigma_0 + it_0$, then $F(s)$ converges absolutely for all complex numbers $s = \sigma + it$ with $\Re(s) = \sigma \geq \sigma_0 = \Re(s_0)$, since

$$\left|\frac{a_n}{n^s}\right| = \frac{|a_n|}{n^\sigma} \leq \frac{|a_n|}{n^{\sigma_0}} = \left|\frac{a_n}{n^{s_0}}\right|.$$

If we let $a_n = 1$ for all $n \geq 1$, we obtain the *Riemann zeta-function*

$$\zeta(s) = \sum_{n=1}^{\infty} \frac{1}{n^s}.$$

This Dirichlet series converges absolutely for all s with $\Re(s) > 1$.

Theorem 6.1 *Let $f(n)$ be a multiplicative function. If the Dirichlet series*

$$F(s) = \sum_{n=1}^{\infty} \frac{f(n)}{n^s}$$

converges absolutely for all complex numbers s with $\Re(s) > \sigma_0$, then $F(s)$ can be represented as the infinite product

$$F(s) = \prod_p \left(1 + \frac{f(p)}{p^s} + \frac{f(p^2)}{p^{2s}} + \cdots \right).$$

If $f(n)$ is completely multiplicative, then

$$F(s) = \prod_p \left(1 - \frac{f(p)}{p^s}\right)^{-1}.$$

This is called the Euler product *for $F(s)$.*

Proof. If $f(n)$ is multiplicative, then so is $f(n)/n^s$. If $f(n)$ is completely multiplicative, then so is $f(n)/n^s$. The result follows immediately from Theorem A.28.

Because the Riemann zeta-function converges absolutely for $\Re(s) > 1$, it follows from Theorem 6.1 that $\zeta(s)$ has the Euler product

$$\zeta(s) = \sum_{n=1}^{\infty} \frac{1}{n^s} = \prod_p \left(1 - \frac{1}{p^s}\right)^{-1}$$

for all s with $\Re(s) > 1$, and so $\zeta(s) \neq 0$ for $\Re(s) > 1$. From the Euler product, we obtain the following analytic proof that there are infinitely many primes.

Theorem 6.2 (Euclid) *There are infinitely many primes.*

Proof. For $0 < x < 1$ we have the Taylor series

$$-\log(1 - x) = \sum_{n=1}^{\infty} \frac{x^n}{n}.$$

If $\sigma > 0$, then $\zeta(1 + \sigma) > 1$ and

$$\log \zeta(1 + \sigma) = \log \prod_p \left(1 - \frac{1}{p^{1+\sigma}}\right)^{-1}$$

$$= -\sum_p \log\left(1 - \frac{1}{p^{1+\sigma}}\right)$$

$$= \sum_p \sum_{n=1}^{\infty} \frac{1}{np^{n(1+\sigma)}}$$

$$= \sum_p \frac{1}{p^{1+\sigma}} + \sum_p \sum_{n=2}^{\infty} \frac{1}{np^{n(1+\sigma)}}.$$

Since

$$0 < \sum_p \sum_{n=2}^\infty \frac{1}{np^{n(1+\sigma)}} < \sum_p \sum_{n=2}^\infty \frac{1}{p^n} = \sum_p \frac{1}{p(p-1)} < \infty, \qquad (6.1)$$

it follows that

$$\log \zeta(1+\sigma) = \sum_p \frac{1}{p^{1+\sigma}} + O(1). \qquad (6.2)$$

Let $0 < \sigma < 1$. Then

$$1 < \frac{1}{\sigma} = \int_1^\infty \frac{1}{x^{1+\sigma}} dx < \zeta(1+\sigma) < 1 + \int_1^\infty \frac{1}{x^{1+\sigma}} dx = \frac{1}{\sigma} + 1$$

and so

$$0 < \log \frac{1}{\sigma} < \log \zeta(1+\sigma)$$

$$< \log \left(\frac{1}{\sigma} + 1 \right) = \log \frac{1}{\sigma} + \log(1+\sigma)$$

$$< \log \frac{1}{\sigma} + \sigma < \log \frac{1}{\sigma} + 1.$$

Therefore,

$$\log \zeta(1+\sigma) = \log \frac{1}{\sigma} + O(1). \qquad (6.3)$$

Combining (6.2) and (6.3), we obtain

$$\log \frac{1}{\sigma} = \sum_p \frac{1}{p^{1+\sigma}} + O(1)$$

for $0 < \sigma < 1$. If there were only finitely many prime numbers, then the sum on the right side of this equation remains bounded as σ tends to 0, but the logarithm on the left side of the equation goes to infinity as σ tends to 0. This is impossible, so there must be infinitely many primes.

6.2   Chebyshev's theorem

The simplest prime-counting functions are

$$\pi(x) = \sum_{p \le x} 1,$$

$$\vartheta(x) = \sum_{p \le x} \log p,$$

and

$$\psi(x) = \sum_{p^k \le x} \log p.$$

$\vartheta(x)$ and $\psi(x)$ are called the *Chebyshev functions*. Chebyschev proved that the functions $\vartheta(x)$ and $\psi(x)$ have order of magnitude x and that $\pi(x)$ has order of magnitude $x/\log x$. Before proving this theorem, we need the following lemma about the unimodality of the sequence of binomial coefficients.

Lemma 6.1 *Let $n \geq 1$ and $1 \leq k \leq n$. Then*

$$\binom{n}{k-1} < \binom{n}{k} \quad \text{if and only if } k < \tfrac{n+1}{2},$$

$$\binom{n}{k-1} > \binom{n}{k} \quad \text{if and only if } k > \tfrac{n+1}{2},$$

$$\binom{n}{k-1} = \binom{n}{k} \quad \text{if and only if } n \text{ is odd and } k = \tfrac{n+1}{2}.$$

Proof. This follows immediately from observing the ratio

$$\frac{\binom{n}{k}}{\binom{n}{k-1}} = \frac{\frac{n!}{k!(n-k)!}}{\frac{n!}{(k-1)!(n-k+1)!}} = \frac{(k-1)!(n-k+1)!}{k!(n-k)!} = \frac{n-k+1}{k}.$$

Lemma 6.2 *Let $n \geq 1$ and $N = \binom{2n}{n}$. Then*

$$N < 2^{2n} \leq 2nN.$$

Proof. Since $\binom{2n}{n}$ is the middle, and hence the largest, binomial coefficient in the expansion of $(1+1)^{2n}$, it follows that

$$N = \binom{2n}{n} < (1+1)^{2n} = 2^{2n}$$

$$= \sum_{k=0}^{2n} \binom{n}{k} = 1 + \sum_{k=1}^{2n-1} \binom{n}{k} + 1$$

$$\leq 2 + (2n-1)\binom{2n}{n} \leq 2n\binom{2n}{n}$$

$$= 2nN.$$

This completes the proof.

For any positive integer n, let $v_p(n)$ denote the highest power of p that divides n. Thus, $v_p(n) = k$ if and only if $p^k \| n$. In this case, $p^k \leq n$ and so $v_p(n) \leq \log n / \log p$.

Lemma 6.3 *For every positive integer n,*

$$v_p(n!) = \sum_{k=1}^{\infty} \left[\frac{n}{p^k} \right] = \sum_{k=1}^{[\log n/\log p]} \left[\frac{n}{p^k} \right]. \tag{6.4}$$

Proof. Since $v_p(mn) = v_p(m)v_p(n)$ for all positive integers m and n, we have

$$v_p(n!) = \sum_{m=1}^{n} v_p(m) = \sum_{m=1}^{n} \sum_{\substack{p^k|m \\ k\geq 1}} 1 = \sum_{k=1}^{\infty} \sum_{\substack{m=1 \\ p^k|m}}^{n} 1 = \sum_{k=1}^{\infty} \left[\frac{n}{p^k}\right].$$

This proves the formula.

Theorem 6.3 (Chebyshev) *There exist positive constants c_1 and c_2 such that*

$$c_1 x \leq \vartheta(x) \leq \psi(x) \leq \pi(x)\log x \leq c_2 x \tag{6.5}$$

for all $x \geq 2$. Moreover,

$$\liminf_{x\to\infty} \frac{\vartheta(x)}{x} = \liminf_{x\to\infty} \frac{\psi(x)}{x} = \liminf_{x\to\infty} \frac{\pi(x)\log x}{x} \geq \log 2$$

and

$$\limsup_{x\to\infty} \frac{\vartheta(x)}{x} = \limsup_{x\to\infty} \frac{\psi(x)}{x} = \limsup_{x\to\infty} \frac{\pi(x)\log x}{x} \leq 4\log 2.$$

Proof. Let $x \geq 2$. If $p^k \leq x$, then $k \leq [\log x/\log p]$, and so

$$\vartheta(x) = \sum_{p\leq x} \log p \leq \psi(x) = \sum_{p^k\leq x} \log p = \sum_{p\leq x} \left[\frac{\log x}{\log p}\right]\log p$$

$$\leq \sum_{p\leq x} \log x = \pi(x)\log x.$$

Therefore,

$$\liminf_{x\to\infty} \frac{\vartheta(x)}{x} \leq \liminf_{x\to\infty} \frac{\psi(x)}{x} \leq \liminf_{x\to\infty} \frac{\pi(x)\log x}{x}$$

and

$$\limsup_{x\to\infty} \frac{\vartheta(x)}{x} \leq \limsup_{x\to\infty} \frac{\psi(x)}{x} \leq \limsup_{x\to\infty} \frac{\pi(x)\log x}{x}.$$

Let

$$0 < \delta < 1.$$

Then

$$\vartheta(x) \geq \sum_{x^{1-\delta}<p\leq x} \log p$$

$$\geq \sum_{x^{1-\delta}<p\leq x} (1-\delta)\log x$$

$$= (1-\delta)\left(\pi(x) - \pi(x^{1-\delta})\right)\log x$$

$$\geq (1-\delta)\pi(x)\log x - x^{1-\delta}\log x,$$

and so

$$\frac{\vartheta(x)}{x} \geq \frac{(1-\delta)\pi(x)\log x}{x} - \frac{\log x}{x^\delta}.$$

It follows that

$$\liminf_{x\to\infty} \frac{\vartheta(x)}{x} \geq (1-\delta)\liminf_{x\to\infty} \frac{\pi(x)\log x}{x}.$$

This holds for all $\delta > 0$, and so

$$\liminf_{x\to\infty} \frac{\vartheta(x)}{x} \geq \liminf_{x\to\infty} \frac{\pi(x)\log x}{x}.$$

Similarly,

$$\limsup_{x\to\infty} \frac{\vartheta(x)}{x} \geq \limsup_{x\to\infty} \frac{\pi(x)\log x}{x}.$$

Therefore,

$$\liminf_{x\to\infty} \frac{\vartheta(x)}{x} = \liminf_{x\to\infty} \frac{\psi(x)}{x} = \liminf_{x\to\infty} \frac{\pi(x)\log x}{x} \qquad (6.6)$$

and

$$\limsup_{x\to\infty} \frac{\vartheta(x)}{x} = \limsup_{x\to\infty} \frac{\psi(x)}{x} = \limsup_{x\to\infty} \frac{\pi(x)\log x}{x}. \qquad (6.7)$$

Let $n \geq 1$, and let

$$N = \binom{2n}{n} = \frac{2n(2n-1)(2n-2)\cdots(n+1)}{n!}.$$

Then N is an integer, since it is a binomial coefficient, and

$$\frac{2^{2n}}{2n} \leq N < 2^{2n}$$

by Lemma 6.2. If p is a prime number such that

$$n < p \leq 2n,$$

then p divides the numerator but not the denominator of N. Therefore, N is divisible by the product of all these primes, and so

$$\prod_{n<p\leq 2n} p \leq N < 2^{2n}.$$

In particular, if $r \geq 1$ and $n = 2^{r-1}$, then

$$\prod_{2^{r-1}<p\leq 2^r} p \leq N < 2^{2^r}.$$

It follows that, for any $R \geq 1$,

$$\prod_{p\leq 2^R} p = \prod_{r=1}^{R}\prod_{2^{r-1}<p\leq 2^r} p < \prod_{r=1}^{R} 2^{2^r} < 2^{2^{R+1}}.$$

For any number $x \geq 2$, there is an integer $R \geq 1$ such that

$$2^{R-1} < x \leq 2^R.$$

Then

$$\prod_{p \leq x} p \leq \prod_{p \leq 2^R} p < 2^{2^{R+1}} < 2^{4x},$$

and so

$$\vartheta(x) = \sum_{p \leq x} \log p = \log \left(\prod_{p \leq x} p \right) < (4 \log 2)x.$$

Thus,

$$\limsup_{x \to \infty} \frac{\vartheta(x)}{x} \leq 4 \log 2.$$

To obtain the lower limit, we use Lemma 6.3 to express N explicitly as a power of primes:

$$N = \binom{2n}{n} = \frac{(2n)!}{n!^2} = \prod_{p \leq 2n} p^{v_p(2n) - 2v_p(n)},$$

where

$$v_p(2n) - 2v_p(n) = \sum_{1 \leq k \leq \frac{\log 2n}{\log p}} \left(\left[\frac{2n}{p^k} \right] - 2 \left[\frac{n}{p^k} \right] \right).$$

Since $[2t] - 2[t] = 0$ or 1 for all real numbers t, it follows that

$$v_p(2n) - 2v_p(n) \leq \frac{\log 2n}{\log p}.$$

By Lemma 6.2,

$$\frac{2^{2n}}{2n} < N = \prod_{p \leq 2n} p^{v_p(2n) - 2v_p(n)} \leq \prod_{p \leq 2n} p^{\frac{\log 2n}{\log p}} \leq \prod_{p \leq 2n} 2n = (2n)^{\pi(2n)}$$

or, equivalently,

$$\pi(2n) \log 2n < 2n \log 2 - \log 2n.$$

Let $n = [x/2]$. Then

$$2n \leq x < 2n + 2$$

and

$$\pi(x) \log x \geq \pi(2n) \log 2n > 2n \log 2 - \log 2n$$
$$> (x - 2) \log 2 - \log x = x \log 2 - \log x - 2 \log 2.$$

It follows that

$$\frac{\pi(x) \log x}{x} > \log 2 - \frac{\log x + 2 \log 2}{x}$$

and so

$$\liminf_{x \to \infty} \frac{\pi(x) \log x}{x} \geq \log 2.$$

Since $\vartheta(2) > 0$, we have $\vartheta(x) \geq c_1 x$ for some $c_1 > 0$ and all $x \geq 2$. This completes the proof.

Theorem 6.4 *Let p_n denote the nth prime number. There exist positive constants c_3 and c_4 such that*

$$c_3 n \log n \leq p_n \leq c_4 n \log n$$

for all $n \geq 2$.

Proof. By Chebyshev's inequality (6.5),

$$\frac{c_1 p_n}{\log p_n} \leq \pi(p_n) = n \leq \frac{c_2 p_n}{\log p_n}$$

and so

$$c_2^{-1} n \log p_n \leq p_n \leq c_1^{-1} n \log p_n.$$

Since

$$\log n \leq \log p_n,$$

we have

$$p_n \geq c_2^{-1} n \log n = c_3 n \log n.$$

For n sufficiently large,

$$\log p_n \leq \log n + \log \log p_n + \log c_1^{-1}$$
$$\leq \log n + 2 \log \log p_n$$
$$\leq \log n + (1/2) \log p_n,$$

so

$$\log p_n \leq 2 \log n$$

and

$$p_n \leq c_1^{-1} n \log p_n \leq 2 c_1^{-1} n \log n.$$

Therefore, there exists a constant c_4 such that $p_n \leq c_4 n \log n$ for all $n \geq 2$. This completes the proof.

6.3   Mertens's theorems

In this section, we derive some important results about the distribution of prime numbers that were originally proved by Mertens.

Lemma 6.4 *For any real number $x \geq 1$ we have*

$$0 \leq \sum_{n \leq x} \log \left(\frac{x}{n}\right) < x.$$

Proof. Since the function $h(t) = \log(x/t)$ is decreasing on the interval $[1, x]$, it follows that

$$\sum_{1 \leq n \leq x} \log\left(\frac{x}{n}\right) < \log x + \int_1^x \log\left(\frac{x}{t}\right) dt$$

$$= x \log x - \int_1^x \log t\, dt$$

$$= x \log x - (x \log x - x + 1)$$

$$< x.$$

This completes the proof.

The function $\Lambda(n)$, called *von Mangoldt's function*, is defined by

$$\Lambda(n) = \begin{cases} \log p & \text{if } n = p^m \text{ is a prime power} \\ 0 & \text{otherwise.} \end{cases}$$

Then

$$\psi(x) = \sum_{1 \leq m \leq x} \Lambda(m).$$

Theorem 6.5 (Mertens) *For any real number $x \geq 1$, we have*

$$\sum_{n \leq x} \frac{\Lambda(n)}{n} = \log x + O(1).$$

Proof. Let $N = [x]$. Then

$$0 \leq \sum_{n \leq x} \log \frac{x}{n} = N \log x - \sum_{n=1}^N \log n = x \log x - \log N! + O(\log x) < x$$

by Lemma 6.4, and so

$$\log N! = x \log x + O(x).$$

It follows from Lemma 6.3 and Theorem 6.3 that

$$\log N! = \sum_{p \leq N} v_p(N) \log p$$

$$= \sum_{p \leq N} \sum_{k=1}^{[\log N / \log p]} \left[\frac{N}{p^k}\right] \log p$$

$$= \sum_{p^k \leq N} \left[\frac{N}{p^k}\right] \log p$$

$$= \sum_{p^k \leq x} \left[\frac{x}{p^k}\right] \log p$$

$$= \sum_{n \leq x} \left[\frac{x}{n}\right] \Lambda(n)$$

$$= \sum_{n \leq x} \left(\frac{x}{n} + O(1) \right) \Lambda(n)$$

$$= x \sum_{n \leq x} \frac{\Lambda(n)}{n} + O \left(\sum_{n \leq x} \Lambda(n) \right)$$

$$= x \sum_{n \leq x} \frac{\Lambda(n)}{n} + O \left(\psi(x) \right)$$

$$= x \sum_{n \leq x} \frac{\Lambda(n)}{n} + O(x).$$

Therefore,

$$x \sum_{n \leq x} \frac{\Lambda(n)}{n} + O(x) = x \log x + O(x)$$

and

$$\sum_{n \leq x} \frac{\Lambda(n)}{n} = \log x + O(1).$$

This completes the proof.

Theorem 6.6 (Mertens)   *For any real number $x \geq 1$, we have*

$$\sum_{p \leq x} \frac{\log p}{p} = \log x + O(1).$$

Proof. Since

$$0 \leq \sum_{n \leq x} \frac{\Lambda(n)}{n} - \sum_{p \leq x} \frac{\log p}{p}$$

$$= \sum_{\substack{p^k \leq x \\ k \geq 2}} \frac{\log p}{p^k}$$

$$\leq \sum_{p \leq x} \log p \sum_{k=2}^{\infty} \frac{1}{p^k}$$

$$\leq \sum_{p \leq x} \frac{\log p}{p(p-1)}$$

$$\leq 2 \sum_{p \leq x} \frac{\log p}{p^2}$$

$$\leq 2 \sum_{n=1}^{\infty} \frac{\log n}{n^2}$$

$$= O(1),$$

it follows from Theorem 6.5 that

$$\sum_{p \leq x} \frac{\log p}{p} = \sum_{n \leq x} \frac{\Lambda(n)}{n} + O(1) = \log x + O(1).$$

This completes the proof.

Theorem 6.7 *There exists a constant $b_1 > 0$ such that*

$$\sum_{p \leq x} \frac{1}{p} = \log \log x + b_1 + O\left(\frac{1}{\log x}\right)$$

for $x \geq 2$.

Proof. We can write

$$\sum_{p \leq x} \frac{1}{p} = \sum_{p \leq x} \frac{\log p}{p} \frac{1}{\log p} = \sum_{n \leq x} u(n) f(n),$$

where

$$u(n) = \begin{cases} \frac{\log p}{p} & \text{if } n = p \\ 0 & \text{otherwise} \end{cases}$$

and

$$f(t) = \frac{1}{\log t}.$$

We define the functions $U(t)$ and $g(t)$ by

$$U(t) = \sum_{n \leq t} u(n) = \sum_{p \leq t} \frac{\log p}{p} = \log t + g(t).$$

Then $U(t) = 0$ for $t < 2$ and $g(t) = O(1)$ by Theorem 6.6. Therefore, the integral $\int_2^\infty g(t)/(t(\log t)^2)dt$ converges absolutely, and

$$\int_x^\infty \frac{g(t)dt}{t(\log t)^2} = O\left(\frac{1}{\log x}\right).$$

Since $f(t)$ is continuous and $U(t)$ is increasing, we can express the sum $\sum_{p \leq x} 1/p$ as a Riemann–Stieltjes integral. Note that $U(t) = 0$ for $t < 2$. By partial summation, we obtain

$$\sum_{p \leq x} \frac{1}{p} = \sum_{n \leq x} u(n) f(n)$$

$$= \frac{1}{2} + \int_2^x f(t)dU(t)$$

$$= f(x)U(x) - \int_2^x U(t)df(t)$$

$$= \frac{\log x + g(x)}{\log x} - \int_2^x U(t)f'(t)dt$$

$$= 1 + O\left(\frac{1}{\log x}\right) + \int_2^x \frac{\log t + g(t)}{t(\log t)^2}dt$$

$$= \int_2^x \frac{1}{t \log t} dt + \int_2^\infty \frac{g(t)}{t(\log t)^2} dt - \int_x^\infty \frac{g(t)}{t(\log t)^2} dt + 1 + O\left(\frac{1}{\log x}\right)$$

$$= \log \log x - \log \log 2 + \int_2^\infty \frac{g(t)}{t(\log t)^2} dt + 1 + O\left(\frac{1}{\log x}\right)$$

$$= \log \log x + b_1 + O\left(\frac{1}{\log x}\right),$$

where

$$b_1 = 1 - \log \log 2 + \int_2^\infty \frac{g(t)}{t(\log t)^2} dt. \tag{6.8}$$

This completes the proof.

From the Taylor series for $\log(1 - x)$, we see that

$$0 < \log\left(1 - \frac{1}{p}\right)^{-1} - \frac{1}{p} = \sum_{n=2}^\infty \frac{1}{np^n} < \sum_{n=2}^\infty \frac{1}{p^n} = \frac{1}{p(p-1)}.$$

It follows from the comparison test that the series

$$b_2 = \sum_p \left(\log\left(1 - \frac{1}{p}\right)^{-1} - \frac{1}{p}\right) = \sum_p \sum_{k=2}^\infty \frac{1}{kp^k} \tag{6.9}$$

converges.

Lemma 6.5 *Let b_1 and b_2 be the positive numbers defined by (6.8) and (6.9). Then*

$$b_1 + b_2 = \gamma,$$

where γ is Euler's constant.

Proof. Let $0 < \sigma < 1$. We define the function $F(\sigma)$ by

$$F(\sigma) = \log \zeta(1 + \sigma) - \sum_p \frac{1}{p^{1+\sigma}}$$

$$= \sum_p \left(\log\left(1 - \frac{1}{p^{1+\sigma}}\right)^{-1} - \frac{1}{p^{1+\sigma}}\right)$$

$$= \sum_p \sum_{n=2}^\infty \frac{1}{np^{n(1+\sigma)}}.$$

By (6.1) and the Weierstrass M-test, the last series converges uniformly for $\sigma \geq 0$ and so represents a continuous function for $\sigma \geq 0$. Therefore,

$$\lim_{\sigma \to 0^+} F(\sigma) = b_2. \tag{6.10}$$

We shall find alternative representations for the functions $\log \zeta(1 + \sigma)$ and $\sum_p p^{-1-\sigma}$. Since

$$1 - \sigma + \frac{\sigma^2}{2e} < e^{-\sigma} < 1 - \sigma + \frac{\sigma^2}{2}$$

for $0 < \sigma < 1$, it follows that

$$1 - \frac{\sigma}{2} < \frac{1 - e^{-\sigma}}{\sigma} < 1 - \frac{\sigma}{2e}$$

and

$$1 + \frac{\sigma}{2e} < 1 + \frac{\sigma}{2e - \sigma} < \frac{\sigma}{1 - e^{-\sigma}} < 1 + \frac{\sigma}{2 - \sigma} < 1 + \sigma.$$

Therefore,

$$0 < \log \sigma + \log \left(1 - e^{-\sigma}\right)^{-1} < \sigma,$$

and so

$$\log \frac{1}{\sigma} = \log(1 - e^{-\sigma})^{-1} + O(\sigma).$$

By (6.3), we have

$$\log \zeta(1 + \sigma) = \log \frac{1}{\sigma} + O(\sigma)$$
$$= \log(1 - e^{-\sigma})^{-1} + O(\sigma)$$
$$= \sum_{n=1}^{\infty} \frac{e^{-\sigma n}}{n} + O(\sigma).$$

By Theorem A.5,

$$L(x) = \sum_{n \leq x} \frac{1}{n} = \log x + \gamma + O\left(\frac{1}{x}\right)$$

for $x \geq 1$. Let $f(x) = e^{-\sigma x}$. By partial summation, we have

$$\log \zeta(1 + \sigma) = \sum_{n=1}^{\infty} \frac{f(n)}{n} + O(\sigma)$$
$$= \int_0^{\infty} f(x) dL(x) + O(\sigma)$$
$$= -\int_0^{\infty} L(x) df(x) + O(\sigma)$$
$$= \sigma \int_0^{\infty} e^{-\sigma x} L(x) dx + O(\sigma).$$

By Theorem 6.7,

$$S(x) = \sum_{p \leq x} \frac{1}{p} = \log \log x + b_1 + O\left(\frac{1}{\log x}\right)$$

for $x \geq 2$. Let $g(x) = x^{-\sigma}$. Again, by partial summation we have

$$\sum_p \frac{1}{p^{1+\sigma}} = \sum_p \frac{g(p)}{p} = \int_1^{\infty} g(x) dS(x) = -\int_1^{\infty} S(x) dg(x)$$

$$= \sigma \int_1^\infty \frac{S(x)dx}{x^{1+\sigma}}$$

$$= \sigma \int_0^\infty e^{-\sigma x} S(e^x) dx.$$

Since

$$S(e^x) = \log x + b_1 + O\left(\frac{1}{x}\right)$$

and

$$L(x) = \log x + \gamma + O\left(\frac{1}{x}\right),$$

it follows that

$$L(x) - S(e^x) = \gamma - b_1 + O\left(\frac{1}{x}\right) = \gamma - b_1 + O\left(\frac{1}{x+1}\right)$$

for $x \geq 1$. We also have

$$L(x) - S(e^x) = \gamma - b_1 + O\left(\frac{1}{x+1}\right)$$

for $0 \leq x \leq 1$. Therefore,

$$F(\sigma) = \log \zeta(1+\sigma) - \sum_p \frac{1}{p^{1+\sigma}}$$

$$= \sigma \int_0^\infty e^{-\sigma x}(L(x) - S(e^x))dx + O(\sigma)$$

$$= \sigma \int_0^\infty e^{-\sigma x}\left(\gamma - b_1 + O\left(\frac{1}{x+1}\right)\right) dx + O(\sigma)$$

$$= (\gamma - b_1)\sigma \int_0^\infty e^{-\sigma x}dx + O\left(\sigma \int_0^\infty \frac{e^{-\sigma x}dx}{x+1}\right) + O(\sigma)$$

$$= \gamma - b_1 + O\left(\sigma \int_0^\infty \frac{e^{-\sigma x}dx}{x+1}\right) + O(\sigma).$$

Since

$$\int_0^\infty \frac{e^{-\sigma x}dx}{x+1} < \int_0^{1/\sigma} \frac{e^{-\sigma x}dx}{x+1} + \int_{1/\sigma}^\infty \frac{e^{-\sigma x}dx}{x}$$

$$< \int_0^{1/\sigma} \frac{dx}{x+1} + \int_1^\infty \frac{e^{-y}dy}{y}$$

$$= \log\left(\frac{1}{\sigma} + 1\right) + O(1)$$

$$\ll \log\left(\frac{1}{\sigma} + 1\right),$$

it follows that

$$F(\sigma) = \gamma - b_1 + O\left(\sigma \log\left(\frac{1}{\sigma} + 1\right)\right).$$

By (6.10), we have

$$b_2 = \lim_{\sigma \to 0^+} F(\sigma) = \gamma - b_1.$$

This completes the proof.

Theorem 6.8 (Mertens's formula)   *For $x \geq 2$,*

$$\prod_{p \leq x}\left(1 - \frac{1}{p}\right)^{-1} = e^\gamma \log x + O(1),$$

where γ is Euler's constant.

   Proof. We begin with two observations. First,

$$\sum_{p>x}\sum_{k=2}^{\infty}\frac{1}{kp^k} < \sum_{p>x}\frac{1}{p(p-1)}$$

$$< \sum_{n>x}\frac{1}{n(n-1)}$$

$$= \sum_{n>x}\left(\frac{1}{n-1} - \frac{1}{n}\right)$$

$$= O\left(\frac{1}{x}\right)$$

$$= O\left(\frac{1}{\log x}\right).$$

Second, since $\exp(t) = 1 + O(t)$ for t in any bounded interval and $O(1/\log x)$ is bounded for $x \geq 2$, it follows that

$$\exp\left(O\left(\frac{1}{\log x}\right)\right) = 1 + O\left(\frac{1}{\log x}\right).$$

Therefore,

$$\log\prod_{p \leq x}\left(1 - \frac{1}{p}\right)^{-1} = \sum_{p \leq x}\log\left(1 - \frac{1}{p}\right)^{-1}$$

$$= \sum_{p \leq x}\sum_{k=1}^{\infty}\frac{1}{kp^k}$$

$$= \sum_{p \leq x}\frac{1}{p} + \sum_{p \leq x}\sum_{k=2}^{\infty}\frac{1}{kp^k}$$

$$= \log\log x + b_1 + O\left(\frac{1}{\log x}\right) + b_2 - \sum_{p>x}\sum_{k=2}^{\infty}\frac{1}{kp^k}$$

$$= \log\log x + \gamma + O\left(\frac{1}{\log x}\right),$$

since $b_1 + b_2 = \gamma$ by Lemma 6.5, and so

$$\prod_{p\leq x}\left(1-\frac{1}{p}\right)^{-1} = e^{\gamma}\log x \exp\left(O\left(\frac{1}{\log x}\right)\right)$$

$$= e^{\gamma}\log x \left(1 + O\left(\frac{1}{\log x}\right)\right)$$

$$= e^{\gamma}\log x + O(1).$$

This is Mertens's formula.

The following result will be used in Chapter 10 in the proof of Chen's theorem.

Theorem 6.9 *For any $\varepsilon > 0$, there exists a number $u_1 = u_1(\varepsilon)$ such that*

$$\prod_{u\leq p<z}\left(1-\frac{1}{p}\right)^{-1} < (1+\varepsilon)\frac{\log z}{\log u}$$

for any $u_1 \leq u < z$.

Proof. Let γ be Euler's constant, and choose $\delta > 0$ such that

$$\frac{\gamma+\delta}{\gamma-\delta} < 1+\varepsilon.$$

By Theorem 6.8, we have

$$\prod_{p<x}\left(1-\frac{1}{p}\right)^{-1} \sim \gamma \log x,$$

and so there exists a number u_1 such that

$$(\gamma-\delta)\log x < \prod_{p<x}\left(1-\frac{1}{p}\right)^{-1} < (\gamma+\delta)\log x$$

for all $x \geq u_1$. Therefore, if $u_1 \leq u < z$, we have

$$\prod_{u\leq p<z}\left(1-\frac{1}{p}\right)^{-1} = \frac{\prod_{p<z}\left(1-\frac{1}{p}\right)^{-1}}{\prod_{p<u}\left(1-\frac{1}{p}\right)^{-1}}$$

$$< \frac{(\gamma+\delta)\log z}{(\gamma-\delta)\log u}$$

$$< (1+\varepsilon)\frac{\log z}{\log u}.$$

This completes the proof.

6.4 Brun's method and twin primes

There is a structural similarity between the twin prime conjecture and the Goldbach conjecture. The twin prime conjecture states that there exist infinitely many prime numbers p such that $p + 2$ is also a prime number or, equivalently, there exist infinitely many integers k such that $k(k + 2)$ has exactly two prime factors. The Goldbach conjecture states that every even integer $n \geq 4$ can be written as the sum of two primes or, equivalently, there exists an integer k such that $1 < k < n - 1$ and $k(n - k)$ has exactly two prime factors. We begin the study of sieve methods with a simple proof of the theorem that the twin primes are sparse in the sense that the sum of the reciprocals of the twin primes converges. This contrasts with the result (Theorem 6.7) that the sum of the reciprocals of all of the primes diverges like $\log \log x$.

Lemma 6.6 *If $\ell \geq 1$ and $0 \leq m \leq \ell$, then*

$$\sum_{k=0}^{m} (-1)^k \binom{\ell}{k} = (-1)^m \binom{\ell - 1}{m}.$$

Proof. This is by induction on m. It is easy to check that the equation is true for $m = 0, 1, 2$. If $1 \leq m \leq \ell$ and the equation holds for $m - 1$, then

$$\sum_{k=0}^{m} (-1)^k \binom{\ell}{k} = \sum_{k=0}^{m-1} (-1)^k \binom{\ell}{k} + (-1)^m \binom{\ell}{m}$$

$$= (-1)^{m-1} \binom{\ell - 1}{m - 1} + (-1)^m \binom{\ell}{m}$$

$$= (-1)^m \left(\binom{\ell}{m} - \binom{\ell - 1}{m - 1} \right)$$

$$= (-1)^m \binom{\ell - 1}{m}.$$

This completes the proof.

The following combinatorial inequality, a version of the principle of inclusion–exclusion, is the simplest form of the Brun sieve.

Theorem 6.10 (The Brun sieve) *Let X be a nonempty, finite set of N objects, and let $P_1, \ldots, P_r$ be r different properties that elements of the set X might have. Let N_0 denote the number of elements of X that have none of these properties. For any subset $I = \{i_1, \ldots, i_k\}$ of $\{1, 2, \ldots, r\}$, let $N(I) = N(i_1, \ldots, i_k)$ denote the number of elements of X that have each of the properties $P_{i_1}, P_{i_2}, \ldots, P_{i_k}$. Let $N(\emptyset) = |X| = N$. If m is a nonnegative even integer, then*

$$N_0 \leq \sum_{k=0}^{m} (-1)^k \sum_{|I|=k} N(I). \tag{6.11}$$

If m is a nonnegative odd integer, then

$$N_0 \geq \sum_{k=0}^{m}(-1)^k \sum_{|I|=k} N(I). \tag{6.12}$$

Proof. Inequalities (6.11) and (6.12) count the elements of X according to the various properties that each element possesses. We shall calculate how much each element of X contributes to the left and right sides of these inequalities.

Let x be an element of the set X, and suppose that x has exactly ℓ properties P_i. If $\ell = 0$, then x is counted once in N_0 and once in $N(\emptyset)$, but is not counted in $N(I)$ if I is nonempty. If $\ell \geq 1$, then x is not counted in N_0. By renumbering the properties, we can assume that x has the properties $P_1, P_2, \ldots, P_\ell$. Let $I \subseteq \{1, 2, \ldots, \ell, \ldots, r\}$. If $i \in I$ for some $i > \ell$, then x is not counted in $N(I)$. If $I \subseteq \{1, 2, \ldots, \ell\}$, then x contributes 1 to $N(I)$. For each $k = 0, 1, \ldots, \ell$, there are exactly $\binom{\ell}{k}$ such subsets with $|I| = k$. If $m \geq \ell$, then the element x contributes

$$\sum_{k=0}^{\ell}(-1)^k \binom{\ell}{k} = 0$$

to the right sides of the inequalities. If $m < \ell$, then x contributes

$$\sum_{k=0}^{m}(-1)^k \binom{\ell}{k}$$

to the right sides of inequalities (6.11) and (6.12). By Lemma 6.6, this contribution is positive if ℓ is even and negative if ℓ is odd. This completes the proof.

Lemma 6.7 *For $x \geq 1$ and for any congruence class a (mod m), the number of positive integers not exceeding x that are congruent to a modulo m is $x/m + \theta$, where $|\theta| < 1$.*

Proof. If $x/m = q \in \mathbf{Z}$, then the set $\{1, \ldots, qm\}$ contains exactly x/m elements in every congruence class modulo m.

Suppose that $x/m \notin \mathbf{Z}$. Let $[x]$ and $\{x\}$ denote the integer and fractional parts of x, respectively, and let $[x] = qm + r$, where $0 \leq r < m$. Then

$$qm < x = qm + r + \{x\} \leq qm + (m-1) + \theta < (q+1)m,$$

and so

$$q < \frac{x}{m} < q+1. \tag{6.13}$$

The positive integers up to x can be partitioned into $q+1$ pairwise disjoint sets such that q of these sets are complete systems of residues modulo m, and the remaining set is a subset of a complete system of residues modulo m. It follows that there are either q or $q+1$ integers in the congruence class a (mod m). The lemma follows from inequality (6.13).

Lemma 6.8 *Let $x \geq 1$, and let $p_{i_1}, \ldots, p_{i_k}$ be distinct odd primes. Let $N(i_1, \ldots, i_k)$ denote the number of positive integers $n \leq x$ such that*

$$n(n+2) \equiv 0 \pmod{p_{i_1} \cdots p_{i_k}}. \tag{6.14}$$

Then

$$N(i_1, \ldots, i_k) = \frac{2^k x}{p_{i_1} \cdots p_{i_k}} + 2^k \theta,$$

where $|\theta| < 1$.

Proof. If p is an odd prime and $n(n+2) \equiv 0 \pmod{p}$, then either

$$n \equiv 0 \pmod{p}$$

or

$$n \equiv -2 \pmod{p}.$$

Moreover, $0 \not\equiv -2 \pmod{p}$ since $p \geq 3$. If the integer n satisfies the congruence (6.14), then there exist unique integers $u_1, \ldots, u_k \in \{0, -2\}$

$$\left. \begin{array}{rcl} n & \equiv & u_1 \pmod{p_1} \\ n & \equiv & u_2 \pmod{p_2} \\ & \vdots & \\ n & \equiv & u_k \pmod{p_k}. \end{array} \right\} \tag{6.15}$$

By the Chinese remainder theorem, for each of the 2^k choices of $u_1, \ldots, u_k$ there exists a unique congruence class $a \pmod{p_1 \cdots p_k}$ such that n is a solution of the system of congruences (6.15) if and only if

$$n \equiv a \pmod{p_1 p_2 \cdots p_k}.$$

By Lemma 6.7, this congruence has

$$\frac{x}{p_1 p_2 \cdots p_k} + \theta(a)$$

solutions in positive integers not exceeding x, where $|\theta(a)| < 1$. Therefore,

$$N(i_1, \ldots, i_k) = \frac{2^k x}{p_{i_1} \cdots p_{i_k}} + 2^k \theta,$$

where $|\theta| < 1$. This completes the proof.

Theorem 6.11 (Brun) *Let $\pi_2(x)$ denote the number of primes p not exceeding x such that $p + 2$ is also prime. Then*

$$\pi_2(x) \ll \frac{x(\log \log x)^2}{(\log x)^2}.$$

Proof. Let $5 \leq y < x$. Let $r = \pi(y) - 1$ denote the number of odd primes not exceeding y. We denote these primes by $p_1, \ldots, p_r$. Let $\pi_2(y, x)$ denote the number of primes p such that $y < p \leq x$ and $p + 2$ is also prime. If $y < n \leq x$ and both n and $n + 2$ are prime numbers, then $n > p_i$ for $i = 1, \ldots, r$, and

$$n(n + 2) \not\equiv 0 \pmod{p_i}$$

for all i. Let $N_0(y, x)$ denote the number of positive integers $n \leq x$ such that

$$n(n + 2) \not\equiv 0 \pmod{p_i}$$

for all $i = 1, \ldots, r$. Then

$$\pi_2(x) \leq y + \pi_2(y, x) \leq y + N_0(y, x).$$

We shall use the Brun sieve to find an upper bound for $N_0(y, x)$.

Let X be the set of positive integers not exceeding x. For each odd prime $p_i \leq y$, we let P_i be the property that $n(n + 2)$ is divisible by p_i. For any subset $I = \{i_1, \ldots, i_k\}$ contained in $\{1, \ldots, r\}$, we let $N(I)$ be the number of integers $n \in X$ such that $n(n + 2)$ is divisible by each of the primes $p_{i_1}, \ldots, p_{i_k}$ or, equivalently, such that $n(n + 2)$ is divisible by $p_{i_1} \cdots p_{i_k}$. By Lemma 6.8, we have

$$N(I) = N(i_1, \ldots, i_k) = \frac{2^k x}{p_{i_1} \cdots p_{i_k}} + 2^k \theta.$$

Let m be an even integer such that $1 \leq m \leq r$. By inequality (6.11), we have

$$N_0(y, x) \leq \sum_{k=0}^{m} (-1)^k \sum_{|I|=k} N(I)$$

$$\leq \sum_{k=0}^{m} (-1)^k \sum_{\{i_1, \ldots, i_k\} \subseteq \{1, \ldots, r\}} \left(\frac{2^k x}{p_{i_1} \cdots p_{i_k}} + O(2^k) \right)$$

$$\leq x \sum_{k=0}^{m} \sum_{\{i_1, \ldots, i_k\} \subseteq \{1, \ldots, r\}} \frac{(-2)^k}{p_{i_1} \cdots p_{i_k}} + \sum_{k=0}^{m} (-1)^k \binom{r}{k} O(2^k)$$

$$\leq x \sum_{k=0}^{r} \sum_{\{i_1, \ldots, i_k\} \subseteq \{1, \ldots, r\}} \frac{(-2)^k}{p_{i_1} \cdots p_{i_k}}$$

$$- x \sum_{k=m+1}^{r} \sum_{\{i_1, \ldots, i_k\} \subseteq \{1, \ldots, r\}} \frac{(-2)^k}{p_{i_1} \cdots p_{i_k}} + O\left(\sum_{k=0}^{m} \binom{r}{k} 2^k \right).$$

We shall estimate these three terms separately. By Theorem 6.8,

$$x \sum_{k=0}^{r} \sum_{\{i_1, \ldots, i_k\} \subseteq \{1, \ldots, r\}} \frac{(-2)^k x}{p_{i_1} \cdots p_{i_k}} = x \prod_{2 < p \leq y} \left(1 - \frac{2}{p} \right)$$

$$< x \prod_{2 < p \leq y} \left(1 - \frac{1}{p} \right)^2$$

$$\ll \frac{x}{(\log y)^2}.$$

Let $s_k(x_1, \ldots, x_r)$ be the elementary symmetric polynomial of degree k in r variables. For any nonnegative real numbers $x_1, \ldots, x_r$ we have

$$
\begin{aligned}
s_k(x_1, \ldots, x_r) &= \sum_{\{i_1, \ldots, i_k\} \subseteq \{1, \ldots, r\}} x_{i_1} \cdots x_{i_k} \\
&\leq \frac{(x_1 + \cdots + x_r)^k}{k!} \\
&= \frac{(s_1(x_1, \ldots, x_r))^k}{k!} \\
&< \left(\frac{e}{k}\right)^k s_1(x_1, \ldots, x_r)^k
\end{aligned}
$$

since $(k/e)^k < k!$. Therefore,

$$
\begin{aligned}
\left| x \sum_{k=m+1}^{r} \sum_{\{i_1, \ldots, i_k\} \subseteq \{1, \ldots, r\}} \frac{(-2)^k}{p_{i_1} \cdots p_{i_k}} \right| \\
\leq x \sum_{k=m+1}^{r} \sum_{\{i_1, \ldots, i_k\} \subseteq \{1, \ldots, r\}} \frac{2^k}{p_{i_1} \cdots p_{i_k}} \\
\leq x \sum_{k=m+1}^{r} \sum_{\{i_1, \ldots, i_k\} \subseteq \{1, \ldots, r\}} \left(\frac{2}{p_{i_1}}\right) \cdots \left(\frac{2}{p_{i_k}}\right) \\
= x \sum_{k=m+1}^{r} s_k \left(\frac{2}{p_1}, \ldots, \frac{2}{p_r}\right) \\
< x \sum_{k=m+1}^{r} \left(\frac{e}{k}\right)^k s_1 \left(\frac{2}{p_1}, \ldots, \frac{2}{p_r}\right)^k \\
= x \sum_{k=m+1}^{r} \left(\frac{e}{k}\right)^k \left(\frac{2}{p_1} + \cdots + \frac{2}{p_r}\right)^k \\
< x \sum_{k=m+1}^{r} \left(\frac{2e}{m}\right)^k \left(\sum_{p \leq y} \frac{1}{p}\right)^k \\
< x \sum_{k=m+1}^{r} \left(\frac{c \log \log y}{m}\right)^k,
\end{aligned}
$$

where c is an absolute positive constant. If we choose the even integer m so that

$$
m > 2c \log \log y,
$$

then

$$
x \sum_{k=m+1}^{r} \left(\frac{c \log \log y}{m}\right)^k \leq x \sum_{k=m+1}^{r} \frac{1}{2^k} < \frac{x}{2^m}.
$$

Since r is the number of odd primes less than or equal to y, it follows that $2r \leq y$, and we get the following estimate for the third term:

$$\sum_{k=0}^{m} \binom{r}{k} 2^k < \sum_{k=0}^{m} (2r)^k \ll (2r)^m \leq y^m.$$

Combining these three estimates, we obtain

$$\pi_2(x) \ll y + \frac{x}{(\log y)^2} + \frac{x}{2^m} + y^m \ll \frac{x}{(\log y)^2} + \frac{x}{2^m} + y^m, \qquad (6.16)$$

where the implied constant is absolute, y is any real number satisfying

$$5 \leq y < x, \qquad (6.17)$$

and m is any even integer such that

$$m > 2c \log \log y. \qquad (6.18)$$

Let $c' = \max\{2c, (\log 2)^{-1}\}$, and let

$$y = \exp\left(\frac{\log x}{3c' \log \log x}\right) = x^{\frac{1}{3c' \log \log x}}$$

and

$$m = 2[c' \log \log x].$$

The number y satisfies conditions (6.17) and (6.18) for x sufficiently large. We estimate the three terms in (6.16) with these values of y and m. Since

$$\log y = \frac{\log x}{3c' \log \log x},$$

we obtain the main term

$$\frac{x}{(\log y)^2} \ll \frac{x(\log \log x)^2}{(\log x)^2}.$$

Next, since $c' \geq (\log 2)^{-1}$ and

$$m = 2[c' \log \log x] > 2c' \log \log x - 2,$$

we obtain

$$\frac{x}{2^m} < \frac{4x}{2^{2c' \log \log x}} = \frac{4x}{(\log x)^{2c' \log 2}} \leq \frac{4x}{(\log x)^2}.$$

Finally,

$$y^m \leq y^{2c' \log \log x} = \exp\left(\frac{2c' \log \log x \log x}{3c' \log \log x}\right) = x^{2/3}.$$

Combining these three estimates, we obtain

$$\pi_2(x) \ll \frac{x(\log \log x)^2}{(\log x)^2}.$$

This completes the proof.

Theorem 6.12 (Brun) *Let $p_1, p_2, \ldots$ be the sequence of prime numbers p such that $p + 2$ is also prime. Then*

$$\sum_{n=1}^{\infty} \left(\frac{1}{p_n} + \frac{1}{p_n + 2} \right)$$

$$= \left(\frac{1}{3} + \frac{1}{5} \right) + \left(\frac{1}{5} + \frac{1}{7} \right) + \left(\frac{1}{11} + \frac{1}{13} \right) + \left(\frac{1}{17} + \frac{1}{19} \right) + \cdots$$

$$< \infty.$$

Proof. Theorem 6.11 implies that

$$\pi_2(x) \ll \frac{x}{(\log x)^{3/2}}$$

for all $x \geq 2$. Therefore,

$$n = \pi_2(p_n) \ll \frac{p_n}{(\log p_n)^{3/2}} \leq \frac{p_n}{(\log n)^{3/2}}$$

for $n \geq 2$, and so

$$\frac{1}{p_n} \ll \frac{1}{n (\log n)^{3/2}}.$$

It follows that the series

$$\sum_{n=1}^{\infty} \frac{1}{p_n} \leq \frac{1}{3} + \sum_{n=2}^{\infty} \frac{1}{p_n} \ll \frac{1}{3} + \sum_{n=2}^{\infty} \frac{1}{n (\log n))^{3/2}}$$

converges. This completes the proof.

6.5 Notes

Dickson [22, vol. I, pp. 421–424] contains a brief account of early results concerning the Goldbach conjecture. Sinisalo [117] has verified the Goldbach conjecture by computer for all even integers up to $4 \cdot 10^{11}$. Wang's book *Goldbach Conjecture* [137] is an anthology of classic papers on this subject.

Brun [7] obtained the first significant result concerning the Goldbach conjecture in 1920. By means of the combinatorial method known today as the Brun sieve, he proved that every sufficiently large even integer can be written as the sum of two integers, each of which is the product of at most nine primes. Brun also obtained the first nontrivial results concerning the twin prime conjecture. In addition to Theorem 6.11 and Theorem 6.12, he also proved that there are infinitely many integers n such that both n and $n + 2$ are the products of at most 9 primes. The application of the Brun sieve to the twin prime conjecture follows Landau [78].

By Theorem 6.12, the sum over the reciprocals of the twin primes converges. The sum of this infinite series is called *Brun's constant*; its value is estimated to be

$1.9021604 \pm 5 \times 10^{-7}$ (see Shanks-Wrench [112] and Brent [5]). It is a difficult computational problem to determine Brun's constant to high precision. In the process of trying to improve the estimates for Brun's constant, Nicely discovered a defect in Intel's Pentium computer chip (see [15]).

A popular game among computational number theorists is to find explicit examples of twin primes. On October 18, 1995, Harvey Dubner announced over the Internet that p and $p + 2$ are prime numbers for

$$p = 570,918,348 \cdot 10^{5120} - 1 = 2^2 \cdot 3^3 \cdot 7 \cdot 11 \cdot 13 \cdot 5281 \cdot 10^{5120} - 1.$$

The prime p has 5129 digits. This established a new record for the largest twin prime.

For other elementary results about the distribution of prime numbers, see Ellison and Ellison [29], Hardy and Wright [51], Ingham [66], and Tenenbaum [121]. Rosen [104] has generalized Mertens's Theorem 6.8 to algebraic number fields.

6.6   Exercises

1. Let n be a positive integer. Prove that

$$\log n = \sum_{d \mid n} \Lambda(d)$$

and

$$\Lambda(n) = - \sum_{d \mid n} \mu(d) \log d.$$

2. Let $\omega(n)$ denote the number of distinct prime divisors of n. Let $n \geq 2$ and $r \geq 0$. Prove that

$$\sum_{\substack{d \mid n \\ \omega(d) \leq 2r+1}} \mu(d) \leq 0 \leq \sum_{\substack{d \mid n \\ \omega(d) \leq 2r}} \mu(d).$$

3. With the notation of Theorem 6.10, prove that

$$N_0 = \sum_{k=0}^{t} (-1)^k \sum_{|I|=k} N(I).$$

This formula is often called the *inclusion–exclusion principle*.

4. Use the inclusion–exclusion principle to prove that

$$\varphi(n) = n \prod_{p \mid n} \left(1 - \frac{1}{p}\right) = n \sum_{d \mid n} \frac{\mu(d)}{d},$$

where $\varphi(n)$ is the Euler φ-function.

5. Let $\Phi(x, y)$ denote the number of positive integers $n \leq x$ that are not divisible by any prime $p \leq y$. Prove that

$$\Phi(x, y) = x \prod_{p \leq y} \left(1 - \frac{1}{p}\right) + 2^{\pi(y)}. \ll \frac{x}{\log y} + 2^{\pi(y)}.$$

6. Prove that

$$\prod_{r < p \leq x} \left(1 - \frac{r}{p}\right) \ll \frac{1}{(\log x)^r}.$$

7. Prove that

$$\sum_{n \leq x} \left(\log \frac{x}{n}\right)^k = k!x + O\left((\log x)^k\right).$$

8. Prove that

$$\exp\left(O\left(\frac{1}{\log x}\right)\right) = 1 + O\left(\frac{1}{\log x}\right).$$

7
The Shnirel'man–Goldbach theorem

Das allgemeine Problem der additiven Zahlentheorie ist die Darstell-
barkeit aller natürlichen Zahlen durch eine beschränkte Anzahl von
Summanden einer gegebenen Folge von natürlichen Zahlen, z. B. der
Primzahlfolge oder der Folge der p-ten Potenzen.[1]

L. G. Shnirel'man [114]

7.1   The Goldbach conjecture

In a letter to Euler in 1742, Goldbach conjectured that every positive even integer
$n > 2$ is the sum of two primes. Euler replied that he believed the conjecture
but could not prove it. It is still unproven, but it has been confirmed by computer
calculations for even integers up to $4 \cdot 10^{11}$.

In 1930, Shnirel'man proved that every integer greater than one is the sum of
a bounded number of primes. This is a great theorem, the first significant result
on the Goldbach conjecture. Shnirel'man used purely combinatorial methods: the
Brun sieve and a theorem about the density of the sum of two sets of integers.
We shall prove Shnirel'man's theorem in this chapter. Instead of the Brun sieve,
however, we shall use a sieve method due to Selberg, which is also completely

---

[1]The general problem in additive number theory is the representation of the natural
numbers as the sum of a bounded number of terms from a given sequence of natural numbers,
e.g. the sequence of prime numbers or the sequence of p-th powers.

elementary but more elegant and in many cases more powerful than Brun's original sieve argument.

7.2   The Selberg sieve

Lemma 7.1 (Cauchy–Schwarz inequality) *Let* $a_1, \ldots, a_n, b_1, \ldots, b_n$ *be real numbers. Then*

$$\left(\sum_{i=1}^{n} a_i b_i\right)^2 \le \left(\sum_{i=1}^{n} a_i^2\right)\left(\sum_{i=1}^{n} b_i^2\right).$$

If $a_j \ne 0$ *for some* j, *then*

$$\left(\sum_{i=1}^{n} a_i b_i\right)^2 = \left(\sum_{i=1}^{n} a_i^2\right)\left(\sum_{i=1}^{n} b_i^2\right)$$

if and only if there is a real number t *such that* $b_i = t a_i$ *for all* $i = 1, \ldots, n$.

Proof. Since

$$
\begin{aligned}
0 &\le \sum_{1 \le i < j \le n} (a_i b_j - a_j b_i)^2 \\
&= \sum_{1 \le i < j \le n} (a_i^2 b_j^2 - 2 a_i a_j b_i b_j + a_j^2 b_i^2) \\
&= \sum_{i=1}^{n} a_i^2 \sum_{j=1}^{n} b_j^2 - \left(\sum_{i=1}^{n} a_i b_i\right)^2,
\end{aligned}
$$

we have

$$\left(\sum_{i=1}^{n} a_i b_i\right)^2 \le \left(\sum_{i=1}^{n} a_i^2\right)\left(\sum_{i=1}^{n} b_i^2\right).$$

Moreover,

$$\left(\sum_{i=1}^{n} a_i b_i\right)^2 = \left(\sum_{i=1}^{n} a_i^2\right)\left(\sum_{i=1}^{n} b_i^2\right)$$

if and only if

$$a_i b_j = a_j b_i$$

for all $i \ne j$. In this case, if $a_j \ne 0$ for some j, let $t = b_j/a_j$. Then

$$b_i = \left(\frac{b_j}{a_j}\right) a_i = t a_i$$

for $i = 1, \ldots, n$. This completes the proof.

Lemma 7.2 *Let $a_1, \ldots, a_n$ be positive real numbers and $b_1, \ldots, b_n$ be any real numbers. The minimum value of the quadratic form*

$$Q(y_1, \ldots, y_n) = a_1 y_1^2 + \cdots + a_n y_n^2$$

subject to the linear constraint

$$b_1 y_1 + \cdots + b_n y_n = 1 \qquad (7.1)$$

is

$$m = \left(\sum_{i=1}^{n} \frac{b_i^2}{a_i} \right)^{-1},$$

and this value is attained if and only if

$$y_i = \frac{m b_i}{a_i}$$

for all $i = 1, \ldots, n$.

Proof. Let $y_1, \ldots, y_n$ be real numbers that satisfy (7.1). By the Cauchy–Schwartz inequality, we have

$$1 = \left(\sum_{i=1}^{n} b_i y_i \right)^2$$

$$= \left(\sum_{i=1}^{n} \left(\frac{b_i}{\sqrt{a_i}} \right) \sqrt{a_i} \, y_i \right)^2$$

$$\leq \left(\sum_{i=1}^{n} \frac{b_i^2}{a_i} \right) \left(\sum_{i=1}^{n} a_i y_i^2 \right),$$

and so

$$\sum_{i=1}^{n} a_i y_i^2 \geq \left(\sum_{i=1}^{n} \frac{b_i^2}{a_i} \right)^{-1} = m.$$

Moreover,

$$\sum_{i=1}^{n} a_i y_i^2 = m$$

if and only if there exists a real number t such that, for all $i = 1, \ldots, n$,

$$\sqrt{a_i} \, y_i = \frac{t b_i}{\sqrt{a_i}},$$

or, equivalently,

$$y_i = \frac{t b_i}{a_i}.$$

This implies that

$$1 = \sum_{i=1}^{n} b_i y_i = t \sum_{i=1}^{n} \frac{b_i^2}{a_i} = \frac{t}{m},$$

and so

$$t = m$$

and

$$y_i = \frac{mb_i}{a_i}.$$

Conversely, if $y_i = mb_i/a_i$ for all i, then $\sum_{i=1}^{n} b_i y_i = 1$ and $Q(y_1, \ldots, y_n) = m$. This completes the proof.

Theorem 7.1 (Selberg sieve) *Let A be a finite sequence of integers, and let $|A|$ denote the number of terms of the sequence. Let $\mathcal{P}$ be a set of primes. For any real number $z \geq 2$, let*

$$P(z) = \prod_{\substack{p < z \\ p \in \mathcal{P}}} p.$$

The "sieving function"

$$S(A, \mathcal{P}, z)$$

denotes the number of terms of the sequence A that are not divisible by any prime $p \in \mathcal{P}$ such that $p < z$. For every square-free positive integer d, let $|A_d|$ denote the number of terms of the sequence A that are divisible by d. Let $g(k)$ be a multiplicative function such that

$$0 < g(p) < 1 \quad \text{for all } p \in \mathcal{P},$$

and let $g_1(m)$ be a completely multiplicative function such that $g_1(p) = g(p)$ for all $p \in \mathcal{P}$. Define the "remainder term" $r(d)$ and the function $G(z)$ by

$$r(d) = |A_d| - g(d)|A|$$

and

$$G(z) = \sum_{\substack{m < z \\ p|m \Rightarrow p \in \mathcal{P}}} g_1(m).$$

Then

$$S(A, \mathcal{P}, z) \leq \frac{|A|}{G(z)} + \sum_{\substack{d < z^2 \\ d|P(z)}} 3^{\omega(d)} |r(d)|, \qquad (7.2)$$

where $\omega(d)$ is the number of distinct prime divisors of d.

Proof. Since g is a multiplicative function, we have, by Theorem A.7,

$$g([d_1, d_2])g((d_1, d_2)) = g(d_1)g(d_2)$$

for all positive integers d_1 and d_2.

Let $z \geq 2$. For every divisor d of $P(z)$, we shall choose a real number $\lambda(d)$ subject only to the conditions that

$$\lambda(1) = 1$$

and

$$\lambda(d) = 0 \quad \text{for all} \quad d \geq z.$$

Since

$$\left(\sum_{d \mid (a, P(z))} \lambda(d) \right)^2 \geq 0$$

for all nonnegative integers a and

$$\left(\sum_{d \mid (a, P(z))} \lambda(d) \right)^2 = 1 \quad \text{if } (a, P(z)) = 1,$$

it follows that

$$S(A, P, z) = \sum_{\substack{a \in A \\ (a, P(z))=1}} 1$$

$$\leq \sum_{a \in A} \left(\sum_{d \mid (a, P(z))} \lambda(d) \right)^2$$

$$= \sum_{a \in A} \sum_{\substack{d_1 \mid a \\ d_1 \mid P(z)}} \sum_{\substack{d_2 \mid a \\ d_2 \mid P(z)}} \lambda(d_1)\lambda(d_2)$$

$$= \sum_{d_1, d_2 \mid P(z)} \lambda(d_1)\lambda(d_2) \sum_{\substack{a \in A \\ [d_1, d_2] \mid a}} 1$$

$$= \sum_{d_1, d_2 \mid P(z)} \lambda(d_1)\lambda(d_2) |A_{[d_1, d_2]}|$$

$$= \sum_{d_1, d_2 \mid P(z)} \lambda(d_1)\lambda(d_2) \left(g([d_1, d_2])|A| + r([d_1, d_2]) \right)$$

$$= |A| \sum_{d_1, d_2 \mid P(z)} g([d_1, d_2])\lambda(d_1)\lambda(d_2) + \sum_{d_1, d_2 \mid P(z)} \lambda(d_1)\lambda(d_2) r([d_1, d_2])$$

$$= |A| \sum_{\substack{d_1, d_2 < z \\ d_1, d_2 \mid P(z)}} \frac{1}{g((d_1, d_2))} g(d_1)\lambda(d_1) g(d_2)\lambda(d_2)$$

$$+ \sum_{\substack{d_1, d_2 < z \\ d_1, d_2 \mid P(z)}} \lambda(d_1)\lambda(d_2) r([d_1, d_2])$$

$$= |A| Q + R,$$

where

$$Q = \sum_{\substack{d_1, d_2 < z \\ d_1, d_2 \mid P(z)}} \frac{1}{g((d_1, d_2))} g(d_1)\lambda(d_1) g(d_2)\lambda(d_2)$$

and

$$R = \sum_{\substack{d_1, d_2 < z \\ d_1, d_2 | P(z)}} \lambda(d_1)\lambda(d_2) r([d_1, d_2]).$$

Let $\mathcal{D}$ be the set of all positive divisors of $P(z)$ that are strictly less than z, that is,

$$\mathcal{D} = \{k | P(z) : 1 \le k < z\}.$$

Then $\mathcal{D}$ is a divisor-closed set of square-free integers. If $k \in \mathcal{D}$, then $0 < g(k) \le 1$ since $0 < g(p) < 1$ for all primes $p \in \mathcal{P}$. For $k \in \mathcal{D}$, we define the function $f(k)$ by

$$f(k) = \sum_{d|k} \frac{\mu(d)}{g(k/d)} = \frac{1}{g(k)} \sum_{d|k} \mu(d)g(d) = \frac{1}{g(k)} \prod_{p|k} (1 - g(p)). \tag{7.3}$$

Then $f(k) > 0$ and $f(k_1 k_2) = f(k_1) f(k_2)$ if $k_1, k_2 \in \mathcal{D}$ and $(k_1, k_2) = 1$. By Möbius inversion (Theorem A.19), we have

$$\frac{1}{g(k)} = \sum_{d|k} f(d). \tag{7.4}$$

Then

$$\begin{aligned}
Q &= \sum_{d_1, d_2 \in \mathcal{D}} \frac{1}{g((d_1, d_2))} g(d_1)\lambda(d_1) g(d_2)\lambda(d_2) \\
&= \sum_{d_1, d_2 \in \mathcal{D}} \sum_{\substack{k|d_1 \\ k|d_2}} f(k) g(d_1)\lambda(d_1) g(d_2)\lambda(d_2) \\
&= \sum_{k \in \mathcal{D}} f(k) \sum_{\substack{d_1, d_2 \in \mathcal{D} \\ k|d_1, k|d_2}} g(d_1)\lambda(d_1) g(d_2)\lambda(d_2) \\
&= \sum_{k \in \mathcal{D}} f(k) \left(\sum_{\substack{d \in \mathcal{D} \\ k|d}} g(d)\lambda(d) \right)^2 \\
&= \sum_{k \in \mathcal{D}} f(k) y_k^2,
\end{aligned}$$

where

$$y_k = \sum_{\substack{d \in \mathcal{D} \\ k|d}} g(d)\lambda(d).$$

Thus, Q is a quadratic form in the variables y_k.

The set $\mathcal{D}$ is finite and divisor-closed. By Möbius inversion (Theorem A.22), we have

$$g(d)\lambda(d) = \sum_{\substack{k \in \mathcal{D} \\ d|k}} \mu\left(\frac{k}{d}\right) y_k = \mu(d) \sum_{\substack{k \in \mathcal{D} \\ d|k}} \mu(k) y_k. \tag{7.5}$$

In particular, for $d = 1$ we obtain

$$\sum_{k \in \mathcal{D}} \mu(k) y_k = 1. \tag{7.6}$$

We define

$$F(z) = \sum_{k \in \mathcal{D}} \frac{1}{f(k)}.$$

By Lemma 7.2, the minimum value of the quadratic form

$$Q = \sum_{k \in \mathcal{D}} f(k) y_k^2$$

subject to the linear constraint (7.6) is

$$\left(\sum_{k \in \mathcal{D}} \frac{\mu(k)^2}{f(k)}\right)^{-1} = \left(\sum_{k \in \mathcal{D}} \frac{1}{f(k)}\right)^{-1} = \frac{1}{F(z)},$$

and this minimum is attained when

$$y_k = \frac{\mu(k)}{F(z)f(k)}.$$

We insert these values of y_k into (7.5) to compute $\lambda(d)$ as follows:

$$\begin{aligned}
\lambda(d) &= \frac{\mu(d)}{g(d)} \sum_{\substack{k \in \mathcal{D} \\ d|k}} \mu(k) y_k \\
&= \frac{\mu(d)}{g(d)} \sum_{\substack{d\ell < z \\ d\ell|P(z)}} \mu(d\ell) y_{d\ell} \\
&= \frac{\mu(d)}{g(d)} \sum_{\substack{\ell < z/d \\ d\ell|P(z)}} \mu(d\ell) \left(\frac{\mu(d\ell)}{F(z)f(d\ell)}\right) \\
&= \frac{\mu(d)}{f(d)g(d)F(z)} \sum_{\substack{\ell < z/d \\ d\ell|P(z)}} \frac{1}{f(\ell)} \\
&= \frac{\mu(d)F_d(z)}{f(d)g(d)F(z)},
\end{aligned}$$

where

$$F_d(z) = \sum_{\substack{\ell < z/d \\ d\ell|P(z)}} \frac{1}{f(\ell)}.$$

In the preceding calculation, we used the fact that if $d\ell$ divides $P(z)$, then d and ℓ are relatively prime since $P(z)$ is square-free. We shall use this fact again to prove that $|\lambda(d)| \leq 1$. Let d be any positive divisor of $P(z)$. Then

$$F(z) = \sum_{k \in D} \frac{1}{f(k)}$$

$$= \sum_{\ell | d} \sum_{\substack{k \in D \\ (k,d)=\ell}} \frac{1}{f(k)}$$

$$= \sum_{\ell | d} \sum_{\substack{\ell m < z \\ \ell m | P(z) \\ (\ell m, d)=\ell}} \frac{1}{f(\ell m)}$$

$$= \sum_{\ell | d} \frac{1}{f(\ell)} \sum_{\substack{m < z/\ell \\ \ell m | P(z) \\ (m, d/\ell)=1}} \frac{1}{f(m)}$$

$$= \sum_{\ell | d} \frac{1}{f(\ell)} \sum_{\substack{m < z/\ell \\ m | P(z) \\ (m,d)=1}} \frac{1}{f(m)}$$

$$= \sum_{\ell | d} \frac{1}{f(\ell)} \sum_{\substack{m < z/\ell \\ dm | P(z)}} \frac{1}{f(m)}$$

$$\geq \sum_{\ell | d} \frac{1}{f(\ell)} \sum_{\substack{m < z/d \\ dm | P(z)}} \frac{1}{f(m)}$$

$$= F_d(z) \sum_{\ell | d} \frac{1}{f(\ell)}$$

$$= \frac{F_d(z)}{f(d)} \sum_{\ell | d} f(d/\ell)$$

$$= \frac{F_d(z)}{f(d) g(d)}$$

by (7.4), and so

$$|\lambda(d)| = \frac{F_d(z)}{f(d) g(d) F(z)} \leq 1.$$

By Exercise 1, for any square-free integer d there are exactly $3^{\omega(d)}$ ordered pairs of positive integers d_1, d_2 such that $[d_1, d_2] = d$. If $d_1, d_2 < z$, then $d = [d_1, d_2] < z^2$. If d_1 and d_2 divide $P(z)$, then $d = [d_1, d_2]$ is a square-free number that also divides $P(z)$. Therefore,

$$|R| = \left| \sum_{\substack{d_1, d_2 < z \\ d_1, d_2 | P(z)}} \lambda(d_1) \lambda(d_2) r([d_1, d_2]) \right|$$

$$\leq \sum_{\substack{d_1, d_2 < z \\ d_1, d_2 | P(z)}} |r([d_1, d_2])|$$

$$\leq \sum_{\substack{d < z^2 \\ d | P(z)}} 3^{\omega(d)} |r(d)|,$$

and so

$$S(A, \mathcal{P}, z) \leq \frac{|A|}{F(z)} + \sum_{\substack{d < z^2 \\ d | P(z)}} 3^{\omega(d)} |r_d|.$$

To obtain the upper bound (7.2) for the sieving function $S(A, \mathcal{P}, z)$, it is enough to prove that $F(z) \geq G(z)$. Let $g_1(k)$ be a completely multiplicative function such that

$$g_1(p) = g(p) \qquad \text{for all primes } p \in \mathcal{P}.$$

By (7.3),

$$
\begin{aligned}
F(z) &= \sum_{k \in \mathcal{D}} \frac{1}{f(k)} \\
&= \sum_{k \in \mathcal{D}} g(k) \prod_{p | k} (1 - g(p))^{-1} \\
&= \sum_{k \in \mathcal{D}} g_1(k) \prod_{p | k} (1 - g_1(p))^{-1} \\
&= \sum_{k \in \mathcal{D}} g_1(k) \prod_{p | k} \sum_{r=0}^{\infty} g_1(p)^r \\
&= \sum_{k \in \mathcal{D}} g_1(k) \prod_{p | k} \sum_{r=0}^{\infty} g_1(p^r) \\
&= \sum_{k \in \mathcal{D}} g_1(k) \sum_{\substack{\ell = 1 \\ p | \ell \Rightarrow p | k}}^{\infty} g_1(\ell) \\
&= \sum_{k \in \mathcal{D}} \sum_{\substack{\ell = 1 \\ p | \ell \Rightarrow p | k}}^{\infty} g_1(k\ell) \\
&= \sum_{k \in \mathcal{D}} \sum_{\substack{m = 1 \\ k | m \\ p | (m/k) \Rightarrow p | k}}^{\infty} g_1(m) \\
&= \sum_{m=1}^{\infty} g_1(m) \left(\sum_{\substack{k \in \mathcal{D} \\ k | m \\ p | (m/k) \Rightarrow p | k}} 1 \right) \\
&\geq \sum_{\substack{m < z \\ p | m \Rightarrow p \in \mathcal{P}}} g_1(m) \left(\sum_{\substack{k \in \mathcal{D} \\ k | m \\ p | m/k \Rightarrow p | k}} 1 \right)
\end{aligned}
$$

$$\geq \sum_{\substack{m<z \\ p|m \Rightarrow p \in \mathcal{P}}} g_1(m)$$

$$= G(z),$$

since, in the last inner sum, we can always choose k to be the "square-free kernel" of m, that is, the product of the distinct primes dividing m. This completes the proof of the theorem.

7.3   Applications of the sieve

In this section, we shall obtain an upper bound for the number of representations of an even integer as the sum of two primes. We also derive an upper bound for the number of representations of an even integer N as the difference of two primes, that is, an upper bound for the number of primes $p \leq x$ such that $p + N$ is also prime.

Theorem 7.2 *Let N be an even integer, and let $r(N)$ denote the number of representations of N as the sum of two primes. Then*

$$r(N) \ll \frac{N}{(\log N)^2} \prod_{p|N} \left(1 + \frac{1}{p}\right),$$

where the implied constant is absolute.

   Proof. The representation function $r(N)$ counts the number of primes $p \leq N$ such that $N - p$ is also prime. Let

$$a_n = n(N - n).$$

Then

$$A = \{a_n\}_{n=1}^{N}$$

is a finite sequence of integers with $|A| = N$ terms. Let $\mathcal{P}$ be the set of all prime numbers. Let

$$2 < z \leq \sqrt{N}.$$

The sieving function $S(A, \mathcal{P}, z)$ denotes the number of terms of the sequence A that are divisible by no prime $p < z$. If

$$\sqrt{N} < n < N - \sqrt{N},$$

and if $a_n \equiv 0 \pmod{p}$ for some prime $p < z$, then either n or $N-n$ is composite. This implies that

$$r(N) \leq 2\sqrt{N} + S(A, \mathcal{P}, z). \tag{7.7}$$

We shall use the Selberg sieve to obtain an upper bound for $S(A, \mathcal{P}, z)$. We continue to use the notation of Theorem 7.1.

Let $g(m)$ be the completely multiplicative function defined by

$$g(p) = \begin{cases} 2/p & \text{if } p \text{ does not divide } N \\ 1/p & \text{if } p \text{ divides } N. \end{cases} \tag{7.8}$$

Then $g_1(m) = g(m)$ for all m. Since N is even, 2 divides N and

$$0 < g(p) < 1$$

for all primes p. Also,

$$a_n = n(N - n) \equiv 0 \pmod{p}$$

if and only if

$$n \equiv 0 \pmod{p} \quad \text{or} \quad n \equiv N \pmod{p}.$$

If p does not divide N, then $N \not\equiv 0 \pmod{p}$ and these two congruences are distinct. If p divides N, then $N \equiv 0 \pmod{p}$ and these two congruences are the same. Let

$$d = p_1 \cdots p_k q_1 \cdots q_\ell$$

be a square-free integer, where the primes p_i divide N and the primes q_j do not divide N. Then

$$g(d) = \frac{2^\ell}{d}.$$

Since $a_n \equiv 0 \pmod{d}$ if and only if $a_n \equiv 0 \pmod{p}$ for every prime p dividing d, it follows from the Chinese remainder theorem that there are exactly 2^ℓ pairwise distinct congruence classes modulo d such that $a_n \equiv 0 \pmod{d}$ if and only if n belongs to one of these 2^ℓ classes. Therefore,

$$|A_d| = |A| g(d) + r(d),$$

where

$$|r(d)| \le 2^\ell \le 2^{\omega(d)}. \tag{7.9}$$

By the Selberg sieve,

$$S(A, \mathcal{P}, z) \le \frac{|A|}{G(z)} + \sum_{\substack{d < z^2 \\ d|P(z)}} 3^{\omega(d)} |r(d)|,$$

where

$$G(z) = \sum_{m < z} g(m)$$

and $\omega(d)$ is the number of distinct prime divisors of d. Let

$$m = \prod_{i=1}^{k} p_i^{r_i} \prod_{j=1}^{\ell} q_j^{s_j},$$

where the primes p_i divide N and the primes q_j do not divide N. Then

$$g(m) = \prod_{i=1}^{k}\left(\frac{1}{p_i}\right)^{r_i} \prod_{j=1}^{\ell}\left(\frac{2}{q_j}\right)^{s_j} = \frac{2^{s_1+\cdots+s_\ell}}{m}.$$

Let $d_N(m)$ denote the number of positive divisors of m that are relatively prime to N. Then

$$d_N(m) = d\left(\prod_{j=1}^{\ell}q_j^{s_j}\right) = \prod_{j=1}^{\ell}(s_j + 1) \le \prod_{j=1}^{m}2^{s_j} = 2^{s_1+\cdots+s_m}.$$

Therefore,

$$g(m) \ge \frac{d_N(m)}{m},$$

and so

$$G(z) = \sum_{m<z}g(m) \ge \sum_{m<z}\frac{d_N(m)}{m}.$$

Since

$$\prod_{p|N}\left(1 - \frac{1}{p}\right)^{-1} = \sum_{\substack{t=1 \\ p|t \Rightarrow p|N}}^{\infty}\frac{1}{t},$$

it follows that

$$\prod_{p|N}\left(1 - \frac{1}{p}\right)^{-1}G(z) \ge \sum_{m<z}\frac{d_N(m)}{m} \sum_{\substack{t=1 \\ p|t \Rightarrow p|N}}^{\infty}\frac{1}{t}$$

$$= \sum_{m<z}d_N(m) \sum_{\substack{t=1 \\ p|t \Rightarrow p|N}}^{\infty}\frac{1}{mt}$$

$$= \sum_{m<z}d_N(m) \sum_{\substack{w=1 \\ m|w \\ p|(w/m) \Rightarrow p|N}}^{\infty}\frac{1}{w}$$

$$= \sum_{w=1}^{\infty}\frac{1}{w} \sum_{\substack{m<z \\ m|w \\ p|(w/m) \Rightarrow p|N}}d_N(m)$$

$$\ge \sum_{w<z}\frac{1}{w} \sum_{\substack{m|w \\ p|(w/m) \Rightarrow p|N}}d_N(m).$$

Let

$$w = \prod_{i=1}^{k}p_i^{u_i} \prod_{j=1}^{\ell}q_j^{v_j}$$

and

$$m = \prod_{i=1}^{k}p_i^{r_i} \prod_{j=1}^{\ell}q_j^{s_j},$$

where the primes p_i divide N and the primes q_j do not divide N. Since m divides w, it follows that $0 \le r_i \le u_i$ for all i, $0 \le s_j \le v_j$ for all j, and

$$\frac{w}{m} = \prod_{i=1}^{k} p_i^{u_i - r_i} \prod_{j=1}^{\ell} q_j^{v_j - s_j}.$$

Since every prime divisor of w/m divides N, it follows that no prime q_j divides w/m, and so $s_j = v_j$ for all j. Therefore,

$$m = \prod_{i=1}^{k} p_i^{r_i} \prod_{j=1}^{\ell} q_j^{v_j}$$

and

$$d_N(m) = \prod_{j=1}^{\ell} (v_j + 1).$$

For each integer w, the number of such divisors m is

$$\prod_{i=1}^{\ell} (u_i + 1).$$

It follows that for every positive integer $w < z$, we have

$$\sum_{\substack{m|w \\ p|(w/m) \Rightarrow p|N}} d_N(m) = \sum_{\substack{m|w \\ p|(w/m) \Rightarrow p|N}} \prod_{j=1}^{m} (v_j + 1) = \prod_{i=1}^{\ell} (u_i + 1) \prod_{j=1}^{m} (v_j + 1) = d(w),$$

where the divisor function $d(w)$ counts the number of all positive divisors of w. Let

$$z = N^{1/8}.$$

From Theorem A.13 we obtain

$$\prod_{p|N} \left(1 - \frac{1}{p}\right)^{-1} G(z) \ge \sum_{w<z} \frac{d(w)}{w} \gg (\log z)^2 \gg (\log N)^2.$$

Equivalently,

$$\frac{|A|}{G(z)} \ll \frac{N}{(\log N)^2} \prod_{p|N} \left(1 - \frac{1}{p}\right)^{-1}$$

$$= \frac{N}{(\log N)^2} \prod_{p|N} \left(1 - \frac{1}{p^2}\right)^{-1} \prod_{p|N} \left(1 + \frac{1}{p}\right)$$

$$\ll \frac{N}{(\log N)^2} \prod_{p|N} \left(1 + \frac{1}{p}\right)$$

since the infinite product $\prod_{p=2}^{\infty} \left(1 - p^{-2}\right)$ converges.

To find an upper bound for the remainder, we use (7.9) to obtain

$$R = \sum_{\substack{d < z^2 \\ d \mid P(z)}} 3^{\omega(d)} |r(d)| \leq \sum_{\substack{d < z^2 \\ d \mid P(z)}} 3^{\omega(d)} 2^{\omega(d)} \leq \sum_{d < z^2} 6^{\omega(d)}.$$

Since

$$2^{\omega(d)} \leq d$$

and

$$6^{\omega(d)} = \left(2^{\omega(d)}\right)^{\log 6 / \log 2} \leq d^{\log 6 / \log 2} < z^{2 \log 6 / \log 2},$$

it follows that

$$R \leq \sum_{d < z^2} z^{2 \log 6 / \log 2} < z^{2 + 2 \log 6 / \log 2} < z^{7.2} = N^{9/10}$$

since $z = N^{1/8}$. Then

$$S(A, \mathcal{P}, z) \ll \frac{N}{(\log N)^2} \prod_{p \mid N} \left(1 + \frac{1}{p}\right) + N^{9/10} \ll \frac{N}{(\log N)^2} \prod_{p \mid N} \left(1 + \frac{1}{p}\right),$$

and so

$$r(N) \leq 2\sqrt{N} + S(A, \mathcal{P}, z) \ll \frac{N}{(\log N)^2} \prod_{p \mid N} \left(1 + \frac{1}{p}\right).$$

This completes the proof.

Theorem 7.3 *Let N be a positive even integer, and let $\pi_N(x)$ denote the number of primes p up to x such that $p + N$ is also prime. Then*

$$\pi_N(x) \ll \frac{x}{(\log x)^2} \prod_{p \mid N} \left(1 + \frac{1}{p}\right),$$

where the implied constant is absolute.

Proof. The proof is similar to the proof of Theorem 7.2. It starts as follows. Let

$$A = \{a_n : 1 \leq n \leq x\}$$

be the finite sequence of integers

$$a_n = n(n + N).$$

Then $|A| = [x]$. Let $\mathcal{P}$ be the set of all prime numbers. For any z satisfying

$$2 < z \leq \sqrt{x},$$

we let $S(A, \mathcal{P}, z)$ denote the number of terms of the sequence A that are divisible by no prime $p < z$. If

$$n > \sqrt{x}$$

and $a_n \equiv 0 \pmod{p}$ for some prime $p < z$, then either n or $n + N$ is composite. This implies that

$$\pi_N(x) \leq \sqrt{x} + S(A, \mathcal{P}, z).$$

We again use the Selberg sieve to obtain an upper bound for $S(A, \mathcal{P}, z)$. Let

$$d = p_1 \cdots p_k q_1 \cdots q_\ell$$

be a square-free integer, where the primes p_i divide N and the primes q_j do not divide N. Let $|A_d|$ denote the number of terms of the sequence A that are divisible by d. For every square-free integer d,

$$|A_d| = \frac{|A|}{g(d)} + r(d),$$

where $g(d)$ is the completely multiplicative function defined by (7.8), and

$$|r(d)| \leq 2^\ell \leq 2^{\omega(d)}.$$

Then

$$S(A, \mathcal{P}, z) \leq \frac{|A|}{G(z)} + \sum_{\substack{d < z^2 \\ d \mid P(z)}} 3^{\omega(d)} |r(d)|,$$

where

$$G(z) = \sum_{m < z} \frac{1}{g(m)}.$$

The proof continues exactly as above.

In the case where $N = 2$, we obtain the following improvement of Brun's Theorem 6.11.

Theorem 7.4 *Let $\pi_2(x)$ denote the number of twin primes up to x. Then*

$$\pi_2(x) \ll \frac{x}{(\log x)^2}.$$

7.4 Shnirel'man density

Let A be a set of integers. For any real number x, let $A(x)$ denote the number of positive elements of A not exceeding x, that is,

$$A(x) = \sum_{\substack{a \in A \\ 1 \leq a \leq x}} 1.$$

The function $A(x)$ is called the *counting function* of the set A. For $x > 0$ we have

$$0 \leq A(x) \leq [x] \leq x$$

and so

$$0 \leq \frac{A(x)}{x} \leq 1.$$

The *Shnirel'man density* of the set A, denoted $\sigma(A)$, is defined by

$$\sigma(A) = \inf_{n=1,2,3,\ldots} \frac{A(n)}{n}.$$

Clearly,

$$0 \leq \sigma(A) \leq 1$$

for every set A of integers. If $\sigma(A) = \alpha$, then

$$A(n) \geq \alpha n$$

for all $n = 1, 2, 3, \ldots$. If $1 \notin A$, then $A(1) = 0$ and so $\sigma(A) = 0$.

If A contains every positive integer, then $A(n) = n$ for all $n \geq 1$ and so $\sigma(A) = 1$. If $m \notin A$ for some $m \geq 1$, then $A(m) \leq m - 1$ and

$$\sigma(A) \leq \frac{A(m)}{m} \leq 1 - \frac{1}{m} < 1.$$

Thus, $\sigma(A) = 1$ if and only if A contains every positive integer.

If A and B are sets of integers, the *sumset* $A + B$ is the set consisting of all integers of the form $a + b$, where $a \in A$ and $b \in B$. If $A_1, \ldots, A_h$ are h sets of integers, then

$$A_1 + A_2 + \cdots + A_h$$

denotes the set of all integers of the form $a_1 + a_2 + \cdots + a_h$, where $a_i \in A_i$ for $i = 1, 2, \ldots, h$. If $A_i = A$ for $i = 1, 2, \ldots, h$, we let

$$hA = \underbrace{A + \cdots + A}_{h \text{ times}}.$$

The set A is called a *basis of order h* if hA contains every nonnegative integer, that is, if every nonnegative integer can be represented as the sum of h not necessarily distinct elements of A. The set A is called a *basis of finite order* if A is a basis of order h for some $h \geq 1$.

Shnirel'man density is an important additive measure of the size of a set of integers. In particular, the set A is a basis of order h if and only if $\sigma(hA) = 1$, and the set A is a basis of finite order if and only if $\sigma(hA) = 1$ for some $h \geq 1$.

Shnirel'man made the simple but extraordinarily powerful discovery that if A is a set of integers that contains 0 and has positive Shnirel'man density, then A is a basis of finite order.

Lemma 7.3 *Let A and B be sets of integers such that $0 \in A, 0 \in B$. If $n \geq 0$ and $A(n) + B(n) \geq n$, then $n \in A + B$.*

Proof. If $n \in A$, then $n = n + 0 \in A + B$. Similarly, if $n \in B$, then $n = 0 + n \in A + B$.

Suppose that $n \notin A \cup B$. Define sets A' and B' by

$$A' = \{n - a : a \in A, 1 \le a \le n - 1\}$$

and

$$B' = \{b : b \in B, 1 \le b \le n - 1\}.$$

Then $|A'| = A(n)$ since $n \notin A$, and $|B'| = B(n)$ since $n \notin B$. Moreover,

$$A' \cup B' \subseteq [1, n - 1].$$

Since

$$|A'| + |B'| = A(n) + B(n) \ge n,$$

it follows that

$$A' \cap B' \ne \emptyset.$$

Therefore, $n - a = b$ for some $a \in A$ and $b \in B$, and so $n = a + b \in A + B$.

Lemma 7.4   *Let A and B be sets of integers such that $0 \in A$ and $0 \in B$. If $\sigma(A) + \sigma(B) \ge 1$, then $n \in A + B$ for every nonnegative integer n.*

Proof. Let $\sigma(A) = \alpha$ and $\sigma(B) = \beta$. If $n \ge 0$, then

$$A(n) + B(n) \ge (\alpha + \beta)n \ge n,$$

and Lemma 7.3 implies that $n \in A + B$.

Lemma 7.5   *Let A be a set of integers such that $0 \in A$ and $\sigma(A) \ge 1/2$. Then A is a basis of order 2.*

Proof. This follows immediately from Lemma 7.4 with $A = B$.

Theorem 7.5 (Shnirel'man)   *Let A and B be sets of integers such that $0 \in A$ and $0 \in B$. Let $\sigma(A) = \alpha$ and $\sigma(B) = \beta$. Then*

$$\sigma(A + B) \ge \alpha + \beta - \alpha\beta. \tag{7.10}$$

Proof. Let $n \ge 1$. Let $a_0 = 0$ and let

$$1 \le a_1 < \cdots < a_k \le n$$

be the $k = A(n)$ positive elements of A that do not exceed n. Since $0 \in B$, it follows that $a_i = a_i + 0 \in A + B$ for $i = 1, \ldots, k$. For $i = 0, \ldots, k - 1$, let

$$1 \le b_1 < \cdots < b_{r_i} \le a_{i+1} - a_i - 1$$

be the $r_i = B(a_{i+1} - a_i - 1)$ positive elements of B less than $a_{i+1} - a_i$. Then

$$a_i < a_i + b_1 < \cdots < a_i + b_{r_i} < a_{i+1}$$

and
$$a_i + b_j \in A + B$$

for $j = 1, \ldots, r_i$. Let

$$1 \le b_1 < \cdots < b_{r_k} \le n - a_k$$

be the $r_k = B(n - a_k)$ positive elements of B not exceeding $n - a_k$. Then

$$a_k < a_k + b_1 < \cdots < a_k + b_{r_k} \le n$$

and

$$a_k + b_j \in A + B$$

for $j = 1, \ldots, r_k$. It follows that

$$(A + B)(n) \ge A(n) + \sum_{i=0}^{k-1} B(a_{i+1} - a_i - 1) + B(n - a_k)$$

$$\ge A(n) + \beta \sum_{i=0}^{k-1}(a_{i+1} - a_i - 1) + \beta(n - a_k)$$

$$= A(n) + \beta \sum_{i=0}^{k-1}(a_{i+1} - a_i) + \beta(n - a_k) - \beta k$$

$$= A(n) + \beta n - \beta k$$

$$= A(n) + \beta n - \beta A(n)$$

$$= (1 - \beta)A(n) + \beta n$$

$$\ge (1 - \beta)\alpha n + \beta n$$

$$= (\alpha + \beta - \alpha\beta)n$$

and so

$$\frac{(A + B)(n)}{n} \ge \alpha + \beta - \alpha\beta.$$

Therefore,

$$\sigma(A + B) \inf_{n=1,2,\ldots} \frac{(A + B)(n)}{n} \ge \alpha + \beta - \alpha\beta.$$

This completes the proof.

Inequality (7.10) can be expressed as follows:

$$1 - \sigma(A + B) \le (1 - \sigma(A))(1 - \sigma(B)). \tag{7.11}$$

The following theorem generalizes this inequality to the sum of any finite number of sets of integers.

Theorem 7.6 *Let $h \ge 1$, and let $A_1, \ldots, A_h$ be sets of integers such that $0 \in A_i$ for $i = 1, \ldots, h$. Then*

$$1 - \sigma(A_1 + \cdots + A_h) \le \prod_{i=1}^{h}(1 - \sigma(A_i)).$$

Proof. This is by induction on h. Let $\sigma(A_i) = \alpha_i$ for $i = 1, \ldots, h$. For $h = 1$, there is nothing to prove, and for $h = 2$ it is inequality (7.11).

Let $h \geq 3$, and assume that the theorem holds for $h - 1$. Let $A_1, \ldots, A_h$ be h sets of integers such that $0 \in A_i$ for all i. Let $B = A_2 + \cdots + A_h$. It follows from the induction hypothesis that

$$1 - \sigma(B) = 1 - \sigma(A_2 + \cdots + A_h) \leq \prod_{i=2}^{h}(1 - \sigma(A_i)),$$

and so

$$
\begin{aligned}
1 - \sigma(A_1 + \cdots + A_h) &= 1 - \sigma(A_1 + B) \\
&\leq (1 - \sigma(A_1))(1 - \sigma(B)) \\
&\leq (1 - \sigma(A_1)) \prod_{i=2}^{h}(1 - \sigma(A_i)) \\
&= \prod_{i=1}^{h}(1 - \sigma(A_i)).
\end{aligned}
$$

This completes the proof.

Theorem 7.7 (Shnirel'man) *Let A be a set of integers such that $0 \in A$ and $\sigma(A) > 0$. Then A is a basis of finite order.*

Proof. Let $\sigma(A) = \alpha > 0$. Then $0 \leq 1 - \alpha < 1$, and so

$$0 \leq (1 - \alpha)^\ell \leq 1/2$$

for some integer $\ell \geq 1$. By Theorem 7.6,

$$1 - \sigma(\ell A) \leq (1 - \sigma(A))^\ell = (1 - \alpha)^\ell \leq 1/2,$$

and so

$$\sigma(\ell A) \geq 1/2.$$

Let $h = 2\ell$. It follows from Lemma 7.5 that the set ℓA is a basis of order 2, and so A is a basis of order $2\ell = h$. This completes the proof.

7.5 The Shnirel'man–Goldbach theorem

We shall apply Shnirel'man's criterion for a set of integers to be a basis of finite order to prove that every integer greater than one is a sum of a bounded number of primes. We begin by proving that the set consisting of 0, 1, and the numbers that can be represented as the sum of two primes has positive Shnirel'man density. To do this, we need estimates for the average number of representations of an integer as the sum of two primes.

Lemma 7.6 *Let $r(N)$ denote the number of representations of the integer N as the sum of two primes. Then*

$$\sum_{N \le x} r(N) \gg \frac{x^2}{(\log x)^2}.$$

Proof. If p and q are primes such that $p, q \le x/2$, then $p + q \le x$. Therefore,

$$\sum_{N \le x} r(N) \ge \pi(x/2)^2 \gg \frac{(x/2)^2}{(\log(x/2))^2} \gg \frac{x^2}{(\log x)^2}$$

by Chebyshev's theorem (Theorem 6.3).

Lemma 7.7 *Let $r(N)$ denote the number of representations of N as the sum of two primes. Then*

$$\sum_{N \le x} r(N)^2 \ll \frac{x^3}{(\log x)^4}.$$

Proof. By Theorem 7.2, if N is even, then

$$r(N) \ll \frac{N}{(\log N)^2} \prod_{p|N}\left(1 + \frac{1}{p}\right) \le \frac{N}{(\log N)^2} \sum_{d|N} \frac{1}{d}.$$

This inequality also holds for odd integers, since an odd integer N can be written as the sum of two primes if and only if $N - 2$ is prime, in which case $r(N) = 2$.

In the following calculation, we use the fact that

$$[d_1, d_2] = \frac{d_1 d_2}{(d_1, d_2)} \ge (d_1 d_2)^{1/2}.$$

Then

$$\sum_{N \le x} r(N)^2 \ll \sum_{N \le x} \frac{N^2}{(\log N)^4}\left(\sum_{d|N} \frac{1}{d}\right)^2$$

$$\ll \frac{x^2}{(\log x)^4} \sum_{N \le x}\left(\sum_{d|N} \frac{1}{d}\right)^2$$

$$\le \frac{x^2}{(\log x)^4} \sum_{N \le x} \sum_{d_1|N} \sum_{d_2|N} \frac{1}{d_1 d_2}$$

$$\le \frac{x^2}{(\log x)^4} \sum_{d_1, d_2 \le x} \frac{1}{d_1 d_2} \sum_{\substack{N \le x \\ d_1|N, d_2|N}} 1$$

$$= \frac{x^2}{(\log x)^4} \sum_{d_1, d_2 \le x} \frac{1}{d_1 d_2} \sum_{\substack{N \le x \\ [d_1, d_2]|N}} 1$$

$$\leq \frac{x^2}{(\log x)^4} \sum_{d_1, d_2 \leq x} \frac{1}{d_1 d_2} \frac{x}{[d_1, d_2]}$$

$$\leq \frac{x^3}{(\log x)^4} \sum_{d_1, d_2 \leq x} \frac{1}{d_1^{3/2} d_2^{3/2}}$$

$$\leq \frac{x^3}{(\log x)^4} \left(\sum_{d \leq x} \frac{1}{d^{3/2}} \right)^2$$

$$\ll \frac{x^3}{(\log x)^4}.$$

This completes the proof.

Theorem 7.8 *The set*

$$A = \{0, 1\} \cup \{p + q : p, q \quad primes\}$$

has positive Shnirel'man density.

Proof. Let $r(N)$ denote the number of representations of N as the sum of two primes. By the Cauchy–Schwarz inequality, we have

$$\left(\sum_{N \leq x} r(N) \right)^2 \leq \sum_{\substack{N \leq x \\ r(N) \geq 1}} 1 \sum_{N \leq x} r(N)^2 \leq A(x) \sum_{N \leq x} r(N)^2.$$

By Lemma 7.6 and Lemma 7.7,

$$\frac{A(x)}{x} \geq \frac{1}{x} \frac{\left(\sum_{N \leq x} r(N) \right)^2}{\sum_{N \leq x} r(N)^2}$$

$$\gg \frac{1}{x} \frac{\frac{x^4}{(\log x)^4}}{\frac{x^3}{(\log x)^4}}$$

$$\gg 1.$$

This means that there exists a number $c_1 > 0$ such that $A(x) \geq c_1 x$ for all $x \geq x_0$. Since 1 belongs to the set A, it follows that there exists a number $c_2 > 0$ such that $A(x) \geq c_2 x$ for $1 \leq x \leq x_0$. Therefore, $A(x) \geq \min(c_1, c_2)x$ for all $x \geq 1$, and so the Shnirel'man density of A is positive. This completes the proof.

Theorem 7.9 (Goldbach–Shnirel'man) *Every integer greater than one is the sum of a bounded number of primes.*

Proof. We have shown that the set

$$A = \{0, 1\} \cup \{p + q : p, q \quad primes\}$$

has positive Shnirel'man density. By Theorem 7.7, there exists an integer h such that every nonnegative integer is the sum of exactly h elements of A. Let $N \geq 2$. Then $N - 2 \geq 0$, so for some integers k and ℓ with $k + \ell \leq h$ there exist ℓ pairs of primes p_i, q_i such that

$$N - 2 = \underbrace{1 + \cdots + 1}_{k} + (p_1 + q_1) + \cdots + (p_\ell + q_\ell).$$

Let $k = 2m + r$, where $r = 0$ or 1. If $r = 0$, then

$$N = \underbrace{2 + \cdots + 2}_{m+1} + (p_1 + q_1) + \cdots + (p_\ell + q_\ell).$$

If $r = 1$, then

$$N = \underbrace{2 + \cdots + 2}_{m} + 3 + (p_1 + q_1) + \cdots + (p_\ell + q_\ell).$$

In both cases, N is a sum of

$$2\ell + m + 1 \leq 3h$$

primes. This completes the proof.

Theorem 7.10 *Let Q be a set of primes that contains a positive proportion of the primes, that is,*

$$Q(x) > \theta \pi(x)$$

for some $\theta > 0$ and all sufficiently large x. Then every sufficiently large integer is the sum of a bounded number of primes belonging to Q.

Proof. We shall first show that the set

$$A(Q) = \{0, 1\} \cup \{p + q : p, q \in Q\}$$

has positive Shnirel'man density. Let $r(N)$ denote the number of representations of N as the sum of two primes, and let $r_Q(N)$ denote the number of representations of N as the sum of two primes belonging to Q. Then

$$\sum_{N \leq x} r_Q(N) \geq (Q(x/2))^2 \geq (\theta \pi(x))^2 \gg \frac{x^2}{(\log x)^2}.$$

By Lemma 7.7,

$$\sum_{N \leq x} r_Q(N)^2 \leq \sum_{N \leq x} r(N)^2 \ll \frac{x^3}{(\log x)^4}.$$

It follows exactly as in the proof of Theorem 7.8 that the set $A(Q)$ has positive Shnirel'man density. Therefore, $A(Q)$ is a basis of finite order. It follows that there exists a number h_1 such that every nonnegative integer is the sum of h_1 elements of $Q \cup \{0, 1\}$.

Choose two primes $p_1, p_2 \in Q$. By Exercise 3, there exists an integer $n_0 = n_0(p_1, p_2)$ such that every integer $n \geq n_0$ can be written in the form

$$n = \ell_1(n)p_1 + \ell_2(n)p_2,$$

where $\ell_1(n)$ and $\ell_2(n)$ are nonnegative integers. Let

$$h_2 = \max\{\ell_1(n) + \ell_2(n) : n = n_0, \ldots, n_0 + h_1\},$$

and let

$$h = h_1 + h_2.$$

If $N \geq n_0$, then $N - n_0$ can be written as the sum of at most h_1 elements of $Q \cup \{1\}$, that is,

$$N - n_0 = \underbrace{1 + \cdots + 1}_{k} + p_{i_1} + \cdots p_{i_\ell},$$

where

$$k + \ell \leq h_1.$$

Then

$$n_0 + k = \ell_1(n)p_1 + \ell_2(n)p_2,$$

where $\ell_1(n) + \ell_2(n) \leq h_2$, and so

$$N = n_0 + k + p_{i_1} + \cdots p_{i_\ell}$$
$$= \ell_1(n)p_1 + \ell_2(n)p_2 + p_{i_1} + \cdots p_{i_\ell}$$

is a sum of

$$\ell + \ell_1(n) + \ell_2(n) \leq h_1 + h_2 = h$$

primes belonging to the set Q. This completes the proof.

7.6 Romanov's theorem

Let a be an integer, $a \geq 2$. We investigate how many numbers N up to x can be written in the form

$$N = p + a^k, \tag{7.12}$$

where p is a prime and k is a positive integer. Let $r(N)$ be the number of representations of N in this form. Since the number of positive powers of a up to x is $\ll \log x$ and the number of primes up to x is $\pi(x) \ll x/\log x$, it follows that

$$\sum_{N \leq x} r(N) = |\{p + a^k \leq x\}| \ll \log x \left(\frac{x}{\log x}\right) = x.$$

Let

$$A = \{p + a^k : p \text{ prime and } k \geq 1\},$$

and let $A(x)$ be the counting function of the set A. In this section, we shall prove a remarkable theorem of Romanov that the lower asymptotic density of the set A is positive, that is, there exists a constant $c > 0$ such that

$$A(x) \geq cx$$

for all sufficiently large x. This means that a positive proportion of the natural numbers can be represented in the form (7.12).

Lemma 7.8 *Let a be an integer, $a \geq 2$. For every integer $d \geq 1$ such that $(a, d) = 1$, let $e(d)$ denote the exponent of a modulo d, that is, the smallest integer such that*

$$a^{e(d)} \equiv 1 \pmod{d}.$$

Then the series

$$\sum_{\substack{d=1 \\ (a,d)=1 \\ \mu^2(d)=1}}^{\infty} \frac{1}{d e(d)}$$

converges.

Proof. If $(a, d) = 1$ and $e(d) = k$, then

$$a^k \equiv 1 \pmod{d},$$

and so d divides $a^k - 1$. Since $a^k - 1$ has only finitely many divisors, it follows that there are only finitely many numbers d such that $e(d) = k$. For $x \geq 2$, let

$$D = D(x) = \prod_{k \leq x} (a^k - 1),$$

and let $n = \omega(D)$ be the number of distinct prime divisors of D. Let

$$E(x) = \sum_{k \leq x} \sum_{\substack{e(d)=k \\ (a,d)=1 \\ \mu^2(d)=1}} \frac{1}{d}.$$

The number d appears in this double sum at most once, and if d appears, then d divides $a^k - 1$ for some $k \leq x$, so d divides D. It follows that

$$E(x) \leq \sum_{\substack{d \mid D \\ \mu^2(d)=1}} \frac{1}{d} = \prod_{p \mid D} \left(1 + \frac{1}{p}\right) \leq \prod_{i=1}^{n} \left(1 + \frac{1}{p_i}\right),$$

where $p_1, p_2, \ldots, p_n$ are the first n prime numbers. Since

$$2^n = 2^{\omega(D)} \leq D = \prod_{k \leq x} (a^k - 1) < \prod_{k \leq x} a^k \leq a^{x(x+1)/2} < a^{x^2},$$

it follows that

$$n < \left(\frac{\log a}{\log 2}\right) x^2 \ll x^2.$$

By Chebyshev (Theorem 6.4),

$$\log p_n \ll \log n \ll \log x,$$

and so, by Mertens's formula (Theorem 6.8),

$$E(x) \ll \prod_{p \le p_n} \left(1 + \frac{1}{p}\right)$$

$$\ll \prod_{p \le p_n} \left(1 - \frac{1}{p}\right)^{-1}$$

$$\ll \log p_n$$

$$\ll \log x.$$

By partial summation,

$$\sum_{k \le x} \frac{1}{k} \left(\sum_{\substack{e(d)=k \\ (a,d)=1 \\ \mu^2(d)=1}} \frac{1}{d} \right) = \frac{E(x)}{x} + \int_1^x \frac{E(t)}{t^2} dt$$

$$\ll \frac{\log x}{x} + \int_1^x \frac{\log t}{t^2} dt$$

$$\ll 1,$$

and so the series

$$\sum_{k=1}^\infty \frac{1}{k} \left(\sum_{\substack{e(d)=k \\ (a,d)=1 \\ \mu^2(d)=1}} \frac{1}{d} \right) = \sum_{\substack{d=1 \\ (a,d)=1 \\ \mu^2(d)=1}}^\infty \frac{1}{de(d)}$$

converges. This completes the proof.

Lemma 7.9 Let a be an integer, $a \ge 2$, and let $r(N)$ denote the number of solutions of the equation

$$N = p + a^k,$$

where p is a prime and k is a positive integer. Then

$$\sum_{N \le x} r(N)^2 \ll x.$$

Proof. Since $r(N)^2$ is equal to the number of quadruples (p_1, p_2, k_1, k_2) such that

$$p_1 + a^{k_1} = p_2 + a^{k_2} = N,$$

it follows that $\sum_{N \leq x} r(N)^2$ is equal to the number of quadruples (p_1, p_2, k_1, k_2) such that

$$p_1 + a^{k_1} = p_2 + a^{k_2} \leq x.$$

This does not exceed the number of solutions of the equation

$$p_2 - p_1 = a^{k_1} - a^{k_2}$$

with $p_1, p_2 \leq x$ and $k_1, k_2 \leq \log x / \log a$.

Choose positive integers $k_1 \neq k_2$, and let

$$h = a^{k_1} - a^{k_2}.$$

Then h is a nonzero, even integer. The number of solutions of the equation

$$p_2 - p_1 = a^{k_1} - a^{k_2} = h$$

with $p_1, p_2 \leq x$ is at most the number of primes $p_1 \leq x$ such that $p_1 + h$ is also prime. By Theorem 7.3, this is

$$\pi_h(x) \ll \frac{x}{(\log x)^2} \prod_{p \mid h} \left(1 + \frac{1}{p} \right).$$

If $k_2 > k_1$, then

$$h = a^{k_1} \left(a^{k_2 - k_1} - 1 \right)$$

and

$$\prod_{p \mid h} \left(1 + \frac{1}{p} \right) = \prod_{p \mid a^{k_1}} \left(1 + \frac{1}{p} \right) \prod_{p \mid (a^{k_2 - k_1} - 1)} \left(1 + \frac{1}{p} \right)$$

$$= \prod_{p \mid a} \left(1 + \frac{1}{p} \right) \prod_{p \mid (a^{k_2 - k_1} - 1)} \left(1 + \frac{1}{p} \right)$$

$$\ll \prod_{p \mid (a^{k_2 - k_1} - 1)} \left(1 + \frac{1}{p} \right),$$

where the implied constant depends on a. Similarly, if $k_1 > k_2$, then

$$h = -a^{k_2} \left(a^{k_1 - k_2} - 1 \right)$$

and

$$\prod_{p \mid h} \left(1 + \frac{1}{p} \right) \ll \prod_{p \mid (a^{k_1 - k_2} - 1)} \left(1 + \frac{1}{p} \right) = \prod_{p \mid (a^{|k_2 - k_1|} - 1)} \left(1 + \frac{1}{p} \right).$$

Finally, if $k_2 = k_1$, the number of solutions of the equation

$$p_2 - p_1 = a^{k_2} - a^{k_1} = 0$$

with $p_1, p_2 \leq x$ and $1 \leq k_2 \leq \log x / \log a$ is

$$\frac{\pi(x) \log x}{\log a} \ll x.$$

It follows that

$$\sum_{N \leq x} r(N)^2 \ll x + 2 \sum_{1 \leq k_1 < k_2 \leq \frac{\log x}{\log a}} \prod_{p \mid (a^{k_2 - k_1} - 1)} \left(1 + \frac{1}{p}\right)$$

$$\ll x + \log x \sum_{1 \leq k \leq \frac{\log x}{\log a}} \prod_{p \mid (a^k - 1)} \left(1 + \frac{1}{p}\right)$$

$$\ll x + \log x \sum_{1 \leq k \leq \frac{\log x}{\log a}} \sum_{\substack{d \mid (a^k - 1) \\ \mu(d)^2 = 1}} \frac{1}{d}.$$

To estimate the last term, we observe that

$$d \mid (a^k - 1)$$

if and only if

$$a^k \equiv 1 \pmod{d}$$

if and only if

$$e(d) \mid k.$$

Then

$$\sum_{N \leq x} r(N)^2 \ll x + \log x \sum_{1 \leq k \leq \frac{\log x}{\log a}} \sum_{\substack{d \mid (a^k - 1) \\ \mu^2(d) = 1 \\ (a,d) = 1}} \frac{1}{d}$$

$$= x + \log x \sum_{\substack{\mu^2(d) = 1 \\ (a,d) = 1}} \frac{1}{d} \sum_{\substack{k \leq \frac{\log x}{\log a} \\ d \mid (a^k - 1)}} 1$$

$$= x + \log x \sum_{\substack{\mu^2(d) = 1 \\ (a,d) = 1}} \frac{1}{d} \sum_{\substack{k \leq \frac{\log x}{\log a} \\ e(d) \mid k}} 1$$

$$\leq x + \log x \sum_{\substack{\mu^2(d) = 1 \\ (a,d) = 1}} \frac{\log x}{d e(d) \log a}$$

$$\ll x + (\log x)^2 \sum_{\substack{\mu^2(d) = 1 \\ (a,d) = 1}} \frac{1}{d e(d)}$$

$$\ll x$$

since the infinite series converges by Lemma 7.8.

Lemma 7.10 *Let a be an integer, $a \geq 2$, and let $r(N)$ denote the number of solutions of the equation*

$$N = p + a^k,$$

where p is a prime and k is a positive integer. Then

$$\sum_{N \leq x} r(N) \gg x.$$

Proof. If $p \leq x/2$ and $a^k \leq x/2$, then $p + a^k \leq x$, so

$$\sum_{N \leq x} r(N) \gg \pi(x/2) \log(x/2) \gg x.$$

This completes the proof.

Theorem 7.11 (Romanov) *Let a be an integer, $a \geq 2$. Let*

$$A = \{p + a^k : p \text{ prime and } k \geq 1\},$$

and let $A(x)$ be the counting function of the set A. There exists a constant $c > 0$ such that

$$A(x) > cx$$

for all sufficiently large x.

Proof. We use the Cauchy–Schwarz inequality. By Lemma 7.10 and Lemma 7.9, there exist positive numbers c_1 and c_2 such that, for x sufficiently large,

$$(c_1 x)^2 \leq \left(\sum_{N \leq x} r(N) \right)^2$$

$$\leq A(x) \sum_{N \leq x} r(N)^2 \leq c_2 x A(x)$$

and so

$$A(x) \geq cx.$$

7.7   Covering congruences

Choosing $a = 2$ in Romanov's theorem, we see that a positive proportion of the natural numbers can be written in the form $p + 2^k$. The only even numbers of this form are $2 + 2^k$, and they constitute a very sparse subset of the even integers, a subset of density zero, so almost all of the integers of the form $p + 2^k$ are odd. We shall prove that there exists an infinite arithmetic progression of odd natural numbers, none of which can be written in the form $p + 2^k$. To do this, we introduce the concept of covering congruences for the integers.

Let

$$1 < m_1 < m_2 < \cdots < m_\ell$$

be a strictly increasing finite sequence of integers, and let $a_1, \ldots, a_\ell$ be any integers. Then the ℓ congruence classes $a_i \pmod{m_i}$ form a *system of covering congruences* if, for every integer k, there exists at least one i such that

$$k \equiv a_i \pmod{m_i}. \tag{7.13}$$

This means that the congruence classes $a_i \pmod{m_i}$ cover the integers in the sense that

$$\mathbf{Z} = \bigcup_{i=1}^{\ell} \{k \in \mathbf{Z} : k \equiv a_i \pmod{m_i}\}.$$

It is an essential part of the definition of covering congruences that the moduli m_i are pairwise distinct integers greater than one. Here is a simple example of a system of covering congruences.

Lemma 7.11 *The six congruences*

$$0 \pmod 2$$
$$0 \pmod 3$$
$$1 \pmod 4$$
$$3 \pmod 8$$
$$7 \pmod{12}$$
$$23 \pmod{24}$$

form a set of covering congruences.

Proof. First, we show that each of the 24 integers $0, 1, \ldots, 23$ satisfies at least one of these six congruences. Every even integer k satisfies $k \equiv 0 \pmod 2$. For odd integers, we have

$$1 \equiv 1 \pmod 4$$
$$3 \equiv 0 \pmod 3$$
$$5 \equiv 1 \pmod 4$$
$$7 \equiv 7 \pmod{12}$$
$$9 \equiv 0 \pmod 3$$
$$11 \equiv 3 \pmod 8$$
$$13 \equiv 1 \pmod 4$$
$$15 \equiv 0 \pmod 3$$
$$17 \equiv 1 \pmod 4$$
$$19 \equiv 7 \pmod{12}$$
$$21 \equiv 0 \pmod 3$$
$$23 \equiv 23 \pmod{24}.$$

For every integer k, there is a unique integer $r \in \{0, 1, \ldots, 23\}$ such that

$$k \equiv r \pmod{24}.$$

Choose i so that

$$r \equiv a_i \pmod{m_i},$$

where $a_i \pmod{m_i}$ is one of our six congruences. Each of the six moduli 2, 3, 4, 6, 12, and 24 divides 24, so m_i divides 24 and

$$k \equiv r \pmod{m_i}.$$

Therefore,

$$k \equiv a_i \pmod{m_i}.$$

This completes the proof.

Theorem 7.12 (Erdős) *There exists an infinite arithmetic progression of odd positive integers, none of which is of the form $p + 2^k$.*

Proof. We shall use the system of covering congruences $a_i \pmod{m_i}$ constructed in Lemma 7.11. For each of the six moduli m_i in this system, we choose distinct primes p_i such that

$$2^{m_i} \equiv 1 \pmod{p_i},$$

as follows:

$$2^2 \equiv 1 \pmod{3}$$
$$2^3 \equiv 1 \pmod{7}$$
$$2^4 \equiv 1 \pmod{5}$$
$$2^8 \equiv 1 \pmod{17}$$
$$2^{12} \equiv 1 \pmod{13}$$
$$2^{24} \equiv 1 \pmod{241}.$$

Let

$$\ell = \max\{p_i\} = 241$$

and

$$m = 2^\ell \cdot 3 \cdot 7 \cdot 5 \cdot 17 \cdot 13 \cdot 241.$$

By the Chinese remainder theorem, there exists a unique congruence class r $\pmod{m}$ such that $r \equiv 1 \pmod{2^\ell}$ and $r \equiv 2^{a_i} \pmod{p_i}$ for $i = 1, \ldots, 6$. This means that

$$r \equiv 1 \pmod{2^\ell}$$
$$r \equiv 2^0 \pmod{3}$$
$$r \equiv 2^0 \pmod{7}$$

$$r \equiv 2^1 \quad (\text{mod } 5)$$
$$r \equiv 2^3 \quad (\text{mod } 17)$$
$$r \equiv 2^7 \quad (\text{mod } 13)$$
$$r \equiv 2^{23} \quad (\text{mod } 241),$$

where the exponents in the powers of 2 are the least nonnegative residues a_i in the six congruence classes in the system of covering congruences. Since r is odd and the modulus m is even, it follows that every integer in the congruence class r (mod m) is odd.

Let N be an integer in the congruence class r (mod m) such that

$$N > 2^\ell + \ell.$$

Let k be a positive integer such that $2^k < N$. There is a congruence class a_i (mod m_i) in the system of covering congruences such that

$$k \equiv a_i \quad (\text{mod } m_i)$$

so $k = a_i + m_i u_i$ for some integer u_i. Since

$$2^{m_i} \equiv 1 \quad (\text{mod } p_i),$$

we have

$$2^k = 2^{a_i} 2^{m_i u_i} \equiv 2^{a_i} \quad (\text{mod } p_i).$$

Since

$$N \equiv r \quad (\text{mod } p_i)$$

and

$$r \equiv 2^{a_i} \quad (\text{mod } p_i),$$

it follows that

$$N \equiv r \equiv 2^{a_i} \equiv 2^k \quad (\text{mod } p_i),$$

and so

$$N = 2^k + p_i v$$

for some positive integer v. If $k \leq \ell$, then

$$p_i v = N - 2^k \geq N - 2^\ell > \ell = \max\{p_i\} \geq p_i$$

for $i = 1, \ldots, 6$, and so $v > 1$. If $k > \ell$, then

$$N - 2^k \equiv N \equiv 1 \quad (\text{mod } 2^\ell)$$

and so

$$p_i v = N - 2^k = 1 + 2^\ell w > 2^\ell > \ell \geq p_i$$

and $v > 1$. In both cases, $N - 2^k$ is composite. This completes the proof.

7.8   Notes

Shnirel'man's fundamental paper was published first in Russian [113] and then expanded and published in German [114]. By *Shnirel' man's constant* we mean the smallest number h such that every integer greater than one is the sum of at most h primes. Using the Brun sieve, Shnirel'man proved that this constant is finite. The best estimate for Shnirel'man's constant is due to Ramaré [100], who has proved that every even integer is the sum of at most six primes. It follows that Shnirel'man's constant is at most seven. The Goldbach conjecture implies that Shirel'man's constant is three.

In this chapter, I use the Selberg sieve instead of the Brun sieve to prove the Goldbach–Shnirel'man theorem. See Hua [63] for a nice account of this approach. Landau [76, 77] gives Shnirel'man's original method. Theorem 7.10, the generalization of the Goldbach–Shnirel'man theorem to dense subsets of the primes, is due to Nathanson [90].

Selberg introduced his sieve in a beautiful short paper [109]. I use Selberg's original proof of the sieve inequality (7.2). See Selberg's *Collected Papers*[110, 111] for his papers on sieve theory. Prachar [97] contains a nice exposition of the Selberg sieve, with many applications. The standard references on sieve methods are the monographs of Halberstam and Richert [44] and Motohashi [87].

Romanov's theorem appears in the paper [103]. Romanov also proved that, for a fixed exponent k, the set of integers of the form $p + n^k$ has positive density. The proof of Theorem 7.8 of Romanov's theorem was simplified by Erdős and Turán [30] and Erdős [33].

Erdős [32] invented covering congruences and used them to construct the infinite arithmetic progression of odd positive integers not of the form $p + 2^k$, as described in Theorem 7.12. Crocker [16] proved that there exists an infinite set of odd positive integers that cannot be represented as the sum of a prime and two positive powers of 2. Crocker's set is sparse. It is an open problem to determine if there exists an infinite arithmetic progression of odd positive integers not of the form $p + 2^{k_1} + 2^{k_2}$.

There are many unsolved problems concerning covering congruences. It is not known, for example, whether there exists a system of covering congruences all of whose moduli are odd. Nor is it known whether, for any number M, there exists a system of covering congruences all of whose moduli are greater than M. The best result is due to Choi [12], who proved that there exists a system of covering congruences with smallest modulus 20.

7.9   Exercises

1. Prove that for any square-free integer d there are exactly $3^{\omega(d)}$ pairs of positive integers d_1, d_2 such that $[d_1, d_2] = d$.

2. Let $\omega(n)$ denote the number of distinct prime divisors of n. Let $n \geq 2$ and $r \geq 0$. Prove that

$$\sum_{\substack{d\mid n \\ \omega(d)\leq 2r+1}} \mu(d) \leq 0 \leq \sum_{\substack{d\mid n \\ \omega(d)\leq 2r}} \mu(d).$$

3. Let a_1 and a_2 be relatively prime positive integers. Prove that there exists an integer $n_0 = n_0(a_1, a_2)$ such that every integer $n \geq n_0$ can be written in the form

$$n = \ell_1(n)a_1 + \ell_2(n)a_2$$

for some nonnegative integers $\ell_1(n)$, $\ell_2(n)$.

4. Construct a system of covering congruences whose moduli are 2, 3, 4, 6, and 12.

5. Let us call an integer n *exceptional* if $n - 2^k$ is prime for all positive integers $k < \log n / \log 2$. Find all exceptional numbers up to 105. Erdős [32] has written that "it seems likely that 105 is the largest exceptional integer."

6. Let $\{a_i \pmod{m_i} : i = 1, \ldots, k\}$ be a system of covering congruences. Prove that

$$\sum_{i=1}^{k} \frac{1}{m_i} \geq 1.$$

8

Sums of three primes

The method which I discovered in 1937 for estimating sums over primes permits, in the first instance, the evaluation of an estimate for the simplest of such sums, i.e. a sum of the type:

$$\sum_{p \leq N} e^{2\pi i \alpha p}.$$

This estimate in combination with the previously known theorems concerning the distribution of primes in arithmetic progressions ... paved the way for establishing unconditionally the asymptotic formula of Hardy and Littlewood in the Goldbach ternary representation problem.

I. M. Vinogradov [135, page 365]

8.1 Vinogradov's theorem

Vinogradov proved that every sufficiently large odd integer is the sum of three primes. In addition, he obtained an asymptotic formula for the number of representations of an odd integer as the sum of three prime numbers. Vinogradov's theorem is one of the great results in additive prime number theory. The principal ingredients of the proof are the circle method and an estimate of a certain exponential sum over prime numbers.

The counting function for the number of representations of an odd integer N as the sum of three primes is

$$r(N) = \sum_{p_1+p_2+p_3=N} 1.$$

The following is Vinogradov's asymptotic formula for $r(N)$.

Theorem 8.1 (Vinogradov) *There exists an arithmetic function* $\mathfrak{S}(N)$ *and positive constants* c_1 *and* c_2 *such that*

$$c_1 < \mathfrak{S}(N) < c_2$$

for all sufficiently large odd integers N, *and*

$$r(N) = \mathfrak{S}(N)\frac{N^2}{2(\log N)^3}\left(1 + O\left(\frac{\log\log N}{\log N}\right)\right).$$

The arithmetic function $\mathfrak{S}(N)$ is called the *singular series* for the ternary Goldbach problem.

8.2   The singular series

We begin by studying the arithmetic function

$$\mathfrak{S}(N) = \sum_{q=1}^{\infty} \frac{\mu(q)c_q(N)}{\varphi(q)^3}, \tag{8.1}$$

where

$$c_q(N) = \sum_{\substack{a=1 \\ (q,a)=1}}^{q} e(aN/q)$$

is *Ramanujan's sum* (A.2). The function $\mathfrak{S}(N)$ is called the *singular series* for the ternary Goldbach problem.

Theorem 8.2 *The singular series* $\mathfrak{S}(N)$ *converges absolutely and uniformly in* N *and has the Euler product*

$$\mathfrak{S}(N) = \prod_{p}\left(1 + \frac{1}{(p-1)^3}\right)\prod_{p|N}\left(1 - \frac{1}{p^2 - 3p + 3}\right).$$

There exist positive constants c_1 *and* c_2 *such that*

$$c_1 < \mathfrak{S}(N) < c_2$$

for all positive integers N. *Moreover, for any* $\varepsilon > 0$,

$$\mathfrak{S}(N, Q) = \sum_{q \leq Q} \frac{\mu(q)c_q(N)}{\varphi(q)^3} = \mathfrak{S}(N) + O\left(Q^{-(1-\varepsilon)}\right), \tag{8.2}$$

where the implied constant depends only on ε.

Proof. Clearly, $c_q(N) \ll \varphi(q)$. By Theorem A.16,

$$\varphi(q) > q^{1-\varepsilon}$$

for $\varepsilon > 0$ and all sufficiently large integers q, and so

$$\frac{\mu(q)c_q(N)}{\varphi(q)^3} \ll \frac{1}{\varphi(q)^2} \ll \frac{1}{q^{2-\varepsilon}}.$$

Thus, the singular series converges absolutely and uniformly in N. Moreover,

$$\mathfrak{S}(N) - \mathfrak{S}(N, Q) \ll \sum_{q>Q} \frac{1}{\varphi(q)^2} \ll \sum_{q>Q} \frac{1}{q^{2-\varepsilon}} \ll \frac{1}{Q^{1-\varepsilon}}.$$

By Theorem A.24, $c_q(N)$ is a multiplicative function of q and

$$c_p(N) = \begin{cases} p-1 & \text{if } p \text{ divides } N \\ -1 & \text{if } p \text{ does not divide } N. \end{cases}$$

Since the arithmetic function

$$\frac{\mu(q)c_q(N)}{\varphi(q)^3}$$

is multiplicative in q and $\mu(p^j) = 0$ for $j \geq 2$, it follows from Theorem A.28 that the singular series has the Euler product

$$\begin{aligned}
\mathfrak{S}(N) &= \prod_p \left(1 + \sum_{j=1}^{\infty} \frac{\mu(p^j)c_{p^j}(N)}{\varphi(p^j)^3}\right) \\
&= \prod_p \left(1 - \frac{c_p(N)}{\varphi(p)^3}\right) \\
&= \prod_{p \mid N} \left(1 + \frac{1}{(p-1)^3}\right) \prod_{p \nmid N} \left(1 - \frac{1}{(p-1)^2}\right) \\
&= \prod_p \left(1 + \frac{1}{(p-1)^3}\right) \prod_{p \mid N} \left(1 - \frac{1}{p^2 - 3p + 3}\right),
\end{aligned}$$

and so there exist positive constants c_1 and c_2 such that

$$c_1 < \mathfrak{S}(N) < c_2$$

for all positive integers N. This completes the proof.

8.3   Decomposition into major and minor arcs

As in the proof of the Hardy–Littlewood asymptotic formula for Waring's problem, we decompose the unit interval $[0, 1]$ into two disjoint sets: the *major arcs* $\mathfrak{M}$ and the *minor arcs* $\mathfrak{m}$.

Let $B > 0$ and

$$Q = (\log N)^B. \tag{8.3}$$

For

$$1 \le q \le Q,$$

$$0 \le a \le q,$$

and

$$(a, q) = 1,$$

the *major arc* $\mathfrak{M}(q, a)$ is the interval consisting of all real numbers $\alpha \in [0, 1]$ such that

$$\left| \alpha - \frac{a}{q} \right| \le \frac{Q}{N}.$$

If $\alpha \in \mathfrak{M}(q, a) \cap \mathfrak{M}(q', a')$ and $a/q \neq a'/q'$, then $|aq' - a'q| \ge 1$ and

$$\frac{1}{Q^2} \le \frac{1}{qq'} \le \frac{|aq' - a'q|}{qq'} = \left| \frac{a}{q} - \frac{a'}{q'} \right|$$

$$\le \left| \frac{a}{q} - \alpha \right| + \left| \alpha - \frac{a'}{q'} \right| \le \frac{2Q}{N},$$

or, equivalently,

$$N \le 2Q^3 = 2(\log N)^{3B}.$$

This is impossible for N sufficiently large. Therefore, the major arcs $\mathfrak{M}(q, a)$ are pairwise disjoint for large N. The set of major arcs is

$$\mathfrak{M} = \bigcup_{q=1}^{Q} \bigcup_{\substack{a=0 \\ (a,q)=1}}^{q} \mathfrak{M}(q, a) \subseteq [0, 1]$$

and the set of minor arcs is

$$\mathfrak{m} = [0, 1] \setminus \mathfrak{M}.$$

We consider a weighted sum over the representations of N as a sum of three primes:

$$R(N) = \sum_{p_1 + p_2 + p_3 = N} \log p_1 \log p_2 \log p_3.$$

Vinogradov obtained an asymptotic formula for $R(N)$, from which Theorem 8.1 will follow by an elementary argument. We can use the circle method to express the representation function $R(N)$ as the integral of a trigonometric polynomial over the major and minor arcs. Let

$$F(\alpha) = \sum_{p \le N} (\log p) e(p\alpha). \tag{8.4}$$

This exponential sum over primes is the generating function for $R(N)$, and

$$R(N) = \sum_{p_1+p_2+p_3=N} \log p_1 \log p_2 \log p_3 = \int_0^1 F(\alpha)^3 e(-N\alpha) d\alpha$$

$$= \int_{\mathfrak{M}} F(\alpha)^3 e(-N\alpha) d\alpha + \int_{\mathfrak{m}} F(\alpha)^3 e(-N\alpha) d\alpha.$$

The main term in Vinogradov's theorem will come from the integral over the major arcs, and the integral over the minor arcs will be negligible.

8.4   The integral over the major arcs

Just as in the Hardy–Littlewood asymptotic formula, the integral over the major arcs in Vinogradov's theorem is (except for a small error term) the product of the singular series $\mathfrak{S}(N)$ and an integral $J(N)$. In this case, the integral $J(N)$ is very easy to evaluate.

Lemma 8.1 *Let*

$$u(\beta) = \sum_{m=1}^{N} e(m\beta).$$

Then

$$J(N) = \int_{-1/2}^{1/2} u(\beta)^3 e(-N\beta) d\beta = \frac{N^2}{2} + O(N).$$

Proof. By Theorem 5.1, the number of representations of N as the sum of three positive integers is

$$J(N) = \int_{-1/2}^{1/2} u(\beta)^3 e(-N\beta) d\beta$$

$$= \int_{-1/2}^{1/2} \sum_{m_1=1}^{N} \sum_{m_2=1}^{N} \sum_{m_3=1}^{N} e((m_1 + m_2 + m_3 - N)\beta) d\beta$$

$$= \binom{N-1}{2}$$

$$= \frac{N^2}{2} + O(N).$$

This completes the proof.

In the next lemma we shall apply the Siegel–Walfisz theorem on the distribution of prime numbers in arithmetic progressions. A proof can be found in Davenport [19].

Theorem 8.3 (Siegel–Walfisz) *If $q \geq 1$ and $(q, a) = 1$, then, for any $C > 0$,*

$$\vartheta(x; q, a) = \sum_{\substack{p \leq x \\ p \equiv a \pmod q}} \log p = \frac{x}{\varphi(q)} + O\left(\frac{x}{(\log x)^C}\right)$$

for all $x \geq 2$, where the implied constant depends only on C.

Lemma 8.2 *Let*

$$F_x(\alpha) = \sum_{p \leq x} (\log p) e(p\alpha).$$

Let B and C be positive real numbers. If $1 \leq q \leq Q = (\log N)^B$ and $(q, a) = 1$, then

$$F_x(a/q) = \frac{\mu(q)}{\varphi(q)} x + O\left(\frac{QN}{(\log N)^C}\right)$$

for $1 \leq x \leq N$, where the implied constant depends only on B and C.

Proof. Let $p \equiv r \pmod q$. Then p divides q if and only if $(r, q) > 1$, and so

$$\sum_{\substack{r=1 \\ (r,q)>1}}^{q} \sum_{\substack{p \leq x \\ p \equiv r \pmod q}} (\log p) e(pa/q) = \sum_{\substack{p \leq x \\ p \mid q}} (\log p) e(pa/q) \ll \sum_{p \mid q} \log p \leq \log q.$$

Therefore,

$$F_x\left(\frac{a}{q}\right) = \sum_{r=1}^{q} \sum_{\substack{p \leq x \\ p \equiv r \pmod q}} (\log p) e\left(\frac{pa}{q}\right)$$

$$= \sum_{\substack{r=1 \\ (r,q)=1}}^{q} \sum_{\substack{p \leq x \\ p \equiv r \pmod q}} (\log p) e\left(\frac{ra}{q}\right) + O(\log q)$$

$$= \sum_{\substack{r=1 \\ (r,q)=1}}^{q} e\left(\frac{ra}{q}\right) \sum_{\substack{p \leq x \\ p \equiv r \pmod q}} (\log p) + O(\log Q)$$

$$= \sum_{\substack{r=1 \\ (r,q)=1}}^{q} e\left(\frac{ra}{q}\right) \vartheta(x; q, r) + O(\log Q)$$

$$= \sum_{\substack{r=1 \\ (r,q)=1}}^{q} e\left(\frac{ra}{q}\right)\left(\frac{x}{\varphi(q)} + O\left(\frac{x}{(\log x)^C}\right)\right) + O(\log Q)$$

$$= \frac{c_q(a)}{\varphi(q)} x + O\left(\frac{qx}{(\log x)^C}\right) + O(\log Q)$$

$$= \frac{\mu(q)}{\vartheta(q)} x + O\left(\frac{QN}{(\log N)^C}\right),$$

since, by Theorem A.24, $c_q(a) = \mu(a)$ if $(q, a) = 1$.

Lemma 8.3 *Let B and C be positive real numbers with $C > 2B$. If $\alpha \in \mathfrak{M}(q, a)$ and $\beta = \alpha - a/q$, then*

$$F(\alpha) = \frac{\mu(q)}{\varphi(q)} u(\beta) + O\left(\frac{Q^2 N}{(\log N)^C}\right)$$

and

$$F(\alpha)^3 = \frac{\mu(q)}{\varphi(q)^3} u(\beta)^3 + O\left(\frac{Q^2 N^3}{(\log N)^C}\right),$$

where the implied constants depend only on B and C.

Proof. If $\alpha \in \mathfrak{M}(q, a)$, then $\alpha = a/q + \beta$, where $|\beta| \leq Q/N$. Let

$$\lambda(m) = \begin{cases} \log p & \text{if } m = p \text{ is prime} \\ 0 & \text{otherwise.} \end{cases}$$

If $1 \leq x \leq N$, then

$$
\begin{aligned}
F(\alpha) - \frac{\mu(q)}{\varphi(q)} u(\beta) &= \sum_{p \leq N} \log p e(p\alpha) - \frac{\mu(q)}{\varphi(q)} \sum_{m=1}^{N} e(m\beta) \\
&= \sum_{m=1}^{N} \lambda(m)e(m\alpha) - \frac{\mu(q)}{\varphi(q)} \sum_{m=1}^{N} e(m\beta) \\
&= \sum_{m=1}^{N} \lambda(m)e\left(\frac{ma}{q} + m\beta\right) - \sum_{m=1}^{N} \frac{\mu(q)}{\varphi(q)} e(m\beta) \\
&= \sum_{m=1}^{N} \left(\lambda(m)e\left(\frac{ma}{q}\right) - \frac{\mu(q)}{\varphi(q)}\right) e(m\beta).
\end{aligned}
$$

By Lemma 8.2, we have

$$
\begin{aligned}
A(x) &= \sum_{1 \leq m \leq x} \left(\lambda(m)e\left(\frac{ma}{q}\right) - \frac{\mu(q)}{\varphi(q)}\right) \\
&= \sum_{1 \leq m \leq x} \lambda(m)e\left(\frac{ma}{q}\right) - \frac{\mu(q)}{\varphi(q)} x + O\left(\frac{1}{\varphi(q)}\right) \\
&= F_x\left(\frac{a}{q}\right) - \frac{\mu(q)}{\varphi(q)} x + O(1) \\
&= O\left(\frac{QN}{(\log N)^C}\right).
\end{aligned}
$$

By partial summation, we obtain

$$
\begin{aligned}
F(\alpha) - \frac{\mu(q)}{\varphi(q)} u(\beta) &= A(N)e(N\beta) - 2\pi i\beta \int_1^N A(x)e(x\beta)dx \\
&\ll |A(N)| + |\beta| N \max\{A(x) : 1 \leq x \leq N\} \\
&\ll \frac{Q^2 N}{(\log N)^C}.
\end{aligned}
$$

Clearly, $|u(\beta)| \leq N$. Since $C > 2B$, we have

$$\frac{Q^2 N}{(\log N)^C} = \frac{N}{(\log N)^{C-2B}} < N,$$

and the estimate for $F(\alpha)^3$ follows immediately. This completes the proof.

Theorem 8.4 *For any positive numbers $B, C,$ and ε with $C > 2B$, the integral over the major arcs is*

$$\int_{\mathfrak{M}} F(\alpha)^3 e(-N\alpha) d\alpha = \mathfrak{S}(N)\frac{N^2}{2} + O\left(\frac{N^2}{(\log N)^{(1-\varepsilon)B}}\right) + O\left(\frac{N^2}{(\log N)^{C-5B}}\right),$$

where the implied constants depend only on $B, C,$ and ε.

Proof. We note that the length of the major arc $\mathfrak{M}(q, a)$ is Q/N if $q = 1$ and $2Q/N$ if $q \geq 2$. By Lemma 8.3,

$$\int_{\mathfrak{M}} \left(F(\alpha)^3 - \frac{\mu(q)}{\varphi(q)^3}u\left(\alpha - \frac{a}{q}\right)^3\right) e(-N\alpha) d\alpha$$

$$= \sum_{q \leq Q} \sum_{\substack{a=0 \\ (a,q)=1}}^{q} \int_{\mathfrak{M}(q,a)} \left(F(\alpha)^3 - \frac{\mu(q)}{\varphi(q)^3}u\left(\alpha - \frac{a}{q}\right)^3\right) e(-N\alpha) d\alpha$$

$$\ll \sum_{q \leq Q} \sum_{\substack{a=0 \\ (a,q)=1}}^{q} \int_{\mathfrak{M}(q,a)} \frac{Q^2 N^3}{(\log N)^C} d\alpha$$

$$\ll \sum_{q \leq Q} \sum_{\substack{a=0 \\ (a,q)=1}}^{q} \frac{Q^3 N^2}{(\log N)^C}$$

$$\leq \frac{Q^5 N^2}{(\log N)^C}$$

$$\leq \frac{N^2}{(\log N)^{C-5B}}.$$

If $\alpha = a/q + \beta \in \mathfrak{M}(q, a)$, then $|\beta| \leq Q/N$ and

$$\sum_{q \leq Q} \sum_{\substack{a=0 \\ (a,q)=1}}^{q} \frac{\mu(q)}{\varphi(q)^3} \int_{\mathfrak{M}(q,a)} u\left(\alpha - \frac{a}{q}\right)^3 e(-N\alpha) d\alpha$$

$$= \sum_{q \leq Q} \sum_{\substack{a=1 \\ (a,q)=1}}^{q} \frac{\mu(q)}{\varphi(q)^3} \int_{a/q-Q/N}^{a/q+Q/N} u\left(\alpha - \frac{a}{q}\right)^3 e(-N\alpha) d\alpha$$

$$= \sum_{q \leq Q} \frac{\mu(q)}{\varphi(q)^3} \sum_{\substack{a=1 \\ (a,q)=1}}^{q} e(-Na/q) \int_{-Q/N}^{Q/N} u(\beta)^3 e(-N\beta) d\beta$$

$$= \sum_{q \le Q} \frac{\mu(q)c_q(-N)}{\varphi(q)^3} \int_{-Q/N}^{Q/N} u(\beta)^3 e(-N\beta) d\beta$$

$$= \mathfrak{S}(N, Q) \int_{-Q/N}^{Q/N} u(\beta)^3 e(-N\beta) d\beta.$$

By Lemma 4.7, if $|\beta| \le 1/2$, then

$$u(\beta) \ll |\beta|^{-1}$$

and

$$\int_{Q/N}^{1/2} u(\beta)^3 e(-N\beta) d\beta \ll \int_{Q/N}^{1/2} |u(\beta)|^3 d\beta$$

$$\ll \int_{Q/N}^{1/2} \beta^{-3} d\beta$$

$$< \frac{N^2}{Q^2}.$$

Similarly,

$$\int_{-1/2}^{-Q/N} u(\beta)^3 e(-N\beta) d\beta \ll \frac{N^2}{Q^2}.$$

By Lemma 8.1,

$$\int_{-Q/N}^{Q/N} u(\beta)^3 e(-N\beta) d\beta = \int_{-1/2}^{1/2} u(\beta)^3 e(-N\beta) d\beta + O(N^2 Q^{-2})$$

$$= \frac{N^2}{2} + O(N) + O\left(\frac{N^2}{Q^2}\right)$$

$$= \frac{N^2}{2} + O\left(\frac{N^2}{Q^2}\right).$$

By Theorem 8.2,

$$\mathfrak{S}(N, Q) = \mathfrak{S}(N) + O\left(\frac{1}{Q^{1-\varepsilon}}\right).$$

Therefore,

$$\int_{\mathfrak{M}} F(\alpha)^3 e(-N\alpha) d\alpha$$

$$= \mathfrak{S}(N, Q) \int_{-Q/N}^{Q/N} u(\beta)^3 e(-N\beta) d\beta + O\left(\frac{N^2}{(\log N)^{C-5B}}\right)$$

$$= \mathfrak{S}(N)\frac{N^2}{2} + O\left(\frac{N^2}{Q^{1-\varepsilon}}\right) + O\left(\frac{N^2}{(\log N)^{C-5B}}\right)$$

$$= \mathfrak{S}(N)\frac{N^2}{2} + O\left(\frac{N^2}{(\log N)^{(1-\varepsilon)B}}\right) + O\left(\frac{N^2}{(\log N)^{C-5B}}\right).$$

This completes the proof.

8.5   An exponential sum over primes

To estimate the integral over the minor arcs, we shall apply Vinogradov's estimate for the exponential sum $F(\alpha)$. The proof is based on a combinatorial identity of Vaughan.

Theorem 8.5 (Vinogradov) *If*

$$\left| \alpha - \frac{a}{q} \right| \leq \frac{1}{q^2},$$

where a and q are integers such that $1 \leq q \leq N$ and $(a, q) = 1$, then

$$F(\alpha) \ll \left(\frac{N}{q^{1/2}} + N^{4/5} + N^{1/2}q^{1/2} \right) (\log N)^4.$$

The proof is divided into a series of lemmas. The first is an identity involving arithmetic functions of two variables and truncated sums of the Möbius function.

Lemma 8.4 (Vaughan's identity) *For $u \geq 1$, let*

$$M_u(k) = \sum_{\substack{d|k \\ d \leq u}} \mu(d).$$

Let $\Phi(k, \ell)$ be an arithmetic function of two variables. Then

$$\sum_{u < \ell \leq N} \Phi(1, \ell) + \sum_{u < k \leq N} \sum_{u < \ell \leq \frac{N}{k}} M_u(k)\Phi(k, \ell) = \sum_{d \leq u} \sum_{u < \ell \leq \frac{N}{d}} \sum_{m \leq \frac{N}{\ell d}} \mu(d)\Phi(dm, \ell).$$

Proof. We shall evaluate the sum

$$S = \sum_{k=1}^{N} \sum_{u < \ell \leq \frac{N}{k}} M_u(k)\Phi(k, \ell)$$

in two different ways. Since

$$\sum_{d|n} \mu(d) = \begin{cases} 1 & \text{if } n = 1 \\ 0 & \text{otherwise,} \end{cases}$$

it follows that

$$M_u(k) = \begin{cases} 1 & \text{if } k = 1 \\ 0 & \text{if } 1 < k \leq u. \end{cases}$$

Therefore,

$$S = \sum_{u < \ell \leq N} \Phi(1, \ell) + \sum_{u < k \leq N} \sum_{u < \ell \leq \frac{N}{k}} M_u(k)\Phi(k, \ell).$$

On the other hand, interchanging summations and letting $k = dm$, we obtain

$$S = \sum_{k=1}^{N} \sum_{\substack{u < \ell \leq \frac{N}{k} \\ d | k \\ d \leq u}} \sum \mu(d)\Phi(k, \ell)$$

$$= \sum_{d \leq u} \sum_{\substack{k=1 \\ d | k}}^{N} \sum_{u < \ell \leq \frac{N}{k}} \mu(d)\Phi(k, \ell)$$

$$= \sum_{d \leq u} \sum_{m \leq \frac{N}{d}} \sum_{u < \ell \leq \frac{N}{dm}} \mu(d)\Phi(dm, \ell)$$

$$= \sum_{d \leq u} \sum_{u < \ell \leq \frac{N}{d}} \sum_{m \leq \frac{N}{\ell d}} \mu(d)\Phi(dm, \ell).$$

Lemma 8.5 *Let $\Lambda(\ell)$ be the von Mangoldt function. For every real number α,*

$$F(\alpha) = S_1 - S_2 - S_3 + O(N^{1/2}),$$

where

$$S_1 = \sum_{d \leq N^{2/5}} \sum_{\ell \leq \frac{N}{d}} \sum_{m \leq \frac{N}{\ell d}} \mu(d)\Lambda(\ell)e(\alpha d\ell m),$$

$$S_2 = \sum_{d \leq N^{2/5}} \sum_{\ell \leq N^{2/5}} \sum_{m \leq \frac{N}{\ell d}} \mu(d)\Lambda(\ell)e(\alpha d\ell m),$$

and

$$S_3 = \sum_{k > N^{2/5}} \sum_{N^{2/5} < \ell \leq N/k} M_{N^{2/5}}(k)\Lambda(\ell)e(\alpha k\ell).$$

Proof. We apply Vaughan's identity with

$$u = N^{2/5}$$

and

$$\Phi(k, \ell) = \Lambda(\ell)e(\alpha k\ell).$$

The first term in Vaughan's identity is

$$\sum_{u < \ell \leq N} \Phi(1, \ell) = \sum_{N^{2/5} < \ell \leq N} \Lambda(\ell)e(\alpha\ell)$$

$$= \sum_{\ell=1}^{N} \Lambda(\ell)e(\alpha\ell) - \sum_{\ell \leq N^{2/5}} \Lambda(\ell)e(\alpha\ell)$$

$$= \sum_{p^k \leq N} (\log p)e(\alpha p^k) + O\left(N^{2/5}\log N\right)$$

$$= \sum_{p \leq N} (\log p)e(\alpha p) + \sum_{\substack{p^k \leq N \\ k \geq 2}} (\log p)e(\alpha p^k) + O\left(N^{2/5}\log N\right)$$

$$= F(\alpha) + O\left(\sum_{\substack{p^k \leq N \\ k \geq 2}} \log p\right) + O\left(N^{2/5} \log N\right)$$

$$= F(\alpha) + O\left(\sum_{p^2 \leq N} \left[\frac{\log N}{\log p}\right] \log p\right) + O\left(N^{2/5} \log N\right)$$

$$= F(\alpha) + O\left(\pi(N^{1/2}) \log N\right) + O\left(N^{2/5} \log N\right)$$

$$= F(\alpha) + O\left(N^{1/2}\right),$$

since

$$\pi(N^{1/2}) \ll \frac{N^{1/2}}{\log N}$$

by Chebyshev (Theorem 6.3).

The second term in Vaughan's identity is simply

$$\sum_{N^{2/5} < k \leq N} \sum_{N^{2/5} < \ell \leq \frac{N}{k}} M_{N^{2/5}}(k) \Lambda(\ell) e(\alpha k \ell) = S_3.$$

The third term in Vaughan's identity is

$$\sum_{d \leq N^{2/5}} \sum_{N^{2/5} < \ell \leq \frac{N}{d}} \sum_{m \leq \frac{N}{\ell d}} \mu(d) \Lambda(\ell) e(\alpha d \ell m)$$

$$= \sum_{d \leq N^{2/5}} \sum_{\ell \leq \frac{N}{d}} \sum_{m \leq \frac{N}{\ell d}} \mu(d) \Lambda(\ell) e(\alpha d \ell m)$$

$$- \sum_{d \leq N^{2/5}} \sum_{\ell \leq N^{2/5}} \sum_{m \leq \frac{N}{\ell d}} \mu(d) \Lambda(\ell) e(\alpha d \ell m)$$

$$= S_1 - S_2.$$

This completes the proof.

In the next three lemmas, we find upper bounds for the sums S_1, S_2, and S_3.

Lemma 8.6 *If*

$$\left|\alpha - \frac{a}{q}\right| \leq \frac{1}{q^2},$$

where $1 \leq q \leq N$ and $(a, q) = 1$, then

$$|S_1| \ll \left(\frac{N}{q} + N^{2/5} + q\right) (\log N)^2.$$

Proof. Let $u = N^{2/5}$. Since $\sum_{\ell | r} \Lambda(\ell) = \log r$, we have

$$S_1 = \sum_{d \leq u} \sum_{\ell \leq \frac{N}{d}} \sum_{m \leq \frac{N}{\ell d}} \mu(d) \Lambda(\ell) e(\alpha d \ell m)$$

$$= \sum_{d \leq u} \sum_{\ell m \leq N/d} \mu(d) \Lambda(\ell) e(\alpha d\ell m)$$

$$= \sum_{d \leq u} \sum_{r \leq N/d} \mu(d) e(\alpha dr) \sum_{\ell \mid r} \Lambda(\ell)$$

$$= \sum_{d \leq u} \mu(d) \sum_{r \leq N/d} e(\alpha dr) \log r$$

$$\ll \sum_{d \leq u} \left| \sum_{r \leq N/d} e(\alpha dr) \log r \right|.$$

We compute the inner sum by writing the logarithm as an integral and interchanging summations:

$$\sum_{r \leq N/d} e(\alpha dr) \log r = \sum_{r \leq N/d} e(\alpha dr) \int_1^r \frac{dx}{x}$$

$$= \sum_{r=2}^{[N/d]} e(\alpha dr) \sum_{s=2}^{r} \int_{s-1}^{s} \frac{dx}{x}$$

$$= \sum_{s=2}^{[N/d]} \sum_{r=s}^{[N/d]} \int_{s-1}^{s} e(\alpha dr) \frac{dx}{x}$$

$$= \sum_{s=2}^{[N/d]} \int_{s-1}^{s} \left(\sum_{r=s}^{[N/d]} e(\alpha dr) \right) \frac{dx}{x}.$$

By Lemma 4.7, the geometric progression inside the integral sign is bounded above by

$$\sum_{r=s}^{[N/d]} e(\alpha dr) \ll \min \left\{ \frac{N}{d}, \|\alpha d\|^{-1} \right\},$$

and so

$$\sum_{r \leq N/d} e(\alpha dr) \log r \ll \min \left(\frac{N}{d}, \|\alpha d\|^{-1} \right) \log N.$$

By Lemma 4.10, we have

$$\sum_{d \leq u} \min \left(\frac{N}{d}, \|\alpha d\|^{-1} \right) \ll \left(\frac{N}{q} + N^{2/5} + q \right) \log N.$$

Therefore,

$$S_1 \ll \sum_{d \leq u} \min \left(\frac{N}{d}, \|\alpha d\|^{-1} \right) \log N$$

$$\ll \left(\frac{N}{q} + N^{2/5} + q \right) (\log N)^2.$$

This completes the proof.

Lemma 8.7 *If*

$$\left| \alpha - \frac{a}{q} \right| \le \frac{1}{q^2},$$

where $1 \le q \le N$ *and* $(a, q) = 1$, *then*

$$|S_2| \ll \left(\frac{N}{q} + N^{4/5} + q \right) (\log N)^2.$$

Proof. If $d \le N^{2/5}$ and $\ell \le N^{2/5}$, then $d\ell \le N^{4/5}$. Making the substitution $k = d\ell$, we obtain

$$S_2 = \sum_{d \le N^{2/5}} \sum_{\ell \le N^{2/5}} \sum_{m \le \frac{N}{d\ell}} \mu(d)\Lambda(\ell)e(\alpha d\ell m)$$

$$= \sum_{k \le N^{4/5}} \left(\sum_{m \le N/k} e(\alpha km) \right) \left(\sum_{\substack{k = d\ell \\ d, \ell \le N^{2/5}}} \mu(d)\Lambda(\ell) \right).$$

Since

$$\sum_{\substack{k = d\ell \\ d, \ell \le N^{2/5}}} \mu(d)\Lambda(\ell) \ll \sum_{\substack{k = d\ell \\ d, \ell \le N^{2/5}}} \Lambda(\ell) \le \sum_{\ell \mid k} \Lambda(\ell) = \log k \ll \log N,$$

it follows again from Lemma 4.10 that

$$S_2 \ll \log N \sum_{k \le N^{4/5}} \sum_{m \le N/k} e(\alpha km)$$

$$\ll \sum_{k \le N^{4/5}} \min \left(\frac{N}{k}, \|\alpha k\|^{-1} \right) \log N$$

$$\ll \left(\frac{N}{q} + N^{4/5} + q \right) (\log N)^2.$$

This completes the proof.

Lemma 8.8 *If*

$$\left| \alpha - \frac{a}{q} \right| \le \frac{1}{q^2},$$

where $1 \le q \le N$ *and* $(a, q) = 1$, *then*

$$|S_3| \ll \left(\frac{N}{q^{1/2}} + N^{4/5} + N^{1/2}q^{1/2} \right) (\log N)^4.$$

Proof. Let $u = N^{2/5}$ and

$$h = \left[\frac{\log N}{5 \log 2} \right] + 1.$$

Then $N^{1/5} < 2^h \leq 2N^{1/5}$ and $h \ll \log N$. If $i \leq h$, then $2^i u \leq 2N^{3/5} \ll N$. If $N^{2/5} < \ell \leq N/k$, then

$$k \leq N/\ell < N^{3/5} = N^{1/5}u < 2^h u,$$

and so

$$S_3 = \sum_{k > N^{2/5}} \sum_{N^{2/5} < \ell \leq N/k} M_u(k) \Lambda(\ell) e(\alpha k \ell)$$

$$= \sum_{i=1}^{h} \sum_{2^{i-1}u < k \leq 2^i u} M_u(k) \sum_{u < \ell \leq N/k} \Lambda(\ell) e(\alpha k \ell)$$

$$= \sum_{i=1}^{h} S_{3,i},$$

where

$$S_{3,i} = \sum_{2^{i-1}u < k \leq 2^i u} M_u(k) \sum_{u < \ell \leq N/k} \Lambda(\ell) e(\alpha k \ell).$$

By the Cauchy–Schwarz inequality,

$$|S_{3,i}|^2 \leq \sum_{2^{i-1}u < k \leq 2^i u} |M_u(k)|^2 \cdot \sum_{2^{i-1}u < k \leq 2^i u} \left| \sum_{u < \ell \leq N/k} \Lambda(\ell) e(\alpha k \ell) \right|^2. \tag{8.5}$$

We shall estimate these sums separately.

To estimate the first sum in (8.5), we observe that

$$|M_u(k)| = \left| \sum_{\substack{d|k \\ d \leq u}} \mu(d) \right| \leq \sum_{\substack{d|k \\ d \leq u}} 1 \leq d(k),$$

where $d(k)$ is the divisor function. It follows from Theorem A.14 that

$$\sum_{2^{i-1}u < k \leq 2^i u} |M_u(k)|^2 \leq \sum_{2^{i-1}u < k \leq 2^i u} d(k)^2$$

$$\ll 2^i u \left(\log 2^i u \right)^3$$

$$\ll 2^i u \left(\log N \right)^3.$$

Next, we estimate the second sum in (8.5). We have

$$\sum_{2^{i-1}u < k \leq 2^i u} \left| \sum_{u < \ell \leq N/k} \Lambda(\ell) e(\alpha k \ell) \right|^2$$

$$= \sum_{2^{i-1}u < k \leq 2^i u} \sum_{u < \ell \leq N/k} \sum_{u < m \leq N/k} \Lambda(\ell) \Lambda(m) e(\alpha k (\ell - m))$$

$$= \sum_{u < \ell < \frac{N}{2^{i-1}u}} \sum_{u < m < \frac{N}{2^{i-1}u}} \Lambda(\ell) \Lambda(m) \sum_{k \in I(\ell,m)} e(\alpha k (\ell - m)),$$

where $I(\ell, m)$ is the interval of consecutive integers k such that

$$2^{i-1}u < k \leq \min\left(2^i u, \frac{N}{\ell}, \frac{N}{m}\right).$$

Clearly,

$$|I(\ell, m)| \leq 2^{i-1}u,$$

and so

$$\sum_{k \in I(\ell,m)} e(\alpha k(\ell - m)) \ll \min\left(2^{i-1}u, \|\alpha(\ell - m)\|^{-1}\right).$$

Since $0 \leq \Lambda(\ell), \Lambda(m) \leq \log N$ for all integers $\ell, m \in [1, N]$, we have

$$\sum_{2^{i-1}u < k \leq 2^i u} \left|\sum_{u < \ell \leq N/k} \Lambda(\ell) e(\alpha k \ell)\right|^2$$

$$\ll \sum_{u < \ell < N/(2^{i-1}u)} \sum_{u < m < N/(2^{i-1}u)} \Lambda(\ell)\Lambda(m) \min\left(2^{i-1}u, \|\alpha(\ell - m)\|^{-1}\right)$$

$$\ll (\log N)^2 \sum_{u < \ell < N/(2^{i-1}u)} \sum_{u < m < N/(2^{i-1}u)} \min\left(2^{i-1}u, \|\alpha(\ell - m)\|^{-1}\right).$$

Let $j = \ell - m$ with $u < \ell, m < N/(2^{i-1}u)$. Then $|j| < N/2^{i-1}u$, and the number of representations of an integer j in this form is at most $N/2^{i-1}u$. By Lemma 4.10, we have

$$\sum_{2^{i-1}u < k \leq 2^i u} \left|\sum_{u < \ell \leq N/k} \Lambda(\ell) e(\alpha k \ell)\right|^2$$

$$\ll (\log N)^2 \frac{N}{2^{i-1}u} \sum_{1 \leq j \leq N/2^{i-1}u} \min\left(2^{i-1}u, \|\alpha j\|^{-1}\right)$$

$$\ll (\log N)^2 \frac{N}{2^{i-1}u} \sum_{1 \leq j \leq N/2^{i-1}u} \min\left(\frac{N}{j}, \|\alpha j\|^{-1}\right)$$

$$\ll \frac{N}{2^{i-1}u}\left(\frac{N}{q} + \frac{N}{2^{i-1}u} + q\right)(\log N)^3.$$

Inserting this into inequality (8.5), we obtain

$$|S_{3,i}|^2 \ll (2^i u (\log N)^3) \frac{N}{2^{i-1}u}\left(\frac{N}{q} + \frac{N}{2^{i-1}u} + q\right)(\log N)^3$$

$$\ll N^2 (\log N)^6 \left(\frac{1}{q} + \frac{1}{u} + \frac{q}{N}\right).$$

Therefore,

$$|S_{3,i}| \ll N(\log N)^3 \left(\frac{1}{q^{1/2}} + \frac{1}{N^{1/5}} + \frac{q^{1/2}}{N^{1/2}}\right).$$

Since $h \ll \log N$, we have

$$S_3 = \sum_{i=1}^{h} S_{3,i} \ll (\log N)^4 \left(\frac{N}{q^{1/2}} + N^{4/5} + q^{1/2} N^{1/2} \right).$$

This completes the proof.

Finally, we obtain Vinogradov's estimate for the exponential sum $F(\alpha)$ by inserting our estimates for the sums S_1, S_2, and S_3 into Lemma 8.5. This completes the proof of Theorem 8.5.

8.6 Proof of the asymptotic formula

We can now estimate the integral over the minor arcs.

Theorem 8.6 *For any $B > 0$, we have*

$$\int_{\mathfrak{m}} F(\alpha)^3 e(-\alpha N) d\alpha \ll \frac{N^2}{(\log N)^{(B/2)-5}},$$

where the implied constant depends only on B.

Proof. Let $\alpha \in \mathfrak{m} = [0, 1] \setminus \mathfrak{M}$. By Dirichlet's theorem (Theorem 4.1), for any real number α there exists a fraction $a/q \in [0, 1]$ with $1 \le q \le N/Q$ and $(a, q) = 1$ such that

$$\left| \alpha - \frac{a}{q} \right| \le \frac{Q}{qN} \le \min \left(\frac{Q}{N}, \frac{1}{q^2} \right).$$

If $q \le Q$, then $\alpha \in \mathfrak{M}(q, a) \subseteq \mathfrak{M}$, which is false. Therefore,

$$Q < q \le \frac{N}{Q}.$$

By Theorem 8.5,

$$F(\alpha) \ll \left(\frac{N}{q^{1/2}} + N^{4/5} + N^{1/2} q^{1/2} \right) (\log N)^4$$

$$\ll \left(\frac{N}{(\log N)^{B/2}} + N^{4/5} + N^{1/2} \left(\frac{N}{(\log N)^B} \right)^{1/2} \right) (\log N)^4$$

$$\ll \frac{N}{(\log N)^{(B/2)-4}}.$$

Since $\vartheta(N) = \sum_{p \le N} \log p \ll N$ by Theorem 6.3, we have

$$\int_0^1 |F(\alpha)|^2 d\alpha = \sum_{p \le N} (\log p)^2 \le \log N \sum_{p \le N} \log p \ll N \log N,$$

and so

$$\int_{\mathfrak{m}} |F(\alpha)|^3 d\alpha \ll \sup\{|F(\alpha)| : \alpha \in \mathfrak{m}\} \int_{\mathfrak{m}} |F(\alpha)|^2 d\alpha$$

$$\ll \frac{N}{(\log N)^{(B/2)-4}} \int_0^1 |F(\alpha)|^2 d\alpha$$

$$\ll \frac{N^2}{(\log N)^{(B/2)-5}}.$$

This completes the proof.

Theorem 8.7 (Vinogradov) *Let $\mathfrak{S}(N)$ be the singular series for the ternary Goldbach problem. For all suffciently large odd integers N and for every $A > 0$,*

$$R(N) = \mathfrak{S}(N)\frac{N^2}{2} + O\left(\frac{N^2}{(\log N)^A}\right),$$

where the implied constant depends only on A.

Proof. It follows from Theorem 8.4 and Theorem 8.6 that, for any positive numbers $B, C,$ and ε with $C > 2B$,

$$R(N) = \int_0^1 F(\alpha)^3 e(-N\alpha) d\alpha$$

$$= \int_{\mathfrak{M}} F(\alpha)^3 e(-N\alpha) d\alpha + \int_{\mathfrak{m}} F(\alpha)^3 e(-N\alpha) d\alpha$$

$$= \mathfrak{S}(N)\frac{N^2}{2} + O\left(\frac{N^2}{(\log N)^{(1-\varepsilon)B}}\right)$$

$$+ O\left(\frac{N^2}{(\log N)^{C-5B}}\right) + O\left(\frac{N^2}{(\log N)^{(B/2)-5}}\right),$$

where the implied constants depend only on $B, C,$ and ε. For any $A > 0$, let $B = 2A + 10$ and $C = A + 5B$. Let $\varepsilon = 1/2$. Then

$$\min((1 - \varepsilon)B, C - 5B, (B/2) - 5) = A,$$

and so

$$R(N) = \mathfrak{S}(N)\frac{N^2}{2} + O\left(\frac{N^2}{(\log N)^A}\right).$$

This completes the proof.

We can now derive Vinogradov's asymptotic formula for $r(N)$.

Proof of Theorem 8.1. We get an upper bound for $R(N)$ as follows:

$$R(N) = \sum_{p_1+p_2+p_3=N} \log p_1 \log p_2 \log p_3$$

$$\le (\log N)^3 \sum_{p_1+p_2+p_3=N} 1$$

$$= (\log N)^3 r(N).$$

For $0 < \delta < 1/2$, let $r_\delta(N)$ denote the number of representations of N in the form $N = p_1 + p_2 + p_3$ such that $p_i \leq N^{1-\delta}$ for some i. Then

$$r_\delta(N) \leq 3 \sum_{\substack{p_1+p_2+p_3=N \\ p_1 \leq N^{1-\delta}}} 1$$

$$\ll \sum_{p_1 \leq N^{1-\delta}} \left(\sum_{p_2+p_3=N-p_1} 1 \right)$$

$$\leq \sum_{p_1 \leq N^{1-\delta}} \left(\sum_{p_2 < N} 1 \right)$$

$$\leq \pi(N^{1-\delta})\pi(N)$$

$$\ll \frac{N^{2-\delta}}{(\log N)^2}.$$

We can now get a lower bound for $R(N)$:

$$R(N) \geq \sum_{\substack{p_1+p_2+p_3=N \\ p_1 \cdot p_2 \cdot p_3 > N^{1-\delta}}} \log p_1 \log p_2 \log p_3$$

$$\geq (1-\delta)^3(\log N)^3 \sum_{\substack{p_1+p_2+p_3=N \\ p_1 \cdot p_2 \cdot p_3 > N^{1-\delta}}} 1$$

$$\geq (1-\delta)^3(\log N)^3(r(N) - r_\delta(N))$$

$$\gg (1-\delta)^3(\log N)^3 \left(r(N) - \frac{N^{2-\delta}}{(\log N)^2} \right).$$

Therefore,

$$(\log N)^3 r(N) \leq (1-\delta)^{-3} R(N) + (\log N)N^{2-\delta}.$$

If $0 < \delta < 1/2$, then $1/2 < 1 - \delta < 1$ and

$$0 < (1-\delta)^{-3} - 1 = \frac{1 - (1-\delta)^3}{(1-\delta)^3} \leq 8\left(1 - (1-\delta)^3\right) < 24\delta.$$

By Theorem 8.7, $R(N) \ll N^2$ and so

$$0 \leq (\log N)^3 r(N) - R(N) \leq \left((1-\delta)^{-3} - 1\right) R(N) + (\log N)N^{2-\delta}$$

$$\ll \delta R(N) + (\log N)N^{2-\delta}$$

$$\ll \delta N^2 + (\log N)N^{2-\delta}$$

$$= N^2 \left(\delta + \frac{\log N}{N^\delta} \right).$$

This inequality holds for all $\delta \in (0, 1/2)$, and the implied constant does not depend on δ. Let

$$\delta = \frac{2 \log \log N}{\log N}.$$

Then
$$\delta + \frac{\log N}{N^\delta} = \frac{2 \log \log N}{\log N} + \frac{\log N}{(\log N)^2} \ll \frac{\log \log N}{\log N}$$

and so
$$0 \le (\log N)^3 r(N) - R(N) \ll \frac{N^2 \log \log N}{\log N}.$$

Let $A \ge 1$. By Theorem 8.7,

$$(\log N)^3 r(N) = R(N) + O\left(\frac{N^2 \log \log N}{\log N}\right)$$

$$= \mathfrak{S}(N)\frac{N^2}{2} + O\left(\frac{N^2}{(\log N)^A}\right) + O\left(\frac{N^2 \log \log N}{\log N}\right)$$

$$= \mathfrak{S}(N)\frac{N^2}{2}\left(1 + O\left(\frac{\log \log N}{\log N}\right)\right).$$

Dividing by $(\log N)^3$, we obtain

$$r(N) = \mathfrak{S}(N)\frac{N^2}{2(\log N)^3}\left(1 + O\left(\frac{\log \log N}{\log N}\right)\right).$$

This completes the proof.

8.7   Notes

For Vinogradov's original papers, see [132, 133]. Vaughan [124] greatly simplified Vinogradov's estimate for the exponential sum $F(\alpha)$ (Theorem 8.5), and it is Vaughan's proof that is given in this book. There are many good expositions of Vinogradov's theorem. See, for example, the books of Davenport [19], Ellison [29], Estermann [38], Hua [64], Vaughan [125], and Vinogradov [135].

Vinogradov's theorem implies that almost all positive even integers can be written as the sum of two primes. This was observed independently by Chudakov [14], van der Corput [123], and Estermann [37]. Let E denote the set of even integers greater than two that cannot be written as the sum of two primes. The set E is called the *exceptional set* for the Goldbach conjecture. Let $E(x)$ denote the number of integers in E not exceeding x. The theorem of Chudakov, van der Corput, and Estermann states that $E(x) \ll_A x/(\log x)^A$ for every $A > 0$. Montgomery and Vaughan [84] proved that there exists $\delta < 1$ such that $E(x) \ll x^\delta$. Of course, if the Goldbach conjecture is true, then $E(x) = 0$ for all x.

8.8   Exercise

1. Let $h \ge 3$. Find an asymptotic formula for the number of representations of a positive integer $N \equiv h \pmod 2$ as a sum of h prime numbers.

9

The linear sieve

We often apply, consciously or not, some kind of sieve procedure
whenever the subject of investigation is not directly recognizable. We
begin by making a long list of suspects, and then we sort it out gradu-
ally by excluding obvious cases with respect to available information.
The process of exclusion itself may yield new data which influences
our decision about what to exclude or include in the next run. When no
clue is provided to drive us further, the process terminates and we are
left with objects which can be examined by other means to determine
their exact identity. These universal ideas were formalized in the con-
text of arithmetic back in the second century B.C. by Eratosthenes,
and are still used today.

H. Iwaniec [68]

9.1 A general sieve

In the next chapter, we shall prove Chen's theorem that every sufficiently large
even integer can be written as the sum of a prime and a number that is the product
of at most two primes. The proof will require more sophisticated sieve estimates
than those obtained from the Selberg sieve in Chapter 7.

We begin by generalizing our concept of a sieve. Let $A = \{a(n)\}_{n=1}^{\infty}$ be an
arithmetic function such that

$$a(n) \geq 0 \qquad \text{for all } n \tag{9.1}$$

and

$$|A| = \sum_{n=1}^{\infty} a(n) < \infty. \tag{9.2}$$

Let $\mathcal{P}$ be a set of prime numbers and let z be a real number, $z \geq 2$. The set $\mathcal{P}$ is called the *sieving range*, and the number z is called the *sieving level*. Let

$$P(z) = \prod_{\substack{p \in \mathcal{P} \\ p < z}} p.$$

The *sieving function* is

$$S(A, \mathcal{P}, z) = \sum_{(n, P(z))=1} a(n).$$

The goal of sieve theory is to obtain "good" upper and lower bounds for this function.

For example, let A be the characteristic function of a finite set of positive integers, that is, $a(n) = 1$ if n is in the set and $a(n) = 0$ if n is not in the set. Then $|A|$ is the cardinality of the set. The sieving function $S(A, \mathcal{P}, z)$ counts the number of integers in the set that are not divisible by any prime $p \in \mathcal{P}, p < z$. This special case is exactly the sieving function for which we obtained, in Chapter 7, an upper bound by means of the Selberg sieve.

Using the fundamental property of the Möbius function, that

$$(1 * \mu)(m) = \sum_{d|m} \mu(d) = \begin{cases} 1 & \text{if } m = 1 \\ 0 & \text{if } m > 1, \end{cases}$$

where 1 denotes the arithmetic function such that $1(n) = 1$ for all $n \geq 1$, we obtain *Legendre's formula*

$$S(A, \mathcal{P}, z) = \sum_{(n, P(z))=1} a(n)$$

$$= \sum_{n} a(n) \sum_{d|(n, P(z))} \mu(d)$$

$$= \sum_{d|P(z)} \mu(d) \sum_{d|n} a(n)$$

$$= \sum_{d|P(z)} \mu(d) |A_d|,$$

where the series

$$|A_d| = \sum_{d|n} a(n)$$

converges because of (9.1) and (9.2).

We shall assume that, for every $n \geq 1$, we have a multiplicative function $g_n(d)$ such that

$$0 \leq g_n(p) < 1$$

for every prime $p \in \mathcal{P}$. Then

$$0 \le g_n(d) \le 1$$

for every integer d that is the product of distinct primes $p \in \mathcal{P}$. For such integers d, the series

$$\sum_n a(n) g_n(d)$$

converges, and we can define the *remainder $r(d)$* by

$$|A_d| = \sum_n a(n) g_n(d) + r(d).$$

Inserting this into Legendre's formula, we obtain

$$S(A, \mathcal{P}, z) = \sum_{d | P(z)} \mu(d) |A_d|$$

$$= \sum_{d | P(z)} \mu(d) \left(\sum_n a(n) g_n(d) + r(d) \right)$$

$$= \sum_n a(n) \sum_{d | P(z)} \mu(d) g_n(d) + \sum_{d | P(z)} \mu(d) r(d)$$

$$= \sum_n a(n) \prod_{p | P(z)} (1 - g_n(p)) + \sum_{d | P(z)} \mu(d) r(d)$$

$$= \sum_n a(n) V_n(z) + R(z),$$

where

$$V_n(z) = \prod_{p | P(z)} (1 - g_n(p))$$

and

$$R(z) = \sum_{d | P(z)} \mu(d) r(d).$$

If $P(z)$ has a large number of divisors, the remainder term $R(z)$ in Legendre's formula may be too large to give useful estimates for $S(A, \mathcal{P}, z)$. For example, let A be the characteristic function of the set of all positive integers not exceeding x, and let $\mathcal{P}$ be the set of all prime numbers. Let

$$g_n(d) = \frac{1}{d}$$

for all n. Then

$$V_n(z) = \prod_{p < z} \left(1 - \frac{1}{p} \right)$$

for all $n \ge 1$. Moreover. for all $d \ge 1$,

$$0 \le |r(d)| = \frac{|A|}{d} - |A_d| = \frac{[x]}{d} - \left[\frac{x}{d} \right] < 1,$$

and so

$$|R(z)| \le \sum_{d|P(z)} |r(d)| \le 2^{\pi(z)}.$$

It follows from Legendre's formula that the number of integers up to x divisible by no prime less than z is

$$S(A, \mathcal{P}, z) = [x] \prod_{p<z} \left(1 - \frac{1}{p}\right) + O\left(2^{\pi(z)}\right).$$

By Mertens's formula (Theorem 6.8),

$$\prod_{p<z} \left(1 - \frac{1}{p}\right) = \frac{e^{-\gamma}}{\log z} \left(1 + O\left(\frac{1}{\log z}\right)\right), \tag{9.3}$$

and so the remainder term will be larger than the main term unless z is very small compared to x.

The sieve idea is to reduce the size of the error term by replacing the Möbius function with carefully constructed arithmetic functions $\lambda^+(d)$ and $\lambda^-(d)$ such that

$$\lambda^+(1) = \lambda^-(1) = 1 \tag{9.4}$$

and, for every $m \ge 2$,

$$(1 * \lambda^+)(m) = \sum_{d|m} \lambda^+(d) \ge 0 \tag{9.5}$$

and

$$(1 * \lambda^-)(m) = \sum_{d|m} \lambda^-(d) \le 0, \tag{9.6}$$

Let $\lambda^+(d)$ and $\lambda^-(d)$ be arithmetic functions that satisfy (9.4), (9.5), and (9.6). If D is a positive number such that $\lambda^+(d) = 0$ for all $d \ge D$, then the arithmetic function $\lambda^+(d)$ is called an *upper bound sieve* with *support level D* . Similarly, if D is a positive number such that $\lambda^-(d) = 0$ for all $d \ge D$, then the arithmetic function $\lambda^-(d)$ is called a *lower bound sieve* with *support level D*.

If $\mathcal{P}$ is a set of primes such that $\lambda^+(d) = 0$ whenever d is divisible by a prime not in $\mathcal{P}$, then $\lambda^+(d)$ is called an *upper bound sieve* with *sieving range* $\mathcal{P}$. Similarly, if $\lambda^-(d) = 0$ whenever d is divisible by a prime not in $\mathcal{P}$, then $\lambda^-(d)$ is called a *lower bound sieve* with *sieving range* $\mathcal{P}$.

The following result is the basic sieve inequality.

Theorem 9.1 *Let $\lambda^+(d)$ be an upper bound sieve with sieving range $\mathcal{P}$ and support level D, and let $\lambda^-(d)$ be a lower bound sieve with sieving range $\mathcal{P}$ and support level D. Then*

$$\sum_{n=1}^{\infty} a(n)G_n(z, \lambda^-) + R^- \le S(A, \mathcal{P}, z) \le \sum_{n=1}^{\infty} a(n)G_n(z, \lambda^+) + R^+,$$

where

$$G_n(z, \lambda^\pm) = \sum_{d \mid P(z)} \lambda^\pm(d) g_n(d)$$

and

$$R^\pm = \sum_{\substack{d \mid P(z) \\ d < D}} \lambda^\pm(d) r(d).$$

Proof. Since the arithmetic function $\lambda^+(d)$ is supported on the finite set of integers $1 \le d < D$, it follows that the series

$$\sum_n a(n) \sum_{d \mid (n, P(z))} \lambda^+(d)$$

converges. By conditions (9.4) and (9.5), the inner sum is 1 if $(n, P(z)) = 1$ and nonnegative for all n. Therefore,

$$
\begin{aligned}
S(A, P, z) &= \sum_{(n, P(z))=1} a(n) \\
&\le \sum_n a(n) \sum_{d \mid (n, P(z))} \lambda^+(d) \\
&= \sum_{d \mid P(z)} \lambda^+(d) \sum_{d \mid n} a(n) \\
&= \sum_{d \mid P(z)} \lambda^+(d) |A_d| \\
&= \sum_{d \mid P(z)} \lambda^+(d) \left(\sum_n a(n) g_n(d) + r(d) \right) \\
&= \sum_{d \mid P(z)} \lambda^+(d) \sum_n a(n) g_n(d) + \sum_{d \mid P(z)} \lambda^+(d) r(d) \\
&= \sum_n a(n) \sum_{d \mid P(z)} \lambda^+(d) g_n(d) + \sum_{\substack{d \mid P(z) \\ d < D}} \lambda^+(d) r(d) \\
&= \sum_n a(n) G_n(z, \lambda^+) + R^+.
\end{aligned}
$$

The proof of the lower bound is similar.

The following result shows how to extend the sieving range of upper and lower bound sieves by any finite set of primes.

Lemma 9.1 *Let $\lambda_1^\pm(d)$ be upper and lower bound sieves with sieving range P_1 and support level D. Let Q be a finite set of primes disjoint from P_1, and let Q be the product of all primes in Q. Every positive integer d can be written uniquely in the form*

$$d = d_1 d_2,$$

where d_1 is relatively prime to Q and d_2 is a product of primes in Q. Define

$$\lambda^\pm(d) = \lambda_1^\pm(d_1) \mu(d_2). \tag{9.7}$$

Then the function $\lambda^+(d)$ (resp. $\lambda^-(d)$) is an upper bound sieve (resp. lower bound sieve) with sieving range

$$\mathcal{P} = \mathcal{P}_1 \cup \mathcal{Q}$$

and support level DQ.

Let g be a multiplicative function, and let

$$G(z, \lambda^\pm) = \sum_{d|P(z)} \lambda^\pm(d)g(d)$$

and

$$G(z, \lambda_1^\pm) = \sum_{d_1|P_1(z)} \lambda_1^\pm(d_1)g(d_1).$$

Then

$$G(z, \lambda^\pm) = G(z, \lambda_1^\pm) \prod_{q|Q(z)} (1 - g(q)).$$

Proof. Clearly, $\lambda^+(1) = \lambda^-(1) = 1$. Every positive integer m factors uniquely into a product $m = m_1 m_2$, where m_1 is relatively prime to Q and m_2 is a product of primes in $\mathcal{Q}$. We have

$$\sum_{d|m} \lambda^+(d) = \sum_{d_1|m_1} \sum_{d_2|m_2} \lambda^+(d_1 d_2)$$

$$= \sum_{d_1|m_1} \lambda_1^+(d_1) \sum_{d_2|m_2} \mu(d_2) \geq 0$$

since

$$\sum_{d_2|m_2} \mu(d_2) = \begin{cases} 1 & \text{if } m_2 = 1 \\ 0 & \text{if } m_2 \geq 2. \end{cases}$$

Similarly, if $m = m_1 m_2 > 1$, then

$$\sum_{d|m} \lambda^-(d) = \sum_{d_1|m_1} \lambda_1^-(d_1) \sum_{d_2|m_2} \mu(d_2) \leq 0$$

since either $m_2 > 1$ and

$$\sum_{d_2|m_2} \mu(d_2) = 0,$$

or $m_2 = 1$, which implies that $m_1 > 1$, and so

$$\sum_{d_1|m_1} \lambda_1(d_1) \leq 0.$$

Thus, the arithmetic functions $\lambda^\pm(d)$ satisfy conditions (9.4), (9.5), and (9.6).

Since $\lambda^\pm(d) = 0$ if d is divisible by some prime not in $\mathcal{P}$, it follows that the functions $\lambda^\pm$ have sieving range $\mathcal{P}$.

Let $d = d_1 d_2$, where d_1 is relatively prime to Q and d_2 is a product of primes in $\mathcal{Q}$. If $d = d_1 d_2 \geq DQ$, then either $d_1 \geq D$ and $\lambda_1^\pm(d_1) = 0$, or $d_2 > Q$, which implies that d_2 is divisible by the square of some prime $q \in Q$, and so $\mu(d_2) = 0$. In both cases, $\lambda^{\pm}(d) = 0$. Therefore, the functions $\lambda^{\pm}(d) = 0$ have support level DQ.

Finally, since $P(z) = P_1(z)Q(z)$,

$$
\begin{aligned}
G(z, \lambda^{\pm}) &= \sum_{d \mid P(z)} \lambda^{\pm}(d)g(d) \\
&= \sum_{d_1 \mid P_1(z)} \sum_{d_2 \mid Q(z)} \lambda^{\pm}(d_1 d_2)g(d_1 d_2) \\
&= \sum_{d_1 \mid P_1(z)} \sum_{d_2 \mid Q(z)} \lambda_1^{\pm}(d_1)g(d_1)\mu(d_2)g(d_2) \\
&= \sum_{d_1 \mid P_1(z)} \lambda_1^{\pm}(d_1)g(d_1) \sum_{d_2 \mid Q(z)} \mu(d_2)g(d_2) \\
&= G(z, \lambda_1^{\pm}) \prod_{q \mid Q(z)} (1 - g(q)).
\end{aligned}
$$

This completes the proof.

Combining Theorem 9.1 and Lemma 9.1, we obtain the following result, which is an important refinement of the basic sieve inequality.

Theorem 9.2 *Let $\lambda_1^{\pm}(d)$ be upper and lower bound sieves with sieving range $\mathcal{P}_1$ and support level D. Let $|\lambda_1^{\pm}(d)| \leq 1$ for all $d \geq 1$. Let Q be a finite set of primes disjoint from $\mathcal{P}_1$, and let Q be the product of the primes in Q. Let $\mathcal{P} = \mathcal{P}_1 \cup Q$. For each $n \geq 1$, let $g_n(d)$ be a multiplicative function such that*

$$
0 \leq g_n(p) < 1 \qquad \text{for all } p \in \mathcal{P}.
$$

Let

$$
G_n(z, \lambda_1^{\pm}) = \sum_{d \mid P_1(z)} \lambda_1^{\pm}(d)g_n(d).
$$

Then

$$
S(A, \mathcal{P}, z) \leq \sum_{n=1}^{\infty} a(n)G_n(z, \lambda_1^{+}) \prod_{q \mid Q(z)} (1 - g_n(q)) + R(DQ, \mathcal{P}, z)
$$

and

$$
S(A, \mathcal{P}, z) \geq \sum_{n=1}^{\infty} a(n)G_n(z, \lambda_1^{-}) \prod_{q \mid Q(z)} (1 - g_n(q)) - R(DQ, \mathcal{P}, z),
$$

where

$$
R(DQ, \mathcal{P}, z) = \sum_{\substack{d \mid P(z) \\ d < DQ}} |r(d)|.
$$

It often happens in applications that the arithmetic functions $g_n(d)$ satisfy one-sided inequalities of the form

$$\prod_{\substack{p \in P \\ u \le p < z}} (1 - g_n(p))^{-1} \le K \left(\frac{\log z}{\log u} \right)^\kappa ,$$

where $K > 1$ and $\kappa > 0$ are constants that are independent of n, and the inequality holds for all n and $1 < u < z$. In this case we say the sieve has *dimension* κ. The case $\kappa = 1$ is called the *linear sieve*. The goal of this chapter is to obtain upper and lower bounds for the linear sieve that were first proved by Jurkat and Richert (Theorem 9.7). This is the only sieve inequality that is needed for Chen's theorem.

9.2   Construction of a combinatorial sieve

In a combinatorial sieve, we reduce the size of the error term in Legendre's formula by replacing the Möbius function with its truncation to a finite set of positive integers. This idea goes back to Viggo Brun [7]. We construct these truncated functions in the following theorem.

Theorem 9.3 *Let $\beta > 1$ and $D > 0$ be real numbers. Let $\mathcal{D}^+$ be the set consisting of 1 and all square-free numbers*

$$d = p_1 p_2 \cdots p_k$$

such that

$$p_k < \cdots < p_2 < p_1 < D$$

and

$$p_m < \left(\frac{D}{p_1 p_2 \cdots p_m} \right)^{1/\beta}$$

for all odd integers m. Let $\mathcal{D}^-$ be the set consisting of 1 and all square-free numbers

$$d = p_1 p_2 \cdots p_k$$

such that

$$p_k < \cdots < p_2 < p_1 < D$$

and

$$p_m < \left(\frac{D}{p_1 p_2 \cdots p_m} \right)^{1/\beta}$$

for all even integers m. Then the sets $\mathcal{D}^+$ and $\mathcal{D}^-$ are finite sets of square-free positive integers $d < D$. Let $\mathcal{P}$ be a set of primes, and let $P(D)$ denote the product of all of the primes in $\mathcal{P}$ that are less than D. Define the arithmetic functions $\lambda^+(d)$ and $\lambda^-(d)$ as follows:

$$\lambda^+(d) = \begin{cases} \mu(d) & \text{if } d \in \mathcal{D}^+ \text{ and } d \mid P(D) \\ 0 & \text{otherwise} \end{cases}$$

and

$$\lambda^-(d) = \begin{cases} \mu(d) & \text{if } d \in \mathcal{D}^- \text{ and } d \mid P(D) \\ 0 & \text{otherwise.} \end{cases}$$

Then $\lambda^+(d)$ and $\lambda^-(d)$ are upper and lower bound sieves with sieving range $\mathcal{P}$ and support level D.

Proof. The condition

$$p_m < \left(\frac{D}{p_1 p_2 \cdots p_m} \right)^{1/\beta}$$

is equivalent to

$$p_1 p_2 \cdots p_{m-1} p_m^{1+\beta} < D.$$

Let $d = p_1 \cdots p_k \in \mathcal{D}^+$. If k is odd, then

$$d = p_1 \cdots p_{k-1} p_k < p_1 \cdots p_{k-1} p_k^{1+\beta} < D.$$

If k is even, then $k - 1$ is odd. Since $p_k < p_{k-1}$ and $\beta > 1$, we have

$$d = p_1 \cdots p_{k-1} p_k < p_1 \cdots p_{k-1}^2 < p_1 \cdots p_{k-1}^{1+\beta} < D.$$

Therefore, $1 \le d < D$ for all $d \in \mathcal{D}^+$.

Similarly, if $d = p_1 \cdots p_k \in \mathcal{D}^-$ and $k \ge 2$, then $1 \le d < D$. For $k = 1$, we have $d = p_1 < D$, that is, $\mathcal{D}^-$ contains all primes strictly less than D. Therefore, $1 \le d < D$ for all $d \in \mathcal{D}^-$.

The arithmetic functions $\lambda^+(d)$ and $\lambda^-(d)$ are truncations of the Möbius function $\mu(d)$ to certain subsets of the sets $\mathcal{D}^+$ and $\mathcal{D}^-$, respectively. Since both sets contain 1, we have

$$\lambda^+(1) = \lambda^-(1) = \mu(1) = 1.$$

Let $m \ge 2$. We must prove that

$$\sum_{d \mid m} \lambda^-(d) \le 0 \le \sum_{d \mid m} \lambda^+(d). \tag{9.8}$$

Since the functions $\lambda^{\pm}(d)$ are supported on divisors of $P(D)$, we may assume that m divides $P(D)$. Let $\omega(m)$ denote the number of distinct prime divisors of m. The proof is by induction on $k = \omega(m)$. If $k = 1$, then $m = p < D$ for some prime $p \in \mathcal{P}$, and so $m \in \mathcal{D}^-$. We have

$$\sum_{d \mid m} \lambda^-(d) = \mu(1) + \mu(p) = 0$$

and

$$\sum_{d \mid m} \lambda^+(d) = \mu(1) + \lambda^+(p) \ge 1 - 1 = 0.$$

This proves the lemma in the case $k = 1$.

Now let $k \geq 1$, and assume that inequalities (9.8) hold for all positive integers m with k distinct prime divisors. If $\omega(m) = k + 1$, then we can write m in the form

$$m = q_0 q_1 \cdots q_k,$$

where

$$q_k < q_{k-1} < \cdots < q_1 < q_0 < D,$$

$q_0, q_1, \ldots, q_k$ are prime numbers in $\mathcal{P}$, and q_0 is the greatest prime divisor of m. Let

$$m_1 = \frac{m}{q_0} = q_1 \cdots q_k.$$

Since m_1 is a divisor of $P(z)$ with k prime factors, it follows from the induction hypothesis that

$$\sum_{d \mid m_1} \lambda^-(d) \leq 0 \leq \sum_{d \mid m_1} \lambda^+(d).$$

Every divisor of m is of the form d or $q_0 d$, where d is a divisor of m_1. Therefore,

$$\sum_{d \mid m} \lambda^+(d) = \sum_{d \mid m_1} \lambda^+(d) + \sum_{d \mid m_1} \lambda^+(q_0 d)$$

$$\geq \sum_{d \mid m_1} \lambda^+(q_0 d)$$

$$= \sum_{\substack{d \mid m_1 \\ q_0 d \in \mathcal{D}^+}} \mu(q_0 d)$$

$$= - \sum_{\substack{d \mid m_1 \\ q_0 d \in \mathcal{D}^+}} \mu(d).$$

Similarly,

$$\sum_{d \mid m} \lambda^-(d) \leq - \sum_{\substack{d \mid m_1 \\ q_0 d \in \mathcal{D}^-}} \mu(d).$$

If d is a divisor of m_1, then

$$d = p_1 \cdots p_j,$$

where $p_1, \ldots, p_j$ are primes in $\mathcal{P}$ such that

$$p_j < \cdots < p_1 \leq q_1 < q_0 < D.$$

Let $D_1 = D/q_0 > 0$, and let $\mathcal{D}_1^+$ and $\mathcal{D}_1^-$ be the sets of integers constructed from β and D_1. Let $\lambda_1^+(d)$ and $\lambda_1^-(d)$ be the Möbius function truncated to the sets $\mathcal{D}_1^+$ and $\mathcal{D}_1^-$, respectively. Then $q_0 d \in \mathcal{D}^+$ if and only if

$$q_0 < \left(\frac{D}{q_0} \right)^{1/\beta}$$

and

$$p_m < \left(\frac{D}{q_0 p_1 \cdots p_m} \right)^{1/\beta} = \left(\frac{D_1}{p_1 \cdots p_m} \right)^{1/\beta}$$

for all *even* integers m. If

$$q_0 \geq \left(\frac{D}{q_0} \right)^{1/\beta},$$

then $q_0 d \notin \mathcal{D}^+$ and so

$$\sum_{\substack{d | m_1 \\ q_0 d \in \mathcal{D}^+}} \mu(d) = 0$$

since the sum is empty. If

$$q_0 < \left(\frac{D}{q_0} \right)^{1/\beta},$$

then $q_0 d \in \mathcal{D}^+$ if and only if $d \in \mathcal{D}_1^-$, and

$$\sum_{\substack{d | m_1 \\ q_0 d \in \mathcal{D}^+}} \mu(d) = \sum_{\substack{d | m_1 \\ d \in \mathcal{D}_1^-}} \mu(d) = \sum_{d | m_1} \lambda_1^-(d) \leq 0$$

by the induction hypothesis. Therefore,

$$\sum_{d | m} \lambda^+(d) \geq 0.$$

Similarly, $q_0 d \in \mathcal{D}^-$ if and only if $d \in \mathcal{D}_1^+$, and so

$$\sum_{\substack{d | m_1 \\ q_0 d \in \mathcal{D}^-}} \mu(d) = \sum_{\substack{d | m_1 \\ d \in \mathcal{D}_1^+}} \mu(d) = \sum_{d | m_1} \lambda_1^+(d) \geq 0.$$

This proves that $\lambda^+(d)$ and $\lambda^-(d)$ are upper and lower bound sieves with sieving range $\mathcal{P}$ and support level D.

Lemma 9.2 *Let $\mathcal{P}$ be a set of primes, and let $g(d)$ be a multiplicative function such that*

$$0 \leq g(p) < 1 \qquad \text{for all } p \in \mathcal{P}.$$

Let

$$V(z) = \prod_{\substack{p \in \mathcal{P} \\ p < z}} (1 - g(p)) = \sum_{p | P(z)} \mu(d) g(d).$$

Then $V(z)$ is a decreasing function of z,

$$0 < V(z) \leq 1$$

for all z, and

$$\sum_{\substack{p \in \mathcal{P} \\ w \leq p < z}} g(p) V(p) = V(w) - V(z) \qquad (9.9)$$

for all $1 \leq w < z$.

Proof. It follows immediately from the definition that $V(z)$ is decreasing and $V(z) \in (0, 1]$ for all z.

The proof of the combinatorial identity (9.9) is by induction on the number k of primes $p \in \mathcal{P}$ that lie in the interval $[w, z)$. If $k = 0$, then $V(w) = V(z)$ and

$$\sum_{\substack{p \in \mathcal{P} \\ w \leq p < z}} g(p)V(p) = 0.$$

If $k \geq 1$, let p_1 be the largest prime in the interval. Then

$$\sum_{\substack{p \in \mathcal{P} \\ w \leq p < z}} g(p)V(p) = \sum_{\substack{p \in \mathcal{P} \\ w \leq p < p_1}} g(p)V(p) + g(p_1)V(p_1)$$

$$= V(w) - V(p_1) + g(p_1)V(p_1)$$

$$= V(w) - (1 - g(p_1))V(p_1)$$

$$= V(w) - V(z).$$

Lemma 9.3 *Let $\mathcal{P}$ be a set of primes. For $\beta > 1$ and $2 \leq z \leq D$, let*

$$y_m = y_m(\beta, D, p_1, \ldots, p_m) = \left(\frac{D}{p_1 \cdots p_m}\right)^{1/\beta}.$$

Let $\lambda^{\pm}(d)$ be the upper and lower bound sieves constructed in Theorem 9.3, and let

$$G(z, \lambda^{\pm}) = \sum_{d \mid P(z)} \lambda^{\pm}(d)g(d).$$

Let

$$T_n(D, z) = \sum_{\substack{p_1, \ldots, p_n \in \mathcal{P} \\ y_n \leq p_n < \cdots < p_1 < z \\ p_m < y_m \forall m < n, m \equiv n \pmod 2}} g(p_1 \cdots p_n)V(p_n).$$

Then

$$G(z, \lambda^+) = V(z) + \sum_{\substack{n=1 \\ n \equiv 1 \pmod 2}}^{\infty} T_n(D, z) \tag{9.10}$$

and

$$G(z, \lambda^-) = V(z) - \sum_{\substack{n=1 \\ n \equiv 0 \pmod 2}}^{\infty} T_n(D, z). \tag{9.11}$$

Moreover,

$$T_n(D, z) \geq 0$$

for all $n \geq 1$, and

$$G(z, \lambda^-) \leq V(z) \leq G(z, \lambda^+).$$

If

$$\beta \leq \frac{\log D}{\log z} = s,$$

then

$$T_n(D, z) = 0 \qquad for\ n \leq s - \beta.$$

Proof. It follows from the construction of the sets $\mathcal{D}^{\pm}$ and the sieves $\lambda^{\pm}(d)$ that

$$G(z, \lambda^+) = \sum_{\substack{d \mid P(z) \\ d \in \mathcal{D}^+}} \mu(d) g_n(d)$$

$$= \sum_{\substack{p_k < \cdots < p_1 < z, \, p_i \in P \\ p_m < y_m \, \forall m \equiv 1 \ (\mathrm{mod}\ 2)}} (-1)^k g_n(p_1 \cdots p_k)$$

and

$$G(z, \lambda^-) = \sum_{\substack{d \mid P(z) \\ d \in \mathcal{D}^-}} \mu(d) g_n(d)$$

$$= \sum_{\substack{p_k < \cdots < p_1 < z, \, p_i \in P \\ p_m < y_m \, \forall m \equiv 0 \ (\mathrm{mod}\ 2)}} (-1)^k g_n(p_1 \cdots p_k).$$

We expand the function $V(z)$ to obtain a partition of $G(z, \lambda^+)$ as a sum of nonnegative functions:

$$V(z) = \sum_{d \mid P(z)} \mu(d) g(d)$$

$$= \sum_{\substack{p_k < \cdots < p_1 < z \\ p_i \in P}} (-1)^k g(p_1 \cdots p_k)$$

$$= \sum_{\substack{p_k < \cdots < p_1 < z, \, p_i \in P \\ p_m < y_m \, \forall m \equiv 1 \ (\mathrm{mod}\ 2)}} (-1)^k g(p_1 \cdots p_k)$$

$$+ \sum_{\substack{p_k < \cdots < p_1 < z, \, p_i \in P \\ \exists m \equiv 1 \ (\mathrm{mod}\ 2): y_m \le p_m}} (-1)^k g(p_1 \cdots p_k)$$

$$= G(z, \lambda^+) + \sum_{\substack{p_k < \cdots < p_1 < z, \, p_i \in P \\ \exists m \equiv 1 \ (\mathrm{mod}\ 2): y_m \le p_m}} (-1)^k g(p_1 \cdots p_k)$$

$$= G(z, \lambda^+) + \sum_{\substack{n=1 \\ n \equiv 1 \ (\mathrm{mod}\ 2)}}^{\infty} \sum_{\substack{p_k < \cdots < p_1 < z, \, p_i \in P \\ p_m < y_m \, \forall m < n, \, m \equiv 1 \ (\mathrm{mod}\ 2) \\ y_n \le p_n}} (-1)^k g(p_1 \cdots p_k)$$

$$= G(z, \lambda^+)$$

$$+ \sum_{\substack{n=1 \\ n \equiv 1 \ (\mathrm{mod}\ 2)}}^{\infty} \left(\sum_{\substack{y_n \le p_n < \cdots < p_1 < z \\ p_i \in P \\ p_m < y_m \, \forall m < n, \\ m \equiv 1 \ (\mathrm{mod}\ 2)}} (-1)^n g(p_1 \cdots p_n) \sum_{\substack{p_k < \cdots < p_{n+1} < p_n \\ p_i \in P}} (-1)^{k-n} g(p_k \cdots p_{n+1}) \right)$$

$$= G(z, \lambda^+) - \sum_{\substack{n=1 \\ n \equiv 1 \ (\mathrm{mod}\ 2)}}^{\infty} \sum_{\substack{y_n \le p_n < \cdots < p_1 < z, \, p_i \in P \\ p_m < y_m \, \forall m < n, \\ m \equiv 1 \ (\mathrm{mod}\ 2)}} g(p_1 \cdots p_n) V(p_n)$$

$$= G(z, \lambda^+) - \sum_{\substack{n=1 \\ n \equiv 1 \ (\mathrm{mod}\ 2)}}^{\infty} T_n(D, z),$$

where

$$T_n(D, z) = \sum_{\substack{y_n \le p_n < \cdots < p_1 < z, \, p_j \in \mathcal{P} \\ p_m < y_m \, \forall m < n, \, m \equiv n \pmod 2}} g(p_1 \cdots p_n) V(p_n) \ge 0.$$

Therefore,

$$G(z, \lambda^+) = V(z) + \sum_{\substack{n=1 \\ n \equiv 1 \pmod 2}}^{\infty} T_n(D, z) \ge V(z).$$

Similarly,

$$G(z, \lambda^-) = V(z) - \sum_{\substack{n=1 \\ n \equiv 0 \pmod 2}}^{\infty} T_n(D, z) \le V(z).$$

If

$$y_n \le p_n < \cdots < p_1 < z, \tag{9.12}$$

then

$$D \le p_1 \cdots p_n p_n^{\beta} < z^{n+\beta}.$$

Let $D = z^s$. Since $T_n(D, z)$ is a sum over integers $p_1 \cdots p_n$ that satisfy inequality (9.12), it follows that $T_n(D, z) = 0$ unless $s < n + \beta$. This completes the proof.

9.3   Approximations

For the rest of this chapter, we shall consider only the case

$$\beta = 2$$

in the construction of the sets $\mathcal{D}^\pm$ and the upper and lower bound sieves $\lambda^\pm(d)$. Then

$$y_m = \left(\frac{D}{p_1 \cdots p_m} \right)^{1/2},$$

and the functions $T_n(D, z)$ satisfy the following recursion relation.

Lemma 9.4 *Let $z \ge 2$ and D be real numbers such that*

$$s = \frac{\log D}{\log z} \ge \begin{cases} 1 & \text{if } n \text{ is odd}, \\ 2 & \text{if } n \text{ is even}. \end{cases}$$

Then

$$T_1(D, z) = V(D^{1/3}) - V(z). \tag{9.13}$$

Let $n \ge 2$. If n is even, or if n is odd and $s \ge 3$, then

$$T_n(D, z) = \sum_{\substack{p \in \mathcal{P} \\ p < z}} g(p) T_{n-1}\left(\frac{D}{p}, p \right). \tag{9.14}$$

If n is odd and $1 \leq s \leq 3$, then

$$T_n(D, z) = \sum_{\substack{p \in P \\ p < D^{1/3}}} g(p)T_{n-1}\left(\frac{D}{p}, p\right). \tag{9.15}$$

Proof. Since $y_1 = (D/p_1)^{1/2}$, it follows from Lemma 9.2 that

$$T_1(D, z) = \sum_{\substack{p_1 \in P \\ y_1 \leq p_1 < z}} g(p_1)V(p_1)$$

$$= \sum_{\substack{p_1 \in P \\ D^{1/3} \leq p_1 < z}} g(p_1)V(p_1)$$

$$= V(D^{1/3}) - V(z).$$

If n is even, then

$$T_n(D, z) = \sum_{\substack{p_n < \cdots < p_1 < z, p_i \in P \\ p_1 \cdots p_n p_n^2 \geq D \\ p_1 \cdots p_m p_m^2 < D \\ \forall m < n, m \equiv n \pmod 2}} g(p_1 \cdots p_n)V(p_n)$$

$$= \sum_{\substack{p_1 \in P \\ p_1 < z}} g(p_1) \sum_{\substack{p_n < \cdots < p_1, p_i \in P \\ p_2 \cdots p_n p_n^2 \geq D/p_1 \\ p_2 \cdots p_m p_m^2 < D/p_1 \forall 2 \leq m < n, \\ m-1 \equiv n-1 \pmod 2}} g(p_2 \cdots p_n)V(p_n)$$

$$= \sum_{\substack{p_1 \in P \\ p_1 < z}} g(p_1)T_{n-1}\left(\frac{D}{p_1}, p_1\right).$$

Let n be odd, $n \geq 3$. If $p_1 < y_1 = (D/p_1)^{1/2}$ and $p_1 < z = D^{1/s}$, then

$$p_1 < \min\left(D^{1/3}, D^{1/s}\right) = \begin{cases} D^{1/3} & \text{if } 1 \leq s \leq 3 \\ z & \text{if } s \geq 3 \end{cases}$$

and the argument proceeds exactly as in the case of even integers n. This completes the proof.

We shall now construct a sequence of continuous functions $f_n(s)$ that will be used later to approximate the discrete functions $T_n(D, z)$. For $s \geq 1$, let $\mathcal{R}_n(s)$ be the open convex region of Euclidean space consisting of all points $(t_1, \ldots, t_n) \in \mathbf{R}^n$ such that

$$0 < t_n < \cdots < t_1 < \frac{1}{s},$$

$$t_1 + \cdots + t_n + 2t_n > 1,$$

and

$$t_1 + \cdots + t_m + 2t_m < 1 \qquad \text{if } m < n \text{ and } m \equiv n \pmod 2. \tag{9.16}$$

For $n \geq 1$ and $s \geq 1$, we define the function $f_n(s)$ by the multiple integral

$$sf_n(s) = \int \cdots \int_{\mathcal{R}_n(s)} \frac{dt_1 \cdots dt_n}{(t_1 \cdots t_n)t_n}. \tag{9.17}$$

The function $f_n(s)$ is nonnegative, continuous, and decreasing, since $\mathcal{R}_n(s_2) \subseteq \mathcal{R}_n(s_1)$ for $s_1 \leq s_2$. If $f_n(s) > 0$, then $\mathcal{R}_n(s)$ is nonempty, so $\mathcal{R}_n(s)$ contains a point $(t_1, \ldots, t_n)$. This point satisfies

$$1 < t_1 + \cdots + t_n + 2t_n \leq (n+2)t_1 < \frac{n+2}{s},$$

and so

$$\frac{1}{n+2} < t_1 < \frac{1}{s}. \tag{9.18}$$

It follows that

$$f_n(s) = 0 \qquad \text{for } s \geq n+2. \tag{9.19}$$

It is easy to compute $f_1(s)$ and $f_2(s)$. We have $f_1(s) = 0$ for $s \geq 3$. For $1 \leq s \leq 3$, we have

$$\mathcal{R}_1(s) = (1/3, 1/s)$$

and so

$$sf_1(s) = \int_{1/3}^{1/s} \frac{dt_1}{t_1^2} = 3 - s. \tag{9.20}$$

Similarly, $f_2(s) = 0$ for $s \geq 4$. For $2 \leq s \leq 4$, we have

$$\mathcal{R}_2(s) = \left\{ (t_1, t_2) : \frac{1}{4} < t_1 < \frac{1}{s} \quad \text{and} \quad \frac{1-t_1}{3} < t_2 < t_1 \right\}$$

and so

$$sf_2(s) = \int_{1/4}^{1/s} \int_{(1-t_1)/3}^{t_1} \frac{dt_2\, dt_1}{t_2^2\, t_1}$$

$$= \int_{1/4}^{1/s} \left(\frac{3}{1-t_1} - \frac{1}{t_1} \right) \frac{dt_1}{t_1}$$

$$= \int_{1/4}^{1/s} \left(\frac{3}{1-t_1} + \frac{3}{t_1} - \frac{1}{t_1^2} \right) dt_1$$

$$= s - 3\log(s-1) + 3\log 3 - 4.$$

The functions $f_n(s)$ satisfy the following recursion relation.

Lemma 9.5 *Let $n \geq 2$. If n is even and $s \geq 2$, or if n is odd and $s \geq 3$, then*

$$sf_n(s) = \int_s^\infty f_{n-1}(t-1)dt. \tag{9.21}$$

If n is odd and $1 \leq s \leq 3$, then

$$sf_n(s) = 3f_n(3) = \int_3^\infty f_{n-1}(t-1)dt. \tag{9.22}$$

Proof. If n is even and $s \geq 2$, or if n is odd and $s \geq 3$, then, from (9.18), we have

$$sf_n(s) = \int \cdots \int_{\mathcal{R}_n(s)} \frac{dt_1 \cdots dt_n}{(t_1 \cdots t_n)t_n}$$

$$= \int_{1/(n+2)}^{1/s} \left(\int_{\substack{0 < t_n < \cdots < t_2 < t_1 \\ t_2 + \cdots + t_n + 2t_n > 1 - t_1 \\ t_2 + \cdots + t_m + 2t_m < 1 - t_1 \\ \forall 1 < m < n, m \equiv n \pmod 2}} \frac{dt_2 \cdots dt_n}{(t_2 \cdots t_n)t_n} \right) \frac{dt_1}{t_1}.$$

In the inner integral, we make the change of variables

$$t_i = (1 - t_1)u_{i-1}$$

for $i = 2, \ldots, n$. Let

$$s_1 = \frac{1 - t_1}{t_1} = \frac{1}{t_1} - 1.$$

Since $t_1 < 1/s$, it follows that $s_1 > 1$ if n is even and $s \geq 2$, and $s_1 > 2$ if n is odd and $s \geq 3$. We obtain

$$\int_{\substack{0 < t_n < \cdots < t_2 < t_1 \\ t_2 + \cdots + t_n + 2t_n > 1 - t_1 \\ t_2 + \cdots + t_m + 2t_m < 1 - t_1 \\ \forall 1 < m < n, m \equiv n \pmod 2}} \frac{dt_2 \cdots dt_n}{(t_2 \cdots t_n)t_n}$$

$$= \int_{\substack{0 < u_{n-1} < \cdots < u_1 < t_1/(1-t_1) \\ u_1 + \cdots + u_{n-1} + 2u_{n-1} > 1 \\ u_1 + \cdots + u_{m-1} + 2u_{m-1} < 1 \\ \forall 1 < m < n, m-1 \equiv n-1 \pmod 2}} \frac{du_1 \cdots du_{n-1}}{(1 - t_1)(u_1 \cdots u_{n-1})u_{n-1}}$$

$$= \frac{1}{1 - t_1} \int_{\substack{0 < u_{n-1} < \cdots < u_1 < 1/s_1 \\ u_1 + \cdots + u_{n-1} + 2u_{n-1} > 1 \\ u_1 + \cdots + u_{m-1} + 2u_{m-1} < 1 \\ \forall 1 < m < n, m-1 \equiv n-1 \pmod 2}} \frac{du_1 \cdots du_{n-1}}{(u_1 \cdots u_{n-1})u_{n-1}}$$

$$= \frac{1}{1 - t_1} \int_{\mathcal{R}_{n-1}(s_1)} \frac{du_1 \cdots du_{n-1}}{(u_1 \cdots u_{n-1})u_{n-1}}$$

$$= \frac{s_1}{1 - t_1} f_{n-1}(s_1)$$

$$= \frac{1}{t_1} f_{n-1}\left(\frac{1}{t_1} - 1\right).$$

Setting $t = 1/t_1$, we obtain

$$sf_n(s) = \int_{1/(n+2)}^{1/s} \frac{1}{t_1} f_{n-1}\left(\frac{1}{t_1} - 1\right) \frac{dt_1}{t_1}$$

$$= -\int_{n+2}^{s} f_{n-1}(t - 1)dt$$

$$= \int_{s}^{n+2} f_{n-1}(t - 1)dt$$

$$= \int_{s}^{\infty} f_{n-1}(t - 1)dt,$$

since $f_{n-1}(t-1) = 0$ for $t - 1 \geq (n-1) + 2$ by (9.19).

Let $n \geq 3$ be an odd integer. If $(t_1, \ldots, t_n) \in \mathcal{R}_n(s)$, then $t_1 < 1/s$. Also, it follows from inequality (9.16) with $m = 1$ that $t_1 < 1/3$, and so $t_1 < 1/\max(s, 3)$. Therefore, if $1 \leq s \leq 3$, then

$$\mathcal{R}_n(s) = \mathcal{R}_n(3)$$

and

$$sf_n(s) = \int \cdots \int_{\mathcal{R}_n(s)} (t_1 \cdots t_n)^{-1} t_n^{-1} dt_1 \cdots dt_n$$

$$= \int \cdots \int_{\mathcal{R}_n(3)} (t_1 \cdots t_n)^{-1} t_n^{-1} dt_1 \cdots dt_n$$

$$= 3 f_n(3).$$

This completes the proof.

We construct the function $h(s)$ for $s \geq 1$ as follows:

$$h(s) = \begin{cases} e^{-2} & \text{for } 1 \leq s \leq 2 \\ e^{-s} & \text{for } 2 \leq s \leq 3 \\ 3s^{-1}e^{-s} & \text{for } s \geq 3. \end{cases} \qquad (9.23)$$

It is easy to check (Exercise 8) that

$$h(s-1) < 4h(s) \qquad \text{for } s \geq 2.$$

For $s \geq 2$, let

$$H(s) = \int_s^\infty h(t-1) dt.$$

Both $h(s)$ and $H(s)$ are continuous, positive, and decreasing functions on their domains. Let

$$\alpha = \frac{H(2)}{2h(2)} = \frac{e^2 H(2)}{2} = 1 - \frac{1}{2e} + \frac{3e^2}{2} \int_3^\infty t^{-1} e^{-t} dt.$$

We can express α in terms of the exponential integral

$$\text{Ei}(x) = \int_{-\infty}^x e^t t^{-1} dt$$

since

$$\int_3^\infty e^{-t} t^{-1} dt = -\text{Ei}(-3) = 0.013048 \ldots.$$

We can obtain this number with technology, such as Maple, or without technology, either by estimating the integral directly or by looking it up in old books, such as Dwight's *Mathematical Tables*[26, page 107]. We find that

$$\alpha = 0.96068 \ldots. \qquad (9.24)$$

Lemma 9.6

$$H(s) \le \alpha sh(s) \qquad \text{for } s \ge 2 \qquad (9.25)$$

and

$$H(3) \le \alpha sh(s) \qquad \text{for } 1 \le s \le 3. \qquad (9.26)$$

Proof. If $s \ge 3$, then $h(s-1) \le e^{1-s}$ and

$$H(s) \le \int_s^\infty e^{1-t}dt = e^{1-s} = \frac{esh(s)}{3} < \alpha sh(s).$$

For $2 \le s \le 3$, let

$$H_0(s) = \alpha sh(s) - H(s).$$

We have

$$s - 1 = 1 + (s-2) \le e^{s-2},$$

and so

$$(1-s)e^{-s} \ge -e^{-2}.$$

Then

$$\begin{aligned}
H_0'(s) &= \alpha h(s) + \alpha sh'(s) - H'(s) \\
&= \alpha(1-s)e^{-s} + h(s-1) \\
&\ge (1-\alpha)e^{-2} \\
&> 0,
\end{aligned}$$

and so $H_0(s)$ is increasing for $2 \le s \le 3$. Since

$$H_0(2) = 0$$

by the definition of α, it follows that

$$H(3) \le H(s) \le \alpha sh(s) \qquad \text{for } 2 \le s \le 3.$$

Let $1 \le s \le 2$. Since $\alpha < 1$, it follows that $h(2) > H(2)/2$ and

$$H(3) = H(2) - e^{-2} = H(2) - h(2) < \frac{H(2)}{2} = \alpha h(2) \le \alpha sh(2) = \alpha sh(s).$$

This completes the proof.

Lemma 9.7 *If n is odd and $s \ge 1$, or if n is even and $s \ge 2$, then*

$$f_n(s) \le 2e^2\alpha^{n-1}h(s).$$

Proof. This is by induction on n. For $n = 1$, we shall show that

$$sf_1(s) \le 2e^2 sh(s).$$

For $1 \leq s \leq 3$, we have $sf_1(s) = 3 - s$ by (9.20). If $1 \leq s \leq 2$, then $h(s) = e^{-2}$ and

$$sf_1(s) = 3 - s \leq 2 = 2e^2h(s) \leq 2e^2sh(s).$$

If $2 \leq s \leq 3$, then $h(s) = e^{-s}$ and

$$sf_1(s) = 3 - s \leq 1 < 4e^2h(s) \leq 2e^2sh(s).$$

If $s \geq 3$, then $f_1(s) = 0$ and

$$sf_1(s) = 0 \leq 2e^2sh(s).$$

This proves the case $n = 1$.

Now let $n \geq 2$, and assume that the lemma holds for $n - 1$. By (9.21) and (9.25), if n is even and $s \geq 2$, or if n is odd and $s \geq 3$, then

$$
\begin{aligned}
sf_n(s) &= \int_s^\infty f_{n-1}(t-1)dt \\
&\leq 2e^2\alpha^{n-2}\int_s^\infty h(t-1)dt \\
&= 2e^2\alpha^{n-2}H(s) \\
&\leq 2e^2\alpha^{n-2}\alpha sh(s) \\
&\leq 2e^2\alpha^{n-1}sh(s).
\end{aligned}
$$

By (9.22) and (9.26), if n is odd and $1 \leq s \leq 3$, then

$$
\begin{aligned}
sf_n(s) &= \int_3^\infty f_{n-1}(t-1)dt \\
&\leq 2e^2\alpha^{n-2}\int_3^\infty h(t-1)dt \\
&\leq 2e^2\alpha^{n-2}H(3) \\
&\leq 2e^2\alpha^{n-2}\alpha sh(s) \\
&\leq 2e^2\alpha^{n-1}sh(s).
\end{aligned}
$$

This completes the proof.

Theorem 9.4 *For $s \geq 1$, the function*

$$F(s) = 1 + \sum_{\substack{n=1 \\ n \equiv 1 \pmod 2}}^\infty f_n(s) \tag{9.27}$$

is continuous and differentiable, and

$$F(s) = 1 + O\left(e^{-s}\right).$$

For $s \geq 2$, the function

$$f(s) = 1 - \sum_{\substack{n=2 \\ n \equiv 0 \pmod 2}}^{\infty} f_n(s) \tag{9.28}$$

is continuous and differentiable, and

$$f(s) = 1 + O\left(e^{-s}\right).$$

Proof. By Lemma 9.7,

$$0 \leq f_n(s) \leq 2e^2 \alpha^{n-1} h(s) \leq 2e^2 \alpha^{n-1} e^{-s}$$

for $s \geq (3 + (-1)^n)/2$. Therefore,

$$\sum_{n=1}^{\infty} f_n(s) \ll e^{-s}.$$

The theorem follows immediately from this inequality.

9.4   The Jurkat–Richert theorem

From now on, we shall consider only arithmetic functions $g(d)$ that satisfy the linear sieve inequality (9.29).

Lemma 9.8 *Let $z \geq 2$ and $1 < w < z$. Let $\mathcal{P}$ be a set of primes, and let $g(d)$ be a multiplicative function such that*

$$0 \leq g(p) < 1 \qquad \text{for all } p \in \mathcal{P}$$

and

$$\prod_{\substack{p \in \mathcal{P} \\ u \leq p < z}} (1 - g(p))^{-1} \leq K \frac{\log z}{\log u} \tag{9.29}$$

for some $K > 1$ and all u such that $1 < u < z$. Let

$$V(z) = \prod_{\substack{p \in \mathcal{P} \\ p < z}} (1 - g(p)),$$

and let Φ be a continuous, increasing function on the interval $[w, z]$. Then

$$\sum_{\substack{p \in \mathcal{P} \\ w \leq p < z}} g(p) V(p) \Phi(p) \leq (K - 1) V(z) \Phi(z) - K V(z) \int_w^z \Phi(u) d\left(\frac{\log z}{\log u}\right).$$

Proof. The step function

$$S(u) = \sum_{\substack{p \in \mathcal{P} \\ u \le p < z}} g(p) V(p)$$

is nonnegative and decreasing. By Lemma 9.2 and inequality (9.29),

$$
\begin{aligned}
S(u) &= V(u) - V(z) \\
&= \left(\frac{V(u)}{V(z)} - 1 \right) V(z) \\
&= \left(\prod_{\substack{p \in \mathcal{P} \\ u \le p < z}} (1 - g(p))^{-1} - 1 \right) V(z) \\
&\le \left(K \frac{\log z}{\log u} - 1 \right) V(z).
\end{aligned}
$$

Let

$$w \le p_k < p_{k-1} < \cdots < p_1 < z$$

be all the primes in $\mathcal{P}$ that lie in the interval $[w, z)$. Then $S(p_k) = S(w)$, $S(p_1) = g(p_1) V(p_1)$, and $S(u) = 0$ for $p_1 < u \le z$. By partial summation and integration by parts of the Riemann–Stieltjes integral,

$$
\begin{aligned}
\sum_{\substack{p \in \mathcal{P} \\ w \le p < z}} g(p) V(p) \Phi(p) &= \sum_{i=1}^{k} g(p_i) V(p_i) \Phi(p_i) \\
&= \sum_{i=2}^{k} (S(p_i) - S(p_{i-1})) \Phi(p_i) + S(p_1) \Phi(p_1) \\
&= \sum_{i=1}^{k} S(p_i) \Phi(p_i) - \sum_{i=1}^{k-1} S(p_i) \Phi(p_{i+1}) \\
&= S(p_k) \Phi(p_k) + \sum_{i=1}^{k-1} S(p_i) (\Phi(p_i) - \Phi(p_{i+1})) \\
&= S(w) \Phi(w) + S(p_k) (\Phi(p_k) - \Phi(w)) \\
&\quad + \sum_{i=1}^{k-1} S(p_i) (\Phi(p_i) - \Phi(p_{i+1})) \\
&= S(w) \Phi(w) + \int_{w}^{p_1} S(u) d\Phi(u) \\
&= S(w) \Phi(w) + \int_{w}^{z} S(u) d\Phi(u) \\
&= S(z) \Phi(z) - \int_{w}^{z} \Phi(u) dS(u) \\
&\le (K - 1) V(z) \Phi(z) - K V(z) \int_{w}^{z} \Phi(u) d\left(\frac{\log z}{\log u} \right).
\end{aligned}
$$

This completes the proof.

Theorem 9.5 *Let $z \geq 2$, and let D be a real number such that $D \geq z$ for n odd and $D \geq z^2$ for n even, that is,*

$$s = \frac{\log D}{\log z} \geq \left\{ \begin{array}{ll} 1 & \text{if n is odd} \\ 2 & \text{if n is even.} \end{array} \right.$$

Let $\mathcal{P}$ be a set of primes, and let $g(d)$ be a multiplicative function such that

$$0 \leq g(p) < 1 \qquad \text{for all } p \in \mathcal{P}$$

and

$$\prod_{\substack{p \in \mathcal{P} \\ u \leq p < z}} (1 - g(p))^{-1} \leq K \frac{\log z}{\log u}$$

for all u such that $1 < u < z$, where the constant K satisfies

$$1 < K < 1 + \frac{1}{200}.$$

Then

$$T_n(D, z) < Vz) \left(f_n(s) + (K - 1) \left(\frac{99}{100} \right)^n e^{10-s} \right). \tag{9.30}$$

Proof. We define the number

$$\tau = \alpha + 5(K - 1) + 11e^{-8}$$

and the functions

$$h_n(s) = (K - 1)\tau^n e^{10} h(s) \tag{9.31}$$

for $n \geq 1$. Note that

$$\alpha < \tau < 0.9607 + 0.0250 + 0.0037 = 0.9894 < \frac{99}{100}.$$

We shall prove that

$$T_n(D, z) < V(z)(f_n(s) + h_n(s)). \tag{9.32}$$

This immediately implies (9.30) since $h(s) \leq e^{-s}$ for all $s \geq 1$.

The proof of (9.32) is by induction on n. Let $n = 1$. By Lemma 9.3 with $\beta = 2$, we have $T_1(D, z) = 0$ for $s > 3$. Since the right side of inequality (9.32) is positive, it follows that the inequality holds for $s > 3$. If $1 \leq s \leq 3$, then $f_1(s) = (3/s) - 1$ and

$$T_1(D, z) = V(D^{1/3}) - V(z)$$

by (9.13). It follows that

$$\frac{T_1(D, z)}{V(z)} = \frac{V(D^{1/3})}{V(z)} - 1$$

$$= \prod_{D^{1/3} \leq p < z} (1 - g(p))^{-1} - 1$$

$$\leq 3K \frac{\log z}{\log D} - 1$$

$$= \frac{3K}{s} - 1$$

$$= \left(\frac{3}{s} - 1\right) + \frac{3}{s}(K - 1)$$

$$\leq f_1(s) + 3(K - 1)$$

$$< f_1(s) + h_1(s)$$

since $h(s) \geq e^{-3}$ and $\tau > 11e^{-8}$, hence

$$h_1(s) = (K - 1)\tau e^{10} h(s) > (K - 1)11e^{-1} > 3(K - 1).$$

This proves the lemma for $n = 1$.

Let $n \geq 2$, and assume that the lemma holds for $n - 1$. For n even and $s \geq 2$, or for n odd and $s \geq 3$, we define the function

$$\Phi(u) = f_{n-1}\left(\frac{\log D}{\log u} - 1\right) + h_{n-1}\left(\frac{\log D}{\log u} - 1\right)$$

for $1 < u \leq w$. The function $\Phi(u)$ is continuous, positive, and increasing. Moreover,

$$\Phi(z) = f_{n-1}(s - 1) + h_{n-1}(s - 1).$$

It follows from the recursion formula (9.14), the induction hypothesis for $n - 1$, and Lemma 9.8 that

$$T_n(D, z) = \sum_{\substack{p \in P \\ p < z}} g(p) T_{n-1}\left(\frac{D}{p}, p\right)$$

$$< \sum_{\substack{p \in P \\ p < z}} g(p) V(p) \left(f_{n-1}\left(\frac{\log D}{\log p} - 1\right) + h_{n-1}\left(\frac{\log D}{\log p} - 1\right) \right)$$

$$= \sum_{\substack{p \in P \\ p < z}} g(p) V(p) \Phi(p)$$

$$= (K - 1)V(z)\Phi(z) - KV(z)\int_1^z \Phi(u) d\left(\frac{\log z}{\log u}\right)$$

$$= (K - 1)V(z)(f_{n-1}(s - 1) + h_{n-1}(s - 1))$$

$$- \frac{KV(z)}{s} \int_1^z \Phi(u) d\left(\frac{\log D}{\log u}\right)$$

$$= (K-1)V(z)\left(f_{n-1}(s-1) + h_{n-1}(s-1)\right)$$
$$+ \frac{KV(z)}{s}\int_s^\infty (f_{n-1}(t-1) + h_{n-1}(t-1))\,dt,$$

where the last equation comes from substituting $t = \log D/\log u$ in the integral. By (9.21), we have

$$\frac{K}{s}\int_s^\infty f_{n-1}(t-1)\,dt = Kf_n(s).$$

Similarly, from the definition of $H(s)$ and (9.25), we have

$$\int_s^\infty h(t-1)\,dt = H(s) \le \alpha s h(s)$$

and so

$$\frac{K}{s}\int_s^\infty h_{n-1}(t-1)\,dt \le \alpha K h_{n-1}(s).$$

Since $h(s-1) < 4h(s)$ for $s \ge 2$, we have

$$(K-1)h_{n-1}(s-1) < 4(K-1)h_{n-1}(s)$$

and

$$(K-1)f_{n-1}(s-1) \le (K-1)2e^2\alpha^{n-2}h(s-1)$$
$$< 8e^2(K-1)\alpha^{n-2}h(s)$$
$$= 8e^{-8}\left(\frac{\alpha}{\tau}\right)^{n-1}\alpha^{-1}(K-1)e^{10}\tau^{n-1}h(s)$$
$$< 9e^{-8}h_{n-1}(s)$$

since $0 < \alpha < \tau$ and $\alpha^{-1} < 9/8$. Therefore,

$$\frac{T_n(D,z)}{V(z)} < Kf_n(s) + \left(\alpha K + 4(K-1) + 9e^{-8}\right)h_{n-1}(s).$$

By Lemma 9.7 and definition (9.31), we have

$$(K-1)f_n(s) \le (K-1)2e^2\alpha^{n-1}h(s) < 2e^{-8}h_{n-1}(s),$$

and so

$$Kf_n(s) < f_n(s) + 2e^{-8}h_{n-1}(s).$$

Since

$$\alpha K = K - (1-\alpha)K < K - (1-\alpha) = (K-1) + \alpha,$$

we have

$$\frac{T_n(D,z)}{V(z)} < f_n(s) + \left(\alpha + 5(K-1) + 11e^{-8}\right)h_{n-1}(s)$$
$$= f_n(s) + \tau h_{n-1}(s)$$
$$= f_n(s) + h_n(s).$$

Let $n \geq 3$ be odd, and let $1 \leq s \leq 3$. If $z = D^{1/3}$, then $\log D / \log z = 3$. By the recursion formula (9.15) and the same argument used above, we obtain

$$T_n(D, z) = \sum_{\substack{p \in \mathcal{P} \\ p < D^{1/3}}} g(p) T_{n-1}\left(\frac{D}{p}, p\right)$$

$$< \sum_{\substack{p \in \mathcal{P} \\ p < D^{1/3}}} g(p) V(p) \Phi(p)$$

$$< (f_n(3) + h_n(3)) V(z)$$

$$\leq (f_n(s) + h_n(s)) V(z)$$

since the functions $f_n(s)$ and $h(s)$ are decreasing. This completes the proof.

Theorem 9.6 *Let $z, D, s, \mathcal{P}, g(d)$, and $K = 1 + \varepsilon$ satisfy the hypotheses of Theorem 9.5. Let*

$$G(z, \lambda^{\pm}) = \sum_{d \mid P(z)} \lambda^{\pm}(d) g(d).$$

Then

$$G(z, \lambda^+) < V(z)\left(F(s) + \varepsilon e^{14-s}\right)$$

and

$$G(z, \lambda^-) > V(z)\left(f(s) - \varepsilon e^{14-s}\right),$$

where $F(s)$ and $f(s)$ are the continuous functions defined by (9.27) and (9.28).

Proof. We note that the sum of the following geometric series satisfies

$$\sum_{\substack{n=0 \\ n \equiv 0 \ (\text{mod } 2)}}^{\infty} \left(\frac{99}{100}\right)^n < 51 < e^4.$$

By (9.10) and Theorem 9.5,

$$G(z, \lambda^+) = V(z) + \sum_{\substack{n=1 \\ n \equiv 1 \ (\text{mod } 2)}}^{\infty} T_n(D, z)$$

$$< V(z)\left(1 + \sum_{\substack{n=1 \\ n \equiv 1 \ (\text{mod } 2)}}^{\infty} f_n(s) + \varepsilon e^{10-s} \sum_{\substack{n=1 \\ n \equiv 1 \ (\text{mod } 2)}}^{\infty} \left(\frac{99}{100}\right)^n\right)$$

$$< V(z)\left(F(s) + \varepsilon e^{14-s}\right).$$

Similarly, by (9.11) and Theorem 9.5,

$$G(z, \lambda^-) = V(z) - \sum_{\substack{n=1 \\ n \equiv 0 \ (\text{mod } 2)}}^{\infty} T_n(D, z)$$

$$> V(z) \left(1 - \sum_{\substack{n=1 \\ n \equiv 0 \ (\mathrm{mod}\ 2)}}^{\infty} f_n(s) - \varepsilon e^{10-s} \sum_{\substack{n=1 \\ n \equiv 0 \ (\mathrm{mod}\ 2)}}^{\infty} \left(\frac{99}{100} \right)^n \right)$$

$$> V(z) \left(f(s) - \varepsilon e^{14-s} \right).$$

This completes the proof.

Theorem 9.7 (Jurkat–Richert) *Let* $A = \{a(n)\}_{n=1}^{\infty}$ *be an arithmetic function such that*

$$a(n) \geq 0 \qquad \text{for all } n$$

and

$$|A| = \sum_{n=1}^{\infty} a(n) < \infty.$$

Let $\mathcal{P}$ *be a set of prime numbers and, for* $z \geq 2$, *let*

$$P(z) = \prod_{\substack{p \in \mathcal{P} \\ p < z}} p.$$

Let

$$S(A, \mathcal{P}, z) = \sum_{\substack{n=1 \\ (n, P(z))=1}}^{\infty} a(n).$$

For every $n \geq 1$, *let* $g_n(d)$ *be a multiplicative function such that*

$$0 \leq g_n(p) < 1 \qquad \text{for all } p \in \mathcal{P}. \tag{9.33}$$

Define $r(d)$ *by*

$$|A_d| = \sum_{\substack{n=1 \\ d|n}}^{\infty} a(n) = \sum_{n=1}^{\infty} a(n) g_n(d) + r(d).$$

Let Q *be a finite subset of* $\mathcal{P}$, *and let* Q *be the product of the primes in* Q. *Suppose that, for some* ε *satisfying* $0 < \varepsilon < 1/200$, *the inequality*

$$\prod_{\substack{p \in \mathcal{P} \backslash Q \\ u \leq p < z}} (1 - g_n(p))^{-1} < (1 + \varepsilon) \frac{\log z}{\log u} \tag{9.34}$$

holds for all n *and* $1 < u < z$. *Then for any* $D \geq z$ *there is the upper bound*

$$S(A, \mathcal{P}, z) < (F(s) + \varepsilon e^{14-s}) X + R, \tag{9.35}$$

and for any $D \geq z^2$ *there is the lower bound*

$$S(A, \mathcal{P}, z) > (f(s) - \varepsilon e^{14-s}) X - R, \tag{9.36}$$

where

$$s = \frac{\log D}{\log z},$$

$f(s)$ and $F(s)$ are the continuous functions defined by (9.27) and (9.28),

$$X = \sum_{n=1}^{\infty} a(n) \prod_{p|P(z)} (1 - g_n(p)), \tag{9.37}$$

and the remainder term is

$$R = \sum_{\substack{d|P(z) \\ d<DQ}} |r(d)|.$$

If there is a multiplicative function $g(d)$ such that $g_n(d) = g(d)$ for all n, then

$$X = V(z)|A|, \tag{9.38}$$

where

$$V(z) = \prod_{p|P(z)} (1 - g(p)).$$

Proof. Let $\mathcal{P}_1 = \mathcal{P} \setminus \mathcal{Q}$. By Theorem 9.3, there exist upper and lower bound sieves $\lambda^{\pm}(d)$ with sieving range $\mathcal{P}_1$ and support level D, and with $|\lambda^{\pm}(d)| \leq 1$ for all $d \geq 1$. We define

$$G_n(z, \lambda^{\pm}) = \sum_{p|P_1(z)} \lambda_1^{\pm}(d)g_n(d)$$

and

$$V_n(z) = \prod_{p|P_1(z)} (1 - g_n(p)).$$

Since $\mathcal{P}_1$ and $\mathcal{Q}$ are disjoint sets of primes, we have

$$\prod_{p|P(z)} (1 - g_n(p)) = V_n(z) \prod_{q|Q(z)} (1 - g_n(q)).$$

By Theorem 9.6,

$$G_n(z, \lambda^+) < V_n(z) \left(F(s) + \varepsilon e^{14-s} \right)$$

and

$$G_n(z, \lambda^-) > V_n(z) \left(F(s) - \varepsilon e^{14-s} \right).$$

It follows from Theorem 9.2 that

$$S(A, \mathcal{P}, z) \leq \sum_{n=1}^{\infty} a(n)G_n(z, \lambda_1^+) \prod_{q|Q(z)} (1 - g_n(q)) + R$$

$$< (F(s) + \varepsilon e^{14-s}) \sum_{n=1}^{\infty} a(n)V_n(z) \prod_{q|Q(z)} (1 - g_n(q)) + R$$

$$= (F(s) + \varepsilon e^{14-s}) \sum_{n=1}^{\infty} a(n) \prod_{p|P(z)} (1 - g_n(p)) + R$$

$$= (F(s) + \varepsilon e^{14-s})X + R.$$

The lower bound is obtained similarly. This completes the proof.

9.5   Differential-difference equations

In this section, we shall compute initial values for the functions

$$F(s) = 1 + \sum_{\substack{n=1 \\ n \equiv 1 \ (\text{mod } 2)}}^{\infty} f_n(s) \qquad \text{for } s \geq 1$$

and

$$f(s) = 1 - \sum_{\substack{n=2 \\ n \equiv 0 \ (\text{mod } 2)}}^{\infty} f_n(s) \qquad \text{for } s \geq 2.$$

We shall prove that

$$F(s) = \frac{2e^{\gamma}}{s} \qquad \text{for } 1 \leq s \leq 3$$

and

$$f(s) = \frac{2e^{\gamma} \log(s-1)}{s} \qquad \text{for } 2 \leq s \leq 4,$$

where γ is Euler's constant. We define $f(s) = 0$ for $1 \leq s \leq 2$.

Lemma 9.9

$$s F(s) = 3F(3)$$

for $1 \leq s \leq 3$.

Proof. Let $1 \leq s \leq 3$. By Lemma 9.5,

$$s f_n(s) = 3 f_n(3) \qquad \text{for all odd } n \geq 3.$$

Since

$$s + s f_1(s) = 3$$

by (9.20), it follows that

$$s F(s) = s + s f_1(s) + \sum_{\substack{n=3 \\ n \equiv 1 \ (\text{mod } 2)}}^{\infty} s f_n(s)$$

$$= 3 + \sum_{\substack{n=3 \\ n \equiv 1 \ (\text{mod } 2)}}^{\infty} 3 f_n(3)$$

$$= 3F(3),$$

which completes the proof.

Define the constants A and B by

$$A = s F(s) \qquad \text{for } 1 \leq s \leq 3$$

and

$$B = 2f(2).$$

Lemma 9.10 *The functions $F(s)$ and $f(s)$ are solutions of the system of differential-difference equations*

$$(sF(s))' = f(s-1) \qquad \text{for } s > 3$$
$$(sf(s))' = F(s-1) \qquad \text{for } s > 2.$$

Proof. Let $n \geq 2$. By Lemma 9.5, for n odd and $s \geq 3$, or for n even and $s \geq 2$, we have

$$sf_n(s) = \int_s^\infty f_{n-1}(t-1)dt$$

and so

$$(sf_n(s))' = -f_{n-1}(s-1).$$

For $s > 3$, we have $sf_1(s) = 0$ and so

$$(sF(s))' = \left(s + \sum_{\substack{n=1 \\ n\equiv 1 \ (\text{mod } 2)}}^{\infty} sf_n(s) \right)'$$

$$= \left(s + \sum_{\substack{n=3 \\ n\equiv 1 \ (\text{mod } 2)}}^{\infty} sf_n(s) \right)'$$

$$= 1 - \sum_{\substack{n=3 \\ n\equiv 1 \ (\text{mod } 2)}}^{\infty} f_{n-1}(s-1)$$

$$= 1 - \sum_{\substack{n=2 \\ n\equiv 0 \ (\text{mod } 2)}}^{\infty} f_n(s-1)$$

$$= f(s-1).$$

Similarly, for $s > 2$ we have

$$(sf(s))' = \left(s - \sum_{\substack{n=2 \\ n\equiv 0 \ (\text{mod } 2)}}^{\infty} sf_n(s) \right)'$$

$$= 1 + \sum_{\substack{n=2 \\ n\equiv 0 \ (\text{mod } 2)}}^{\infty} f_{n-1}(s-1)$$

$$= 1 + \sum_{\substack{n=1 \\ n\equiv 1 \ (\text{mod } 2)}}^{\infty} f_n(s-1)$$

$$= F(s-1).$$

This completes the proof.

Lemma 9.11 *For $s \geq 2$, let*

$$P(s) = F(s) + f(s)$$

and

$$Q(s) = F(s) - f(s).$$

For $s > 3$, the functions $P(s)$ and $Q(s)$ are the unique solutions of the differential-difference equations

$$sP'(s) = -P(s) + P(s - 1) \tag{9.39}$$

and

$$sQ'(s) = -Q(s) - Q(s - 1) \tag{9.40}$$

that satisfy the initial conditions

$$sP(s) = A + B + A \log(s - 1)$$

and

$$sQ(s) = A - B - A \log(s - 1)$$

for $2 \leq s \leq 3$. Moreover,

$$P(s) = 2 + O(e^{-s})$$

and

$$Q(s) = O(e^{-s}).$$

Proof. Since

$$sF(s) = A \qquad \text{for } 1 \leq s \leq 3,$$

it follows that

$$F(s) = \frac{A}{s} \qquad \text{for } 1 \leq s \leq 3$$

or, equivalently, that

$$F(s - 1) = \frac{A}{s - 1} \qquad \text{for } 2 \leq s \leq 4.$$

Since $(sf(s))' = F(s - 1)$ for $s > 2$, it follows that

$$sf(s) = 2f(2) + \int_2^s \frac{A}{t - 1} dt = B + A \log(s - 1)$$

for $2 \leq s \leq 4$. Since

$$sF(s) = A \qquad \text{for } 1 \leq s \leq 3,$$

it follows that

$$sP(s) = A + B + A \log(s - 1) \tag{9.41}$$

and

$$sQ(s) = A - B - A \log(s - 1) \tag{9.42}$$

for $2 \leq s \leq 3$. For $s > 3$, we have

$$(sP(s))' = (sF(s))' + (sf(s))' = f(s-1) + F(s-1) = P(s-1),$$

and so

$$sP'(s) = -P(s) + P(s-1).$$

Similarly,

$$(sQ(s))' = (sF(s))' - (sf(s))' = f(s-1) - F(s-1) = -Q(s-1)$$

and so

$$sQ'(s) = -Q(s) - Q(s-1).$$

By Theorem 9.4, we have $F(s) = 1 + O(e^{-s})$ and $f(s) = 1 + O(e^{-s})$, and so $P(s) = 2 + O(e^{-s})$ and $Q(s) = O(e^{-s})$. This completes the proof.

The differential-difference equations (9.39) and (9.40) are of the form

$$sR'(s) = -aR(s) - bR(s-1). \tag{9.43}$$

Associated with this equation is the *adjoint equation*

$$(sr(s))' = ar(s) + br(s+1). \tag{9.44}$$

To every solution $R(s)$ of equation (9.43) and every solution $r(s)$ of equation (9.44), we associate the function

$$\langle R(s), r(s) \rangle = sR(s)r(s) - b \int_{s-1}^{s} R(x)r(x+1)dx$$

for $s \geq 3$. Differentiating with respect to s, we obtain

$$\frac{d}{ds}\langle R(s), r(s) \rangle$$

$$= R(s)r(s) + sR'(s)r(s) + sR(s)r'(s) - bR(s)r(s+1) + bR(s-1)r(s)$$

$$= (sR'(s) + bR(s-1))r(s) + (r(s) + sr'(s) - br(s+1))R(s)$$

$$= -aR(s)r(s) + aR(s)r(s)$$

$$= 0.$$

Therefore, $\langle R(s), r(s) \rangle$ is constant for $s \geq 3$.

The equation adjoint to (9.40) is

$$(sq(s))' = q(s) + q(s+1)$$

or, equivalently,

$$sq'(s) = q(s+1).$$

This has the solution

$$q(s) = s - 1.$$

Clearly,

$$q(s) \sim s$$

as s tends to infinity, and

$$q(1) = 0.$$

Since $Q(s) = O(e^{-s})$, it follows that

$$s Q(s) q(s) = O\left(s^2 e^{-s}\right) = o(1)$$

and

$$\int_{s-1}^{s} Q(x) q(x+1) dx = o(1).$$

Therefore,

$$\lim_{s \to \infty} \langle Q(s), q(s) \rangle = 0.$$

Since $\langle Q(s), q(s) \rangle$ is constant for $s \geq 3$, it follows that

$$\langle Q(s), q(s) \rangle = 0$$

for $s \geq 3$. This implies that $B = 0$, since $(x Q(x))' = -(x-1)^{-1}$ by (9.42), and

$$\begin{aligned}
0 &= \langle Q(3), q(3) \rangle \\
&= 3 Q(3) q(3) - \int_{2}^{3} Q(x) q(x+1) dx \\
&= 3 Q(3) q(3) - \int_{2}^{3} x Q(x) q'(x) dx \\
&= 3 Q(3) q(3) - [x Q(x) q(x)]_{x=2}^{x=3} + \int_{2}^{3} (x Q(x))' q(x) dx \\
&= 2 Q(2) q(2) - A \int_{2}^{3} \frac{q(x)}{x-1} dx \\
&= (A - B) - A \\
&= B.
\end{aligned}$$

Similarly, the equation adjoint to (9.39) is

$$(sp(s))' = p(s) - p(s+1)$$

or, equivalently,

$$sp'(s) = -p(s+1). \tag{9.45}$$

For $s > 0$, we introduce the function

$$p(s) = \int_{0}^{\infty} \exp\left(-sx - I(x)\right) dx, \tag{9.46}$$

where

$$I(x) = \int_0^x (1 - e^{-t})t^{-1}dt.$$

Since

$$0 < \frac{1 - e^{-t}}{t} < 1 \qquad \text{for } t > 0,$$

we have

$$0 < I(x) < x \qquad \text{for } x > 0,$$

and so

$$\exp(-(s+1)x) < \exp(-sx - I(x)) < \exp(-sx).$$

Therefore, the integral converges for all $s > 0$, and

$$\frac{1}{s+1} = \int_0^\infty \exp(-(s+1)x)\,dx < p(s) < \int_0^\infty \exp(-sx)\,dx = \frac{1}{s}.$$

It follows that

$$sp(s) \sim 1$$

as s tends to infinity. Using integration by parts and the observation that

$$x I'(x) = 1 - e^{-x},$$

we obtain

$$sp'(s) = - \int_0^\infty sx \exp(-sx - I(x))\,dx$$

$$= \int_0^\infty \left(\frac{d}{dx} \exp(-sx) \right) x \exp(-I(x))\,dx$$

$$= \left[x \exp(-sx - I(x)) \right]_{x=0}^\infty - \int_0^\infty \exp(-sx) \left(\frac{d}{dx} x \exp(-I(x)) \right) dx$$

$$= - \int_0^\infty \exp(-sx)(1 - x I'(x)) \exp(-I(x))\,dx$$

$$= - \int_0^\infty \exp(-sx) \exp(-x) \exp(-I(x))\,dx$$

$$= - \int_0^\infty \exp(-(s+1)x - I(x))\,dx$$

$$= -p(s+1).$$

This proves that $p(s)$ is a solution to the adjoint equation (9.45) for all $s > 0$.
   We shall prove that

$$p(1) = e^\gamma.$$

We need the following integral representation for Euler's constant:

$$\gamma = \int_0^1 (1 - e^{-t})t^{-1}dt - \int_1^\infty e^{-t}t^{-1}dt \qquad (9.47)$$

(see Exercise 16 and Gradshteyn and Ryzhik [42, page 956]). Then

$$
\begin{aligned}
I(x) &= \int_0^x (1 - e^{-t})t^{-1}dt \\
&= \int_0^1 (1 - e^{-t})t^{-1}dt + \int_1^x (1 - e^{-t})t^{-1}dt \\
&= \int_0^1 (1 - e^{-t})t^{-1}dt - \int_1^x e^{-t}t^{-1}dt + \log x \\
&= \int_0^1 (1 - e^{-t})t^{-1}dt - \int_1^\infty e^{-t}t^{-1}dt + \int_x^\infty e^{-t}t^{-1}dt + \log x \\
&= \gamma + \int_x^\infty e^{-t}t^{-1}dt + \log x.
\end{aligned}
$$

It follows that

$$
\begin{aligned}
-sp'(s) &= \int_0^\infty sx \exp(-sx - I(x))dx \\
&= e^{-\gamma} \int_0^\infty s \exp\left(-sx - \int_x^\infty e^{-t}t^{-1}dt\right) dx \\
&= e^{-\gamma} \int_0^\infty \exp\left(-u - \int_{u/s}^\infty e^{-t}t^{-1}dt\right) du.
\end{aligned}
$$

For $u > 0$, we have

$$
\lim_{s\to 0^+} \int_{u/s}^\infty e^{-t}t^{-1}dt = 0,
$$

and so

$$
\begin{aligned}
p(1) &= \lim_{s\to 0^+} p(s+1) \\
&= -\lim_{s\to 0^+} sp'(s) \\
&= e^{-\gamma} \lim_{s\to 0^+} \int_0^\infty \exp\left(-u - \int_{u/s}^\infty e^{-t}t^{-1}dt\right) du \\
&= e^{-\gamma} \int_0^\infty \lim_{s\to 0^+} \exp\left(-u - \int_{u/s}^\infty e^{-t}t^{-1}dt\right) du \\
&= e^{-\gamma} \int_0^\infty \exp(-u)du \\
&= e^{-\gamma}.
\end{aligned}
$$

Since $P(s) = 2 + O(e^{-s})$ and $sp(s) \sim 1$, it follows that

$$
\lim_{s\to\infty} \langle P(s), p(s)\rangle = \lim_{s\to\infty} \left(sP(s)p(s) + \int_{s-1}^s P(x)p(x+1)dx\right) = 2.
$$

Since $\langle P(s), p(s) \rangle$ is constant for $s \geq 3$, it follows that

$$\langle P(s), p(s) \rangle = 2$$

for all $s \geq 3$. Letting $B = 0$ in (9.41), we have

$$sP(s) = A + A \log(s - 1)$$

and

$$(sP(s))' = \frac{A}{s - 1}$$

for $2 \leq s \leq 3$. Therefore, $2P(2) = A$ and

$$
\begin{aligned}
2 &= \langle P(3), p(3) \rangle \\
&= 3P(3)p(3) + \int_2^3 P(x)p(x + 1)dx \\
&= 3P(3)p(3) - \int_2^3 xP(x)p'(x)dx \\
&= 3P(3)p(3) - [xP(x)p(x)]_{x=2}^{x=3} + \int_2^3 (xP(x))'p(x)dx \\
&= 2P(2)p(2) + A \int_2^3 \frac{p(x)}{x - 1}dx \\
&= Ap(2) + A \int_2^3 \frac{p(x)}{x - 1}dx \\
&= Ap(2) - A \int_2^3 p'(x - 1)dx \\
&= Ap(2) - Ap(2) + Ap(1) \\
&= Ae^{-\gamma}.
\end{aligned}
$$

This proves that

$$A = 2e^{\gamma}.$$

We can now determine the initial values of $F(s)$ and $f(s)$.

Theorem 9.8

$$F(s) = \frac{2e^{\gamma}}{s} \qquad for\ 1 \leq s \leq 3$$

and

$$f(s) = \frac{2e^{\gamma} \log(s - 1)}{s} \qquad for\ 2 \leq s \leq 4,$$

where γ is Euler's constant.

Proof. Let $2 \leq s \leq 3$, and let $A = 2e^{\gamma}$ and $B = 0$ in (9.41) and (9.42). Then

$$sP(s) = 2e^{\gamma} + 2e^{\gamma} \log(s - 1)$$

and

$$s\,Q(s) = 2e^\gamma - 2e^\gamma \log(s-1).$$

Therefore,

$$s\,F(s) = \frac{s\,P(s) + s\,Q(s)}{2} = 2e^\gamma.$$

By Lemma 9.9, $s\,F(s)$ is constant for $1 \le s \le 3$ and so

$$s\,F(s) = 2e^\gamma \qquad \text{for } 1 \le s \le 3.$$

By Lemma 9.10, we have $(s f(s))' = F(s-1)$ for $s > 2$ and so

$$s f(s) = 2f(2) + \int_2^s F(t-1)\,dt = \int_2^s \frac{2e^\gamma}{t-1}\,dt = 2e^\gamma \log(s-1)$$

for $2 \le s \le 4$. This completes the proof.

9.6 Notes

The material in this chapter is based on unpublished lecture notes of Henryk Iwaniec[68]. See Jurkat and Richert [69] for the original proof of Theorem 9.7. Standard references on sieve methods are the monographs of Halberstam and Richert [44] and Motohashi [87].

9.7 Exercises

1. Let P be the product of the primes up to $\sqrt{x}$. Prove Legendre's formula

$$\pi(x) - \pi(\sqrt{x}) + 1$$

$$= [x] - \sum_{p_1 \le \sqrt{x}} \left[\frac{x}{p_1}\right] + \sum_{p_2 < p_1 \le \sqrt{x}} \left[\frac{x}{p_1 p_2}\right] - \sum_{p_3 < p_2 < p_1 \le \sqrt{x}} \left[\frac{x}{p_1 p_2 p_3}\right] + \cdots$$

$$= \sum_{d \mid P} \mu(d) \left[\frac{x}{d}\right].$$

2. Let P be the product of the primes up to $\sqrt{x}$. Prove Sylvester's formula

$$\sum_{\sqrt{x} < p \le x} p + 1 = \frac{1}{2} \sum_{d \mid P} \mu(d) \left[\frac{x}{d}\right] \left(\left[\frac{x}{d}\right] + 1\right).$$

3. Let $A_1 = \{a_1(n)\}$ and $A_2 = \{a_2(n)\}$ be arithmetic functions such that $a_1(n) \le a_2(n)$ for all $n \ge 1$. Prove that

$$S(A_1, \mathcal{P}, z) \le S(A_2, \mathcal{P}, z).$$

4. Let $A_\ell = \{a_\ell(n)\}$ be a nonnegative arithmetic function for $\ell = 1, \ldots, k$, and let $A = \{a(n)\}$ be the arithmetic function defined by $a(n) = a_1(n) + \cdots + a_k(n)$ for all n. Prove that

$$S(A, \mathcal{P}, z) = \sum_{\ell=1}^{k} S(A_\ell, \mathcal{P}, z).$$

5. Let $2 \le w < z$. Prove Buchstab's identity:

$$S(A, \mathcal{P}, z) = S(A, \mathcal{P}, w) - \sum_{w \le p < z} S(A_p, \mathcal{P}, p).$$

In particular,

$$S(A, \mathcal{P}, z) = |A| - \sum_{p < z} S(A_p, \mathcal{P}, p).$$

6. By iterating the Buchstab identity, prove that, for $z_1 \le z$,

$$S(A, \mathcal{P}, z) \le |A| - \sum_{p_1 < z_1} |A_{p_1}| + \sum_{p_2 < p_1 < z_1} |A_{p_1 p_2}|$$
$$- \sum_{p_3 < p_2 < p_1 < z} S(A_{p_1 p_2 p_3}, \mathcal{P}, p_3).$$

7. Let $\mathcal{P}$ be a set of primes, and let $\lambda^\pm(d)$ be upper and lower bound sieves with sieving range $\mathcal{P}$ and support level D. Let $\mathcal{P}_1$ be a subset of $\mathcal{P}$. We define functions $\lambda_1^\pm(d)$ by $\lambda_1^\pm(d) = \lambda^\pm(d)$ if d is divisible only by primes in $\mathcal{P}_1$, and $\lambda_1^\pm(d) = 0$ otherwise. Prove that $\lambda_1^\pm(d)$ are upper and lower bound sieves with sieving range $\mathcal{P}_1$ and support level D.

8. Let $h(s)$ be the function defined by 9.23. Prove that

$$h(s - 1) < 4h(s) \qquad \text{for } s \ge 2.$$

9. Use the recurrence relation

$$sf_2(s) = \int_s^\infty f_1(t - 1)dt$$

to prove that

$$sf_2(s) = s - 3\log(s - 1) + 3\log 3 - 4$$

for $2 \le s \le 4$.

10. Prove that

$$f(x) = x \log \frac{9x}{9x - 1} \le \log \frac{9}{8}$$

for $x \ge 1$. Hint: Show that the function $f(x)$ is decreasing for $x \ge 1$.

11. Let $Q(s)$ be a continuous function on the interval $[1, 2]$. Prove that there exists a unique continuous function $Q(s)$ defined for all $s \geq 1$ that satisfies this initial condition and that is a solution of the differential-difference equation

$$s Q'(s) = -Q(s) - Q(s - 1)$$

for all $s > 2$. Hint: For $2 < s \leq 3$, we must have

$$s Q(s) = - \int_2^s Q(x - 1)dx + 2Q(2).$$

Similarly, for $3 < s \leq 4$, we must have

$$s Q(s) = - \int_3^s Q(x - 1)dx + 3Q(3).$$

The proof proceeds by induction.

12. Let $Q(s)$ be the function defined in Lemma 9.11. Prove that

$$s(s - 1)Q(s) = \int_{s-1}^s x Q(x)dx$$

for all $s \geq 3$. Prove that

$$0 < s Q(s) \ll s^{-s}.$$

13. Let $\mathcal{P}_1$ and $\mathcal{P}_2$ be disjoint sets of prime numbers, and let f_1 and f_2 be arithmetic functions such that $f_1(d) \neq 0$ only if d is a product of primes belonging to $\mathcal{P}_1$ and $f_2(d) \neq 0$ only if d is a product of primes belonging to $\mathcal{P}_2$. Let $f = f_1 * f_2$. Prove that

$$1 * f = (1 * f_1)(1 * f_2).$$

14. Let $\lambda_1^+(d)$ and $\lambda_2^+(d)$ be upper bound sieves with support levels D_1 and D_2, respectively, and with disjoint sieving ranges $\mathcal{P}_1$ and $\mathcal{P}_2$. Let $\lambda^+ + (d)$ be the convolution of $\lambda_1^+(d)$ and $\lambda_2^+(d)$, that is,

$$\lambda^+(d) = \lambda_1^+ * \lambda_2^+(d) = \sum_{d=d_1 d_2} \lambda_1^+(d_1)\lambda_2^+(d_2).$$

Prove that λ^+ is an upper bound sieve with support level $D = D_1 D_2$ and sieving range $\mathcal{P}_1 \cup \mathcal{P}_2$.

15. Let $\lambda_1^+(d)$ and $\lambda_2^+(d)$ be upper bound sieves with support levels D_1 and D_2, respectively, and with disjoint sieving ranges $\mathcal{P}_1$ and $\mathcal{P}_2$, and let $\lambda_1^-(d)$ and $\lambda_2^-(d)$ be lower bound sieves with support levels D_1 and D_2, respectively, and with disjoint sieving ranges $\mathcal{P}_1$ and $\mathcal{P}_2$. Prove that

$$\lambda^-(d) = \lambda_1^1 * \lambda_2^+(d) - \lambda_1^+ * \lambda_2^+(d) + \lambda_1^+ * \lambda_2^-(d)$$

Prove that λ^- is a lower bound sieve with support level $D = D_1 D_2$ and sieving range $\mathcal{P}_1 \cup \mathcal{P}_2$.

16. In the theory of the Gamma function, it is proved that

$$-\gamma = \Gamma'(1) = \int_0^\infty e^{-x} \log x\, dx.$$

From this formula, use integration by parts to obtain (9.47):

$$\gamma = \int_0^1 (1 - e^{-t})t^{-1}\, dt - \int_1^\infty e^{-t}t^{-1}\, dt.$$

10

Chen's theorem

Is it even true that every even n is the sum of 2 primes? To show this seems to transcend our present mathematical powers.... The prime numbers remain very elusive fellows.

H. Weyl [142]

10.1 Primes and almost primes

In this chapter, we shall prove one of the most famous results in additive prime number theory: Chen's theorem that every sufficiently large even integer can be written as the sum of an odd prime and a number that is either prime or the product of two primes. An integer that is the product of at most r not necessarily distinct prime numbers is called an *almost prime of order* r, denoted P_r, and so Chen's theorem can be written in the form

$$N = p + P_2$$

for every sufficiently large even integer N. We shall prove not only that every large even integer N has at least one representation as the sum of a prime and an almost prime of order two but that there are, in fact, many such representations.

Theorem 10.1 (Chen) *Let $r(N)$ denote the number of representations of N in the form*

$$N = p + n,$$

where p is an odd prime and n is the product of at most two primes. Then

$$r(N) \gg \mathfrak{S}(N) \frac{2N}{(\log N)^2},$$ (10.1)

where

$$\mathfrak{S}(N) = \prod_{p>2} \left(1 - \frac{1}{(p-1)^2}\right) \prod_{\substack{p|N \\ p>2}} \frac{p-1}{p-2}.$$ (10.2)

The number $\mathfrak{S}(N)$ is called the *singular series* for the Goldbach conjecture.

The proof has two ingredients. The first is the Jurkat–Richert theorem (Theorem 9.7), which gives upper and lower bounds for the linear sieve. The second is the Bombieri–Vinogradov theorem, which describes the average distribution of prime numbers in arithmetic progressions. Throughout this chapter, p and q denote prime numbers.

10.2   Weights

Let N be an even integer, $N \geq 4^8$. We begin by assigning a weight $w(n)$ to every positive integer n. Let

$$z = N^{1/8}$$ (10.3)

and

$$y = N^{1/3}.$$ (10.4)

Then $z \geq 4$. We define

$$w(n) = 1 - \frac{1}{2} \sum_{\substack{z \leq q < y \\ q^k \| n}} k - \frac{1}{2} \sum_{\substack{p_1 p_2 p_3 = n \\ z \leq p_1 < y \leq p_2 \leq p_3}} 1.$$ (10.5)

Clearly,

$$w(n) \leq 1$$

for all n, and $w(n) = 1$ if and only if n is divisible by no prime in the interval $[z, y)$.

Let $\mathcal{P}$ be the set of prime numbers that do not divide N. Then $2 \notin \mathcal{P}$ since N is even. Let

$$P(z) = \prod_{\substack{p \in \mathcal{P} \\ p < z}} p.$$

Let n be a positive integer such that

$$n < N \quad \text{and} \quad (n, N) = (n, P(z)) = 1.$$

Then n is divisible only by primes $p \geq z$ that do not divide N. If $n = p_1 p_2 \cdots p_r p_{r+1} \cdots p_{r+s}$, where

$$z \leq p_1 \leq \cdots \leq p_r < y \leq p_{r+1} \leq \cdots \leq p_{r+s},$$

then

$$N^{s/3} = y^s \leq p_{r+1} \cdots p_{r+s} \leq n < N$$

and so $s = 0$, 1, or 2. Suppose that $w(n) > 0$. Since

$$\frac{1}{2} \sum_{\substack{q^k \| n \\ z \leq q < y}} k = \frac{r}{2},$$

it follows that $r = 0$ or 1. If $r = 1$ and $s = 2$, then $n = p_1 p_2 p_3$, where $z \leq p_1 < y \leq p_2 \leq p_3$, and so $w(n) = 0$. Therefore, if $w(n) > 0$, then either $r = 0$ and $s = 0$, 1, or 2, or $r = 1$ and $s = 0$ or 1. In all of these cases, $r + s \leq 2$. Therefore, if $(n, N) = (n, P(z)) = 1$ and $w(n) > 0$, then either $n = 1$ or n is an integer of the form p_1 or $p_1 p_2$, where p_1 and p_2 are primes $\geq z$ that do not divide N.

Consider the set

$$\mathcal{A} = \{N - p : p \leq N, p \in \mathcal{P}\}. \tag{10.6}$$

Then $\mathcal{A}$ is a finite set of positive integers, and $|\mathcal{A}| = \pi(N) - \omega(N)$, where $\omega(N)$ denotes the number of distinct prime divisors of N. If $n = N - p \in \mathcal{A}$ and if $(n, N) > 1$, then p divides N and so $p \notin \mathcal{P}$, which is absurd. Therefore, $(n, N) = 1$ for all $n \in \mathcal{A}$. We obtain a lower bound for $r(N)$ as follows.

$$
\begin{aligned}
r(N) &\geq \sum_{\substack{N=p+n \\ n \in \{1, p_1, p_1 p_2 : p_1, p_2 \geq z\}}} 1 \\
&\geq \sum_{\substack{n \in \mathcal{A} \\ n \in \{1, p_1, p_1 p_2 : p_1, p_2 \geq z\}}} 1 \\
&= \sum_{\substack{n \in \mathcal{A} \\ (n, P(z))=1 \\ n \in \{1, p_1, p_1 p_2 : p_1, p_2 \geq z\}}} 1 \\
&\geq \sum_{\substack{n \in \mathcal{A} \\ (n, P(z))=1 \\ n \in \{1, p_1, p_1 p_2 : p_1, p_2 \geq z\}}} w(n) \\
&\geq \sum_{\substack{n \in \mathcal{A} \\ (n, P(z))=1}} w(n) \\
&= \sum_{\substack{n \in \mathcal{A} \\ (n, P(z))=1}} \left(1 - \frac{1}{2} \sum_{\substack{z \leq q < y \\ q^k \| n}} k - \frac{1}{2} \sum_{\substack{p_1 p_2 p_3 = n \\ z \leq p_1 < y \leq p_2 \leq p_3}} 1 \right) \\
&= \left(\sum_{\substack{n \in \mathcal{A} \\ (n, P(z))=1}} 1 \right) - \frac{1}{2} \left(\sum_{\substack{n \in \mathcal{A} \\ (n, P(z))=1}} \sum_{\substack{z \leq q < y \\ q^k \| n}} k \right) \\
&\quad - \frac{1}{2} \left(\sum_{\substack{n \in \mathcal{A} \\ (n, P(z))=1}} \sum_{\substack{p_1 p_2 p_3 = n \\ z \leq p_1 < y \leq p_2 \leq p_3}} 1 \right).
\end{aligned}
$$

We shall express these three sums as sieving functions. If we let $A = \{a(n)\}_{n=1}^{\infty}$ be the characteristic function of the finite set $\mathcal{A}$, then the first sum becomes simply

$$\sum_{\substack{n \in \mathcal{A} \\ (n,P(z))=1}} 1 = \sum_{(n,P(z))=1} a(n) = S(A, \mathcal{P}, z).$$

We divide the second sum into two pieces:

$$\sum_{\substack{n \in \mathcal{A} \\ (n,P(z))=1}} \sum_{\substack{z \leq q < y \\ q^k \| n}} k = \left(\sum_{\substack{n \in \mathcal{A} \\ (n,P(z))=1}} \sum_{\substack{z \leq q < y \\ q | n}} 1 \right) + \left(\sum_{\substack{n \in \mathcal{A} \\ (n,P(z))=1}} \sum_{\substack{z \leq q < y \\ q^k \| n \\ k \geq 2}} (k-1) \right).$$

The first piece can be expressed as a sieving function as follows: For every prime q, let $A_q = \{a_q(n)\}_{n=1}^{\infty}$ be the arithmetic function defined by

$$a_q(n) = \begin{cases} 1 & \text{if } n \in \mathcal{A} \text{ and } q | n \\ 0 & \text{otherwise.} \end{cases}$$

Since $(n, N) = 1$ for all $n \in \mathcal{A}$, we have $q \in \mathcal{P}$ if $a_q(n) = 1$, and

$$\sum_{\substack{n \in \mathcal{A} \\ (n,P(z))=1}} \sum_{\substack{z \leq q < y \\ q | n}} 1 = \sum_{z \leq q < y} \sum_{(n,P(z))=1} a_q(n)$$

$$= \sum_{z \leq q < y} S(A_q, \mathcal{P}, z).$$

It is easy to estimate the second piece. Since $z = N^{1/8} \geq 4$ and

$$\sum_{k=2}^{\infty} \frac{k-1}{q^k} = \frac{1}{(q-1)^2},$$

we have

$$\sum_{\substack{n \in \mathcal{A} \\ (n,P(z))=1}} \sum_{\substack{z \leq q < y \\ q^k \| n \\ k \geq 2}} (k-1) = \sum_{z \leq q < y} \sum_{k=2}^{\infty} \sum_{\substack{n \in \mathcal{A} \\ (n,P(z))=1 \\ q^k \| n}} (k-1)$$

$$\leq \sum_{z \leq q < y} \sum_{k=2}^{\infty} \sum_{\substack{n < N \\ q^k \| n}} (k-1)$$

$$< N \sum_{z \leq q < y} \sum_{k=2}^{\infty} \frac{k-1}{q^k}$$

$$= N \sum_{z \leq q <} \frac{1}{(q-1)^2}$$

$$< \frac{N}{z-2}$$

$$\leq \frac{2N}{z}$$

$$= 2N^{7/8}.$$

For the third sum, we let B be the set of all positive integers of the form

$$N - p_1 p_2 p_3,$$

where the primes p_1, p_2, p_3 satisfy the conditions

$$z \le p_1 < y \le p_2 \le p_3$$

$$p_1 p_2 p_3 < N$$

$$(p_1 p_2 p_3, N) = 1.$$

Let $B = \{b(n)\}_{n=1}^{\infty}$ be the characteristic function of the finite set B. An element of B is a prime p if and only if $p < N$ and $N - p = p_1 p_2 p_3 \in A$, where $z \le p_1 < y \le p_2 \le p_3$. Therefore,

$$
\begin{aligned}
\sum_{\substack{n \in A \\ (n, P(z))=1}} \sum_{\substack{p_1 p_2 p_3 = n \\ z \le p_1 < y \le p_2 \le p_3}} 1 &= \sum_{\substack{p_1 p_2 p_3 \in A \\ z \le p_1 < y \le p_2 \le p_3}} 1 \\
&= \sum_{p \in B} 1 - \sum_{\substack{p \in B \\ p < y}} 1 + \sum_{\substack{p \in B \\ p \ge y}} 1 \\
&< y + \sum_{\substack{p \in B \\ p \ge y}} 1 \\
&\le y + \sum_{\substack{n \in B \\ (n, P(y))=1}} 1 \\
&= y + \sum_{\substack{(n, P(y))=1}} b(n) \\
&= N^{1/3} + S(B, \mathcal{P}, y).
\end{aligned}
$$

We now have a lower bound for $r(N)$ in terms of sieving functions.

Theorem 10.2

$$r(N) > S(A, \mathcal{P}, z) - \frac{1}{2} \sum_{z \le q < y} S(A_q, \mathcal{P}, z) - \frac{1}{2} S(B, \mathcal{P}, y) - 2N^{7/8} - N^{1/3}.$$

We shall obtain a lower bound for $S(A, \mathcal{P}, z)$ and upper bounds for $\sum_q S(A_q, \mathcal{P}, z)$ and $S(B, \mathcal{P}, y)$.

10.3 Prolegomena to sieving

In applying the linear sieve to estimate the three sieving functions, we choose the multiplicative function

$$g(d) = g_n(d) = \frac{1}{\varphi(d)}$$

for all $n \geq 1$. Since N is even, we have $2 \notin \mathcal{P}$ and

$$0 < g(p) = \frac{1}{p-1} < 1 \qquad \text{for all } p \in \mathcal{P},$$

so the functions $g(d)$ satisfy (9.33). To establish inequality (9.34), we apply Theorem 6.9, which says that there exists a number $u_1(\varepsilon)$ such that

$$\prod_{u \leq p < z} \left(1 - \frac{1}{p}\right)^{-1} < (1 + \varepsilon/3) \frac{\log z}{\log u}$$

for any $u_1(\varepsilon) \leq u < z$. Also, there exists $u_2(\varepsilon)$ such that

$$\prod_{p \geq u_2(\varepsilon)} \frac{(p-1)^2}{p(p-2)} = \prod_{p \geq u_2(\varepsilon)} \left(1 + \frac{1}{p(p-2)}\right) < 1 + \frac{\varepsilon}{3}$$

since the infinite product converges. Therefore, for

$$u \geq u_0(\varepsilon) = \max(u_1(\varepsilon), u_2(\varepsilon))$$

we have

$$\prod_{u \leq p < z} (1 - g(p))^{-1} = \prod_{u \leq p < z} \left(1 - \frac{1}{p-1}\right)^{-1}$$

$$= \prod_{u \leq p < z} \frac{(p-1)^2}{p(p-2)} \prod_{u \leq p < z} \left(1 - \frac{1}{p}\right)^{-1}$$

$$< (1 + \varepsilon/3)^2 \frac{\log z}{\log u}$$

$$< (1 + \varepsilon) \frac{\log z}{\log u}.$$

Let $\mathcal{Q}(\varepsilon)$ be the set of all primes $p < u_0(\varepsilon)$, and let $\mathcal{Q} = \mathcal{P} \cap \mathcal{Q}(\varepsilon)$. This gives (9.34). Let $Q(\varepsilon)$ be the product of the primes in $\mathcal{Q}(\varepsilon)$, and let Q be the product of the primes in $\mathcal{Q}$. Then $Q(\varepsilon)$ depends only on ε, not on N, and so

$$Q \leq Q(\varepsilon) < \log N \tag{10.7}$$

for all sufficiently large integers N.

Theorem 10.3 *Let N be an even positive integer, and let*

$$V(z) = \prod_{p \mid P(z)} (1 - g(p)) = \prod_{\substack{p < z \\ (p,N)=1}} \left(1 - \frac{1}{p-1}\right) \tag{10.8}$$

Then

$$V(z) = \mathfrak{S}(N) \frac{e^{-\gamma}}{\log z} \left(1 + O\left(\frac{1}{\log N}\right)\right),$$

where

$$\mathfrak{S}(N) = \prod_{p>2}\left(1 - \frac{1}{(p-1)^2}\right)\prod_{\substack{p\mid N \\ p>2}}\frac{p-1}{p-2}.$$

Proof. Let

$$W(z) = \prod_{2<p<z}\left(1 - \frac{1}{p-1}\right).$$

Then

$$\frac{V(z)}{W(z)} = \prod_{\substack{2<p<z \\ p\mid N}}\left(1 - \frac{1}{p-1}\right)^{-1}$$

$$= \prod_{\substack{p>2 \\ p\mid N}}\left(1 - \frac{1}{p-1}\right)^{-1}\prod_{\substack{p\geq z \\ p\mid N}}\left(1 - \frac{1}{p-1}\right)$$

$$= \prod_{\substack{p>2 \\ p\mid N}}\frac{p-1}{p-2}\prod_{\substack{p\geq z \\ p\mid N}}\left(1 - \frac{1}{p-1}\right).$$

Since $1 - x > e^{-2x}$ for $0 < x < (\log 2)/2$ and $1 - x < e^{-x}$ for all x, we have

$$\prod_{\substack{p\geq z \\ p\mid N}}\left(1 - \frac{1}{p-1}\right) > \prod_{\substack{p\geq z \\ p\mid N}}\exp\left(-\frac{2}{p-1}\right)$$

$$= \exp\left(-2\sum_{\substack{p\geq z \\ p\mid N}}\frac{1}{p-1}\right)$$

$$\geq \exp\left(\frac{-2\omega(N)}{z-1}\right)$$

$$> \exp\left(\frac{-8\log N}{z}\right)$$

$$= \exp\left(\frac{-8\log N}{N^{1/8}}\right)$$

$$> 1 - \frac{8\log N}{N^{1/8}}.$$

Thus,

$$\frac{V(z)}{W(z)} = \prod_{\substack{p>2 \\ p\mid N}}\frac{p-1}{p-2}\left(1 + O\left(\frac{\log N}{N^{1/8}}\right)\right).$$

To estimate $W(z)$, we see that

$$W(z)\prod_{p<z}\left(1 - \frac{1}{p}\right)^{-1} = \prod_{2<p<z}\left(1 - \frac{1}{p-1}\right)\prod_{p<z}\left(1 - \frac{1}{p}\right)^{-1}$$

$$= 2 \prod_{2 < p < z} \frac{p(p-2)}{(p-1)^2}$$

$$= 2 \prod_{2 < p < z} \left(1 - \frac{1}{(p-1)^2} \right)$$

$$= 2 \prod_{p > 2} \left(1 - \frac{1}{(p-1)^2} \right) \prod_{p \geq z} \left(1 + \frac{1}{p(p-2)} \right).$$

Since $1 + x < e^x < 1 + 2x$ for $0 < x < \log 2$, it follows that

$$\prod_{p \geq z} \left(1 + \frac{1}{p(p-2)} \right) < \exp \left(\sum_{p \geq z} \frac{1}{p(p-2)} \right)$$

$$< \exp \left(\sum_{n \geq z} \frac{1}{n(n-2)} \right)$$

$$\leq \exp \left(\frac{1}{2(z-2)} \right)$$

$$\leq \exp \left(\frac{1}{z} \right)$$

$$< 1 + \frac{2}{z}.$$

By Mertens's formula (Theorem 6.8), we obtain

$$W(z) = 2 \prod_{p > 2} \left(1 - \frac{1}{(p-1)^2} \right) \left(1 + O \left(\frac{1}{z} \right) \right) \prod_{p < z} \left(1 - \frac{1}{p} \right)$$

$$= 2 \prod_{p > 2} \left(1 - \frac{1}{(p-1)^2} \right) \left(1 + O \left(\frac{1}{z} \right) \right) \frac{e^{-\gamma}}{\log z} \left(1 + O \left(\frac{1}{\log z} \right) \right)$$

$$= 2 \prod_{p > 2} \left(1 - \frac{1}{(p-1)^2} \right) \frac{e^{-\gamma}}{\log z} \left(1 + O \left(\frac{1}{\log N} \right) \right).$$

Therefore,

$$V(z) = \frac{V(z)}{W(z)} W(z)$$

$$= \prod_{\substack{p > 2 \\ p \mid N}} \frac{p-1}{p-2} \prod_{p > 2} \left(1 - \frac{1}{(p-1)^2} \right) \frac{e^{-\gamma}}{\log z} \left(1 + O \left(\frac{1}{\log N} \right) \right)$$

$$= \mathfrak{S}(N) \frac{e^{-\gamma}}{\log z} \left(1 + O \left(\frac{1}{\log N} \right) \right).$$

10.4   A lower bound for $S(A, \mathcal{P}, z)$

Theorem 10.4

$$S(A, \mathcal{P}, z) > \left(\frac{e^\gamma \log 3}{2} + O(\varepsilon) \right) \frac{N V(z)}{\log N}.$$

Proof. We shall apply the linear sieve and results about the distribution of prime numbers in arithmetic progressions to obtain a lower bound for the sieving function $S(A, \mathcal{P}, z)$. We use the prime number theorem in the form

$$\pi(N) = \frac{N}{\log N} \left(1 + O \left(\frac{1}{\log N} \right) \right).$$

Then

$$|A| = \sum_{\substack{p < N \\ (p,N)=1}} 1$$

$$= \pi(N) - \omega(N)$$

$$= \pi(N) + O(\log N)$$

$$= \frac{N}{\log N} \left(1 + O \left(\frac{1}{\log N} \right) \right).$$

In the Jurkat–Richert theorem, the main term in the lower bound (9.36) is $f(s)X$, where

$$X = V(z)|A| = V(z) \frac{N}{\log N} \left(1 + O \left(\frac{1}{\log N} \right) \right)$$

and $V(z)$ is defined by (10.8).

The remainder term in the Jurkat–Richert theorem is

$$R = \sum_{\substack{d < QD \\ d|P(z)}} |r(d)|,$$

where

$$r(d) = |A_d| - \sum_n a(n)g(d) = |A_d| - \frac{|A|}{\varphi(d)}. \qquad (10.9)$$

We want to obtain

$$R \ll \frac{N}{(\log N)^3}$$

with $D = D(N)$ as large as possible. We want D large because the function $f(s)$ in the lower bound of the Jurkat–Richert theorem is an increasing function of $s = \log D / \log z$ for $2 \le s \le 4$. We have

$$|A_d| = \sum_{\substack{n=1 \\ d|n}}^{\infty} a(n)$$

$$= \sum_{\substack{N-p \in A \\ N-p \equiv 0 \pmod{d}}}^{\infty} 1$$

$$= \sum_{\substack{p \in P \\ p \leq N \\ p \equiv N \pmod{d}}} 1$$

$$= \sum_{\substack{p \leq N \\ p \equiv N \pmod{d}}} 1 + O(\omega(N))$$

$$= \pi(N; d, N) + O(\log N),$$

where the term $\omega(N)$ appears when we include the primes that divide N. Therefore,

$$r(d) = |A_d| - \frac{|A|}{\varphi(d)}$$

$$= \pi(N; d, N) - \frac{\pi(N)}{\varphi(d)} + O(\log N)$$

$$= \delta(N; d, N) + O(\log N),$$

where

$$\delta(x; d, a) = \pi(x; d, a) - \frac{\pi(x)}{\varphi(d)}$$

for $x \geq 2, d \geq 1$, and $(d, a) = 1$. There are two important results that provide estimates for $\delta(x; d, a)$. The Siegel–Walfisz theorem states that

$$\delta(x; d, a) \ll \frac{x}{(\log x)^A}$$

for any positive number A, where the implied constant depends only on A. This result is useful if the modulus d is not too large, say, $d \ll (\log x)^A$. The Bombieri–Vinogradov theorem tells us about the average distribution of primes in congruence classes over a large set of moduli. It states that, for every $A > 0$, there exists a positive number $B(A)$ such that

$$\sum_{d < D(A)} \max_{(d,a)=1} |\delta(x; d, a)| \ll \frac{x}{(\log x)^A}$$

for

$$D(A) = \frac{x^{1/2}}{(\log x)^{B(A)}},$$

where the implied constant depends only on A.

We shall apply the Bombieri–Vinogradov theorem with $x = a = N$ and $A = 3$. Let

$$D = \frac{D(3)}{\log N} = \frac{N^{1/2}}{(\log N)^{B(3)+1}}.$$

Then $D \geq z^2 = N^{1/4}$. Since $Q \leq Q(\varepsilon) < \log N$ for $N \geq N(\varepsilon)$, we have

$$QD < \frac{N^{1/2}}{(\log N)^{B(3)}} = D(3)$$

and

$$QD \log N < \frac{N^{1/2}}{(\log N)^{B(3)-1}} \ll \frac{N}{(\log N)^3}$$

for N sufficiently large. Therefore,

$$
\begin{aligned}
R &= \sum_{\substack{d < QD \\ d \mid P(z)}} |r(d)| \\
&\leq \sum_{\substack{d < QD \\ (d,N)=1}} |r(d)| \\
&\leq \sum_{\substack{d < QD \\ (d,N)=1}} |\delta(N; d, N)| + QD \log N \\
&\ll \sum_{\substack{d < D(3) \\ (d,N)=1}} |\delta(N; d, N)| + \frac{N}{(\log N)^3} \\
&\ll \frac{N}{(\log N)^3}.
\end{aligned}
$$

Now we apply the Jurkat–Richert theorem (Theorem 9.7) with $z = N^{1/8}$ and N sufficiently large. We have

$$s = \frac{\log D}{\log z} = 4 - \frac{8(B(3) - 1)) \log \log N}{\log N} \in [3, 4]$$

and so

$$f(s) = \frac{2e^\gamma \log(s - 1)}{s} = \frac{e^\gamma \log 3}{2} + O\left(\frac{\log \log N}{\log N}\right) = \frac{e^\gamma \log 3}{2} + O(\varepsilon).$$

Therefore,

$$
\begin{aligned}
S(A, \mathcal{P}, z) &> (f(s) - \varepsilon e^{14-s})X - R \\
&> (f(s) - \varepsilon e^{11})V(z)\frac{N}{\log N}\left(1 + O\left(\frac{1}{\log N}\right)\right) + O\left(\frac{N}{(\log N)^3}\right) \\
&> \left(\frac{e^\gamma \log 3}{2} + O(\varepsilon)\right)\frac{NV(z)}{\log N}.
\end{aligned}
$$

10.5   An upper bound for $S(A_q, \mathcal{P}, z)$

Theorem 10.5

$$\sum_{z \leq q < y} S(A_q, \mathcal{P}, z) < \left(\frac{e^\gamma \log 6}{2} + O(\varepsilon)\right)\frac{NV(z)}{\log N}.$$

Proof. We shall apply the Jurkat–Richert theorem again to get an upper bound for $S(A_q, \mathcal{P}, z)$, where q is a prime number such that $z \leq q < y$. If $n = N - p \in \mathcal{A}$ and q divides both n and N, then $q = p$, which is impossible since the prime p does not divide N. Therefore, $|A_q| = 0$ if q divides N, so we can assume that $(q, N) = 1$.

Again we choose $g(d) = g_n(d) = 1/\varphi(d)$ for all n, so inequalities (9.33) and (9.34) are satisfied. The error term $r_q(d)$ is defined by

$$r_q(d) = |(A_q)_d| - \frac{|A_q|}{\varphi(d)}.$$

Let d divide $P(z)$. Since d is a product of primes strictly less than z, it follows that $(q, d) = 1$ for every prime number $q \geq z$, and so

$$|(A_q)_d| = \sum_{\substack{q|n \\ d|n}} a(n) = \sum_{qd|n} a(n) = |A_{qd}|.$$

Then

$$r_q(d) = |A_{qd}| - \frac{|A_q|}{\varphi(d)}$$
$$= |A_{qd}| - \frac{|A|}{\varphi(qd)} + \frac{|A|}{\varphi(qd)} - \frac{|A_q|}{\varphi(d)}$$
$$= r(qd) - \frac{r(q)}{\varphi(d)},$$

where $r(qd)$ and $r(q)$ are error terms of the form (10.9). Let

$$D = \frac{D(4)}{\log N} = \frac{N^{1/2}}{(\log N)^{B(4)+1}}$$

and

$$D_q = \frac{D}{q}.$$

Then $D_q \geq D/z \geq z$. The remainder term for $S(A_q, \mathcal{P}, z)$ is

$$R_q = \sum_{\substack{d < QD_q \\ d|P(z)}} |r_q(d)| \leq \sum_{\substack{d < QD_q \\ d|P(z)}} |r(qd)| + r(q) \sum_{\substack{d < QD_q \\ d|P(z)}} \frac{1}{\varphi(d)}.$$

From Theorem 9.7, we have the upper bound

$$S(A_q, \mathcal{P}, z) < \left(F(s_q) + \varepsilon e^{14-s_q}\right) |A_q| V(z) + R_q,$$

where

$$s_q = \frac{\log D_q}{\log z}.$$

We do not estimate the main term and the remainder term for individual primes q. Instead, summing over $z \leq q < y$, we obtain

$$\sum_{\substack{z \leq q < y \\ (q,N)=1}} S(A_q, \mathcal{P}, z) < \sum_{\substack{z \leq q < y \\ (q,N)=1}} \left(F(s_q) + \varepsilon e^{14} \right) |A_q| V(z) + R',$$

where

$$R' = \sum_{\substack{z \leq q < y \\ (q,N)=1}} R_q$$

$$\leq \sum_{\substack{z \leq q < y \\ (q,N)=1}} \sum_{\substack{d < QD/q \\ d|P(z)}} |r(qd)| + \sum_{\substack{z \leq q < y \\ (q,N)=1}} r(q) \sum_{\substack{d < QD/q \\ d|P(z)}} \frac{1}{\varphi(d)}$$

$$\leq \sum_{\substack{d' < QD \\ (d',N)=1}} |r(d')| + \sum_{\substack{z \leq q < y \\ (q,N)=1}} |r(q)| \sum_{d < N^{1/2}} \frac{1}{\varphi(d)}$$

and $QD < D(4)$. Applying the Bombieri–Vinogradov theorem as in the previous section, we obtain

$$\sum_{\substack{d' < QD \\ (d',N)=1}} |r(d')| \leq \sum_{\substack{d' < DQ \\ (d',N)=1}} |\delta(N; d', N)| + \sum_{\substack{d' < DQ \\ (d',N)=1}} O(\log N)$$

$$\ll \frac{N}{(\log N)^4}.$$

Since $y = N^{1/3} < D \leq QD$ for sufficiently large N, we also have

$$\sum_{\substack{z \leq q < y \\ (q,N)=1}} |r(q)| \ll \frac{N}{(\log N)^4}.$$

By Theorem A.17,

$$\sum_{d < N} \frac{1}{\varphi(d)} \ll \log N$$

and so

$$R' \ll \frac{N}{(\log N)^3}.$$

Next, we estimate the main term. We have

$$s_q = \frac{\log D/q}{\log z} = \frac{8 \log(N^{1/2}/q)}{\log N} - \frac{8(B(4)+1) \log \log N}{\log N}.$$

Since $N^{1/8} = z \leq q < y = N^{1/3}$, it follows that

$$\frac{4}{3} < \frac{8 \log(N^{1/2}/q)}{\log N} \leq 3, \qquad (10.10)$$

and so $1 \le s_q \le 3$. By Theorem 9.8, $F(s) = 2e^\gamma/s$ for $1 \le s \le 3$. Therefore,

$$F(s_q) = \frac{2e^\gamma}{s_q} = \frac{e^\gamma \log N}{4 \log(N^{1/2}/q)} + O\left(\frac{\log \log N}{\log N}\right)$$

and so

$$F(s_q) + \varepsilon e^{14} = \frac{e^\gamma \log N}{4 \log(N^{1/2}/q)} + O(\varepsilon). \qquad (10.11)$$

Also,

$$|A_q| = \pi(N; q, N) + O(\log N)$$

$$= \frac{\pi(N)}{\varphi(q)} + \delta(N; q, N) + O(\log N)$$

$$= \frac{N}{\varphi(q) \log N}\left(1 + O\left(\frac{1}{\log N}\right)\right) + \delta(N; q, N).$$

Therefore,

$$\sum_{\substack{z \le q < y \\ (q,N)=1}} (F(s_q) + \varepsilon e^{14})|A_q|$$

$$= \sum_{\substack{z \le q < y \\ (q,N)=1}} \left(\frac{e^\gamma \log N}{4 \log(N^{1/2}/q)} + O(\varepsilon)\right) \frac{N}{\varphi(q) \log N}\left(1 + O\left(\frac{1}{\log N}\right)\right)$$

$$+ \sum_{\substack{z \le q < y \\ (q,N)=1}} (F(s_q) + \varepsilon e^{14})\delta(N; q, N)$$

$$= \frac{e^\gamma N}{4} \sum_{\substack{z \le q < y \\ (q,N)=1}} \frac{1}{\varphi(q) \log(N^{1/2}/q)}$$

$$+ O\left(\frac{N}{\log N}\right) \sum_{\substack{z \le q < y \\ (q,N)=1}} \frac{1}{\varphi(q) \log(N^{1/2}/q)}$$

$$+ O\left(\frac{\varepsilon N}{\log N}\right) \sum_{\substack{z \le q < y \\ (q,N)=1}} \frac{1}{\varphi(q)} + O\left(\sum_{\substack{z \le q < y \\ (q,N)=1}} \delta(N; q, N)\right).$$

It is not difficult to evaluate these terms. By the Bombieri–Vinogradov theorem
again, we have

$$\sum_{\substack{z \le q < y \\ (q,N)=1}} \delta(N; q, N) = O\left(\frac{N}{(\log N)^3}\right).$$

By Theorem 6.7, we have

$$\sum_{\substack{z \le q < y \\ (q,N)=1}} \frac{1}{\varphi(q)} = \sum_{\substack{z \le q < y \\ q \in P}} \frac{1}{q - 1}$$

$$\ll \sum_{\substack{z \le q < y \\ q \in \mathcal{P}}} \frac{1}{q}$$

$$= \log \log y - \log \log z + O\left(\frac{1}{\log z}\right)$$

$$= \log(8/3) + O\left(\frac{1}{\log z}\right)$$

$$= O(1).$$

Using this estimate and inequality (10.10), we have

$$\frac{N}{\log N} \sum_{\substack{z \le q < y \\ (q,N)=1}} \frac{1}{\varphi(q) \log(N^{1/2}/q)} - \frac{N}{(\log N)^2} \sum_{\substack{z \le q < y \\ (q,N)=1}} \frac{\log N}{\varphi(q) \log(N^{1/2}/q)}$$

$$\ll \frac{N}{(\log N)^2} \sum_{\substack{z \le q < y \\ (q,N)=1}} \frac{1}{\varphi(q)}$$

$$\ll \frac{N}{(\log N)^2}.$$

Therefore,

$$\sum_{\substack{z \le q < y \\ (q,N)=1}} (F(s_q) + \varepsilon e^{14})|A_q| = \frac{e^\gamma N}{4} \sum_{\substack{z \le q < y \\ (q,N)=1}} \frac{1}{\varphi(q) \log(N^{1/2}/q)} + O\left(\frac{\varepsilon N}{\log N}\right).$$

We note that

$$\frac{1}{\varphi(q)} - \frac{1}{q-1} = \frac{1}{q} + O\left(\frac{1}{q^2}\right)$$

and

$$N \sum_{\substack{z \le q < y \\ (q,N)=1}} \frac{1}{q^2 \log(N^{1/2}/q)} \le N \sum_{z \le q < y} \frac{1}{q^2 \log N^{1/2}/y}$$

$$= \frac{6N}{\log N} \sum_{z \le q < y} \frac{1}{q^2}$$

$$\ll \frac{N}{z \log N}$$

$$= \frac{N^{7/8}}{\log N}.$$

Let

$$S(t) = \sum_{q < t} \frac{1}{q} = \log \log t + B + O\left(\frac{1}{\log t}\right)$$

and

$$f(t) = \frac{1}{\log(N^{1/2}/t)}.$$

The functions $S(t)$ and $f(t)$ are increasing. We shall estimate the sum

$$\sum_{z \leq q < y} \frac{1}{q \log(N^{1/2}/q)}$$

by using integration by parts twice in Riemann–Stieltjes integrals. We have

$$\sum_{z \leq q < y} \frac{1}{q \log(N^{1/2}/q)} = \int_z^y \frac{dS(t)}{\log(N^{1/2}/t)} = \int_z^y f(t) dS(t)$$

$$= f(y)S(y) - f(z)S(z) - \int_z^y S(t) df(t)$$

$$= f(y)(\log \log y + B) - f(z)(\log \log z + B)$$

$$- \int_z^y (\log \log t + B) df(t)$$

$$+ O\left(\frac{f(y)}{\log z}\right) + O\left(\int_z^y \frac{df(t)}{\log t}\right)$$

$$= \int_z^y f(t) d \log \log t + O\left(\frac{1}{(\log N)^2}\right).$$

We compute the integral explicitly by making the change of variable $t = N^\alpha$. Then

$$\int_z^y f(t) d \log \log t = \int_z^y \frac{dt}{t \log t \log(N^{1/2}/t)}$$

$$= \frac{1}{\log N} \int_{1/8}^{1/3} \frac{d\alpha}{\alpha((1/2) - \alpha)}$$

$$= \frac{2 \log 6}{\log N}.$$

Therefore,

$$\sum_{\substack{z \leq q < y \\ (q,N)=1}} (F(s_q) + \varepsilon e^{14})|A_q| = \left(\frac{e^\gamma \log 6}{2} + O(\varepsilon)\right) \frac{N}{\log N}$$

and so

$$\sum_{z \leq q < y} S(A_q, \mathcal{P}, z) < \left(\frac{e^\gamma \log 6}{2} + O(\varepsilon)\right) \frac{NV(z)}{\log N}.$$

10.6   An upper bound for $S(B, \mathcal{P}, y)$

Theorem 10.6

$$S(B, \mathcal{P}, y) < \left(\frac{ce^\gamma}{2} + O(\varepsilon)\right) \frac{NV(z)}{\log N} + O\left(\frac{\varepsilon^{-1}N}{(\log N)^3}\right).$$

Proof. Recall that

$$B = \{N - p_1 p_2 p_3 : z \leq p_1 < y \leq p_2 \leq p_3, \, p_1 p_2 p_3 < N, \, (p_1 p_2 p_3, N) = 1\}.$$

Before estimating the sieving function $S(B, \mathcal{P}, y)$, we shall drop the requirement that $(p_1, N) = 1$ and relax the condition that $p_1 p_2 p_3 < N$ so that the numbers p_1 and $p_2 p_3$ range over intervals independent of each other. This will produce a "bilinear form" in p_1 and $p_2 p_3$. We shall let the prime p_1 vary over pairwise disjoint intervals

$$\ell \leq p_1 < (1 + \varepsilon)\ell,$$

where ℓ is a number of the form

$$\ell = z(1 + \varepsilon)^k$$

such that $z \leq \ell < y$. Then

$$0 \leq k \leq \frac{\log(y/z)}{\log(1 + \varepsilon)} \ll \frac{\log N}{\varepsilon}. \tag{10.12}$$

Let

$$\begin{aligned} B^{(\ell)} = \{N - p_1 p_2 p_3 : z \leq p_1 < y \leq p_2 \leq p_3, \\ \ell \leq p_1 < (1 + \varepsilon)\ell, \, \ell p_2 p_3 < N, \, (p_2 p_3, N) = 1\} \end{aligned} \tag{10.13}$$

and

$$\tilde{B} = \bigcup_\ell B^{(\ell)}.$$

Then

$$B \subseteq \tilde{B} \subseteq \{N - p_1 p_2 p_3 : z \leq p_1 < y \leq p_2 \leq p_3, \, p_1 p_2 p_3 < (1 + \varepsilon)N\}. \tag{10.14}$$

Let $b(n)$, $b^{(\ell)}(n)$, and $\tilde{b}(n)$ be the characteristic functions of the sets B, $B^{(\ell)}$, and $\tilde{B}$, respectively. Since the sets $B^{(\ell)}$ are pairwise disjoint, we have

$$|\tilde{B}| = \sum_\ell |B^{(\ell)}|$$

and

$$S(B, \mathcal{P}, y) \leq S(\tilde{B}, \mathcal{P}, y) = \sum_\ell S(B^{(\ell)}, \mathcal{P}, y).$$

We shall estimate the sieving function $S(B^{(\ell)}, \mathcal{P}, y)$ by using Theorem 9.7 with the functions

$$g(d) = g_n(d) = \frac{1}{\varphi(d)}$$

for all $n \geq 1$, and with support level

$$D = \frac{N^{1/2}}{(\log N)^A}.$$

Then

$$|B_d^{(\ell)}| = \sum_{\substack{p_1 p_2 p_3 \equiv N \ (\mathrm{mod}\ d) \\ z \leq p_1 < y \leq p_2 \leq p_3, \ell \leq p_1 < (1+\varepsilon)\ell \\ \ell p_2 p_3 < N, (p_2 p_3, N)=1}} 1,$$

and the error term $r_d^{(\ell)}$ is defined by

$$|B_d^{(\ell)}| = \frac{|B^{(\ell)}|}{\varphi(d)} + r_d^{(\ell)}.$$

In the next section, we shall prove that

$$R^{(\ell)} = \sum_{\substack{d < D \\ d | P(y)}} |r_d^{(\ell)}| \ll \frac{N}{(\log N)^4}. \tag{10.15}$$

With this estimate for the remainder, Theorem 9.7 gives the upper bound

$$S(B^{(\ell)}, \mathcal{P}, y) < (F(s) + \varepsilon e^{14})|B^{(\ell)}|V(y) + O\left(\frac{N}{(\log N)^4}\right),$$

where

$$s = \frac{\log D}{\log y} = \frac{3}{2} + O\left(\frac{\log \log N}{\log N}\right) \in [1, 3]$$

and so, by Theorem 9.8,

$$F(s) = \frac{4e^\gamma}{3} + O\left(\frac{\log \log N}{\log N}\right).$$

It follows from (10.3) that

$$\frac{V(y)}{V(z)} = \frac{\log z}{\log y}\left(1 + O\left(\frac{1}{\log N}\right)\right) = \frac{3}{8} + O\left(\frac{1}{\log N}\right).$$

This gives

$$S(B^{(\ell)}, \mathcal{P}, y)$$
$$< \left(\frac{4e^\gamma}{3} + O(\varepsilon)\right)\left(\frac{3}{8} + O\left(\frac{1}{\log N}\right)\right)|B^{(\ell)}|V(z) + O\left(\frac{N}{(\log N)^4}\right)$$
$$< \left(\frac{e^\gamma}{2} + O(\varepsilon)\right)|B^{(\ell)}|V(z) + O\left(\frac{N}{(\log N)^4}\right).$$

Summing over the sets $\mathcal{B}^{(\ell)}$, we obtain

$$S(B, \mathcal{P}, y) \leq \sum_\ell S(B^{(\ell)}, \mathcal{P}, y) < \left(\frac{e^\gamma}{2} + O(\varepsilon)\right)|\tilde{B}|V(z) + O\left(\frac{\varepsilon^{-1}N}{(\log N)^3}\right)$$

since the number of sets $\mathcal{B}^{(\ell)}$ is $O(\varepsilon^{-1}\log N)$ by (10.12).

Next, we estimate $|\tilde{B}|$. By the prime number theorem,

$$\pi\left(\frac{(1+\varepsilon)N}{p_1 p_2}\right) < \frac{(1+2\varepsilon)N}{p_1 p_2 \log(N/p_1 p_2)}$$

for $N \geq N(\varepsilon)$. If $p_1 \leq p_2 \leq p_3$, and $p_1 p_2 p_3 < (1+\varepsilon)N$, then $p_1 p_2^2 < (1+\varepsilon)N$ and

$$p_3 < \frac{(1+\varepsilon)N}{p_1 p_2}.$$

It follows from (10.14) that

$$|\tilde{B}| \leq \sum_{\substack{z \leq p_1 < y \leq p_2 \leq p_3 \\ p_1 p_2 p_3 < (1+\varepsilon)N}} 1$$

$$\leq \sum_{\substack{z \leq p_1 < y \leq p_2 \\ p_1 p_2^2 < (1+\varepsilon)N}} \pi\left(\frac{(1+\varepsilon)N}{p_1 p_2}\right)$$

$$< (1+2\varepsilon)N \sum_{z \leq p_1 < y} \frac{1}{p_1} \sum_{y \leq p_2 < ((1+\varepsilon)N/p_1)^{1/2}} \frac{1}{p_2 \log(N/p_1 p_2)}.$$

To estimate the inner sum, we introduce the functions

$$h(t) = \frac{1}{\log(N/p_1 t)}$$

and

$$H(u) = \int_y^{(N/u)^{1/2}} \frac{1}{\log(N/ut)} d \log \log t.$$

The function $h(t)$ is positive and increasing for $0 < t < N_1/p_1$. Since $y = N^{1/3}$, we have $(N/y)^{1/2} = y$ and so $H(y) = 0$. Since $z = N^{1/8}$, we have, with the change of variable $t = N^\alpha$,

$$H(z) = \int_{N^{1/3}}^{N^{7/16}} \frac{1}{\log(N^{7/8}/t)} d \log \log t$$

$$= \frac{1}{\log N} \int_{1/3}^{7/16} \frac{d\alpha}{(7/8) - \alpha}$$

$$= O\left(\frac{1}{\log N}\right).$$

Recall that

$$S(t) = \sum_{p < t} \frac{1}{p} = \log \log t + B + O\left(\frac{1}{\log t}\right).$$

Applying integration by parts to the inner sum, we obtain

$$\sum_{y \leq p_2 < ((1+\varepsilon)N/p_1)^{1/2}} \frac{1}{p_2 \log(N/p_1 p_2)}$$

$$= \sum_{y \le p_2 < ((1+\varepsilon)N/p_1)^{1/2}} \frac{h(p_2)}{p_2}$$

$$= \int_y^{((1+\varepsilon)N/p_1)^{1/2}} h(t) dS(t)$$

$$= \int_y^{((1+\varepsilon)N/p_1)^{1/2}} h(t) d \log\log t + O\left(\frac{h(((1+\varepsilon)N/p_1)^{1/2})}{\log y}\right)$$

$$= \int_y^{(N/p_1)^{1/2}} \frac{1}{\log(N/p_1 t)} d \log\log t$$

$$+ \int_{(N/p_1)^{1/2}}^{((1+\varepsilon)N/p_1)^{1/2}} \frac{1}{\log(N/p_1 t)} d \log\log t + O\left(\frac{1}{(\log N)^2}\right)$$

$$= H(p_1) + O\left(\frac{1}{(\log N)^2}\right).$$

The error term is obtained as follows. First,

$$\frac{h(((1+\varepsilon)N/p_1)^{1/2})}{\log y} = \frac{2}{\log\left(\frac{N}{(1+\varepsilon)p_1}\right) \log y}$$

$$\le \frac{2}{\log\left(\frac{N}{(1+\varepsilon)y}\right) \log y}$$

$$= \frac{2}{\log\left(\frac{N^{2/3}}{(1+\varepsilon)}\right) \log N^{1/3}}$$

$$\ll \frac{1}{(\log N)^2}.$$

Second, with the change of variable $t = (N/p_1)^{1/2} s$,

$$\int_{(N/p_1)^{1/2}}^{((1+\varepsilon)N/p_1)^{1/2}} \frac{1}{\log(N/p_1 t)} d \log\log t$$

$$= \int_{(N/p_1)^{1/2}}^{((1+\varepsilon)N/p_1)^{1/2}} \frac{1}{t \log t \log(N/p_1 t)} dt$$

$$= \int_1^{(1+\varepsilon)^{1/2}} \frac{ds}{s \log\left((N/p_1)^{1/2} s\right) \log\left((N/p_1)^{1/2} s^{-1}\right)}$$

$$= \int_1^{(1+\varepsilon)^{1/2}} \frac{ds}{s(\log\left((N/p_1)^{1/2}\right) + \log s)(\log\left((N/p_1)^{1/2}\right) - \log s)}$$

$$= \int_1^{(1+\varepsilon)^{1/2}} \frac{ds}{s \left((\log(N/p_1)^{1/2})^2 - (\log s)^2\right)}$$

$$\ll \frac{1}{(\log N)^2} \int_1^{(1+\varepsilon)^{1/2}} \frac{ds}{s}$$

$$= O\left(\frac{1}{(\log N)^2}\right).$$

It follows that the outer sum is

$$\sum_{z \le p_1 < y} \frac{H(p_1)}{p_1} + O\left(\sum_{z \le p_1 < y} \frac{1}{p_1 (\log N)^2}\right) = \sum_{z \le p_1 < y} \frac{H(p_1)}{p_1} + O\left(\frac{1}{(\log N)^2}\right),$$

where the error term comes from the fact that

$$\sum_{z \le p_1 < y} \frac{1}{p_1} = \log\log y - \log\log z + O((\log z)^{-1})$$

$$= \log(8/3) + O((\log N)^{-1})$$

$$= O(1).$$

We calculate the main term, as usual, by integration by parts:

$$\sum_{z \le p_1 < y} \frac{H(p_1)}{p_1} = \int_z^y H(u)\, dS(u)$$

$$= \int_z^y H(u)\, d\log\log u + O\left(\frac{\max(H(z), H(y))}{\log y}\right)$$

$$= \int_z^y H(u)\, d\log\log u + O\left(\frac{1}{(\log N)^2}\right).$$

To evaluate the integral, we make the change of variables $t = N^\alpha$ and $u = N^\beta$. This gives

$$\int_z^y H(u)\, d\log\log u = \int_{N^{1/8}}^{N^{1/3}} \int_{N^{1/3}}^{(N/u)^{1/2}} \frac{1}{\log(N/ut)}\, d\log\log t\, d\log\log u$$

$$= \frac{1}{\log N} \int_{1/8}^{1/3} \int_{1/3}^{(1-\beta)/2} \frac{d\alpha\, d\beta}{\alpha\beta(1 - \alpha - \beta)}$$

$$= \frac{1}{\log N} \int_{1/8}^{1/3} \frac{\log(2 - 3\beta)}{\beta(1 - \beta)}\, d\beta$$

$$= \frac{c}{\log N},$$

where

$$c = \int_{1/8}^{1/3} \frac{\log(2 - 3\beta)}{\beta(1 - \beta)}\, d\beta = 0.363\ldots.$$

Therefore,

$$|\bar{B}| < \frac{(1 + O(\varepsilon))cN}{\log N} + O\left(\frac{N}{(\log N)^2}\right)$$

and

$$S(B, \mathcal{P}, y) < \left(\frac{e^\gamma}{2} + O(\varepsilon)\right)|\tilde{B}|V(z) + O\left(\frac{\varepsilon^{-1}N}{(\log N)^3}\right)$$

$$< \left(\frac{ce^\gamma}{2} + O(\varepsilon)\right)\frac{NV(z)}{\log N} + O\left(\frac{\varepsilon^{-1}N}{(\log N)^3}\right).$$

10.7   A bilinear form inequality

We must still prove inequality (10.15) for the remainder $R^{(\ell)}$. This will be a consequence of the following theorem.

Theorem 10.7 *Let $a(n)$ be an arithmetic function such that $|a(n)| \leq 1$ for all n. Let A be a positive number, let $X > (\log Y)^{2A}$, and let*

$$D^* = \frac{(XY)^{1/2}}{(\log Y)^A}.$$

Then

$$\sum_{d < D^*} \max_{(a,d)=1} \left| \sum_{n < X} \sum_{\substack{Z \leq p < Y \\ np \equiv a \pmod{d}}} a(n) - \frac{1}{\varphi(d)} \sum_{n < X} \sum_{\substack{Z \leq p < Y \\ (np,d)=1}} a(n) \right|$$

$$\ll \frac{XY(\log XY)^2}{(\log Y)^A}, \qquad (10.16)$$

where the implied constant depends only on A.

Proof. Let $(a, d) = 1$. By the orthogonality property of Dirichlet characters χ (mod d), we have

$$\sum_{\chi \pmod{d}} \overline{\chi}(a)\chi(np) = \begin{cases} \varphi(d) & \text{if } np \equiv a \pmod{d} \\ 0 & \text{otherwise.} \end{cases}$$

This gives

$$\sum_{n < X} \sum_{\substack{Z \leq p < Y \\ np \equiv a \pmod{d}}} a(n) = \sum_{n < X} \sum_{Z \leq p < Y} \frac{a(n)}{\varphi(d)} \sum_{\chi \pmod{d}} \overline{\chi}(a)\chi(np)$$

$$= \frac{1}{\varphi(d)} \sum_{\chi \pmod{d}} \overline{\chi}(a) \sum_{n < X} a(n)\chi(n) \sum_{Z \leq p < Y} \chi(p).$$

The contribution of the principal character χ_0 (mod d) to this sum is

$$\frac{1}{\varphi(d)} \sum_{n < X} \sum_{\substack{Z \leq p < Y \\ (np,d)=1}} a(n).$$

It follows that the left side of (10.16) is bounded above by

$$\sum_{d < D^*} \frac{1}{\varphi(d)} \sum_{\substack{\chi \pmod d \\ \chi \ne \chi_0}} \left| \sum_{n < X} a(n)\chi(n) \right| \left| \sum_{Z \le p < Y} \chi(p) \right|.$$

Every character χ (mod d) factors uniquely into the product of a primitive character (mod r) and the principal character (mod s), where $rs = d$. Therefore, the sum can be written in the form

$$\sum_{rs < D^*} \frac{1}{\varphi(rs)} \sum_{\substack{\chi \pmod r \\ \chi \ne \chi_0}}^{*} \left| \sum_{\substack{n < X \\ (n,s)=1}} a(n)\chi(n) \right| \left| \sum_{\substack{Z \le p < Y \\ (p,s)=1}} \chi(p) \right| \qquad (10.17)$$

$$\le \sum_{s < D^*} \frac{1}{\varphi(s)} \sum_{r < D^*} \frac{1}{\varphi(r)} \sum_{\substack{\chi \pmod r \\ \chi \ne \chi_0}}^{*} \left| \sum_{\substack{n < X \\ (n,s)=1}} a(n)\chi(n) \right| \left| \sum_{\substack{Z \le p < Y \\ (p,s)=1}} \chi(p) \right|,$$

where $\sum^*$ denotes the sum over primitive characters (mod r). To obtain the last inequality, we used the fact that the Euler φ-function satisfies $\varphi(rs) \ge \varphi(r)\varphi(s)$. We can estimate the character sum $\sum_{p < Y} \chi(p)$ by means of the Siegel–Walfisz theorem. We have

$$\sum_{p < Y} \chi(p) \le \sum_{a \pmod r} \chi(a) \sum_{\substack{p < Y \\ p \equiv a \pmod r}} 1$$

$$= \sum_{a \pmod r} \chi(a)\pi(Y; r, a)$$

$$= \sum_{a \pmod r} \chi(a) \left(\frac{\pi(Y)}{\varphi(r)} + O\left(\frac{Y}{(\log Y)^B} \right) \right)$$

$$\ll \frac{rY}{(\log Y)^B}$$

since

$$\sum_{a \pmod r} \chi(a) = 0$$

for every nonprincipal character χ. Since also

$$\sum_{p < Z} \chi(p) \ll \frac{rZ}{(\log Z)^B} \ll \frac{rY}{(\log Y)^B},$$

it follows by subtraction that

$$\sum_{Z \le p < Y} \chi(p) \ll \frac{rY}{(\log Y)^B}.$$

If we add the condition $(p, s) = 1$, we remove at most $\omega(s) \ll \log s \ll \log D^*$ terms from the character sum and so

$$\sum_{\substack{Z \le p < Y \\ (p,s)=1}} \chi(p) \ll \frac{rY}{(\log Y)^B} + \log D^*.$$

Since $|a(n)| \le 1$, we also have

$$\left| \sum_{\substack{n < X \\ (n,s)=1}} a(n)\chi(n) \right| \le X.$$

Let D_0^* be "small." The inner sum in (10.17), restricted to $r < D_0^*$, is

$$\sum_{r < D^*} \frac{1}{\varphi(r)} {\sum_{\substack{\chi \ (\mathrm{mod}\ r) \\ \chi \ne \chi_0}}}^* \left| \sum_{\substack{n < X \\ (n,s)=1}} a(n)\chi(n) \right| \left| \sum_{\substack{Z \le p < Y \\ (p,s)=1}} \chi(p) \right|$$

$$\ll \sum_{r < D_0^*} \frac{rX}{\varphi(r)} \left(\frac{rY}{(\log Y)^B} + \log D^* \right)$$

$$\ll \frac{D_0^{*3} XY \log D^*}{(\log Y)^B}. \tag{10.18}$$

The rest of the inner sum in (10.17) ranges over $D_0^* \le r < D^*$. We partition this interval into pairwise disjoint subintervals of the form $D_1^* \le r < 2D_1^*$, where $D_1^* = 2^k D_0^*$ and $0 \le k \ll \log D^*$. This produces partial sums of the form

$$\sum_{\substack{D_1^* \le r < 2D_1^* \\ D_0^* \le r < D^*}} \frac{1}{\varphi(r)} {\sum_{\substack{\chi \ (\mathrm{mod}\ r) \\ \chi \ne \chi_0}}}^* \left| \sum_{\substack{n < X \\ (n,s)=1}} a(n)\chi(n) \right| \left| \sum_{\substack{Z \le p < Y \\ (p,s)=1}} \chi(p) \right|$$

$$\le \frac{1}{D_1^*} \sum_{\substack{D_1^* \le r < 2D_1^* \\ D_0^* \le r < D^*}} {\sum_{\substack{\chi \ (\mathrm{mod}\ r) \\ \chi \ne \chi_0}}}^* \left(\left(\frac{r}{\varphi(r)} \right)^{1/2} \left| \sum_{\substack{n < X \\ (n,s)=1}} a(n)\chi(n) \right| \right)$$

$$\times \left(\left(\frac{r}{\varphi(r)} \right)^{1/2} \left| \sum_{\substack{Z \le p < Y \\ (p,s)=1}} \chi(p) \right| \right).$$

By Cauchy's inequality, this sum is bounded above by

$$\frac{1}{D_1^*} \left(\sum_{D_1^* < r < 2D_1^*} \frac{r}{\varphi(r)} {\sum_{\substack{\chi \ (\mathrm{mod}\ r) \\ \chi \ne \chi_0}}}^* \left| \sum_{\substack{n < X \\ (n,s)=1}} a(n)\chi(n) \right|^2 \right)^{1/2}$$

$$\times \left(\sum_{D_1^* \le r < 2D_1^*} \frac{r}{\varphi(r)} {\sum_{\substack{\chi \ (\mathrm{mod}\ r) \\ \chi \ne \chi_0}}}^* \left| \sum_{\substack{Z \le p < Y \\ (p,s)=1}} \chi(p) \right|^2 \right)^{1/2}.$$

The large sieve inequality [19, page 160] states that

$$\sum_{r<R}\frac{r}{\varphi(r)}\sum_{\substack{\chi \ (\mathrm{mod}\ r)\\ \chi\neq\chi_0}}^{*}\left|\sum_{n=L+1}^{L+M}a(n)\chi(n)\right|^2 \ll (R^2+M)\sum_{n=L+1}^{L+M}|a(n)|^2$$

for every arithmetic function $a(n)$. Applying this inequality to each of the factors in the product, and using the condition that $|a(n)|\le 1$, we obtain

$$\sum_{\substack{D_1^*\le r<2D_1^*\\ D_0^*\le r<D^*}}\frac{1}{\varphi(r)}\sum_{\substack{\chi\ (\mathrm{mod}\ r)\\ \chi\neq\chi_0}}^{*}\left|\sum_{\substack{n<X\\ (n,s)=1}}a(n)\chi(n)\right|\left|\sum_{\substack{Z\le p<Y\\ (p,s)=1}}\chi(p)\right|$$

$$\ll \frac{1}{D_1^*}\left((D_1^{*2}+X)X\right)^{1/2}\left((D_1^{*2}+Y)Y\right)^{1/2}$$

$$=\left(\left(D_1^{*2}+X+Y+\frac{XY}{D_1^{*2}}\right)XY\right)^{1/2}$$

$$\ll\left(D_1^*+X^{1/2}+Y^{1/2}+\frac{(XY)^{1/2}}{D_1^*}\right)(XY)^{1/2}$$

$$\ll\left(D^*+X^{1/2}+Y^{1/2}+\frac{(XY)^{1/2}}{D_0^*}\right)(XY)^{1/2}.$$

Multiplying this by the number of partial sums, which is $O(\log D^*)$, and adding (10.18), we obtain the following upper bound for the left side of (10.16):

$$\sum_{d<D^*}\frac{1}{\varphi(d)}\sum_{\substack{\chi\ (\mathrm{mod}\ d)\\ \chi\neq\chi_0}}\left|\sum_{n<X}a(n)\chi(n)\right|\left|\sum_{Z\le p<Y}\chi(p)\right|$$

$$\le\sum_{s<D^*}\frac{1}{\varphi(s)}\sum_{r<D^*}\frac{1}{\varphi(r)}\sum_{\substack{\chi\ (\mathrm{mod}\ r)\\ \chi\neq\chi_0}}^{*}\left|\sum_{\substack{n<X\\ (n,s)=1}}a(n)\chi(n)\right|\left|\sum_{\substack{Z\le p<Y\\ (p,s)=1}}\chi(p)\right|$$

$$\ll\sum_{s<D^*}\frac{1}{\varphi(s)}\frac{D_0^{*3}XY\log D^*}{(\log Y)^B}$$

$$+\sum_{s<D^*}\frac{1}{\varphi(s)}\left(D^*+X^{1/2}+Y^{1/2}+\frac{(XY)^{1/2}}{D_0^*}\right)(XY)^{1/2}\log D^*$$

$$\le\frac{D_0^{*3}XY(\log D^*)^2}{(\log Y)^B}$$

$$+\left(D^*+X^{1/2}+Y^{1/2}+\frac{(XY)^{1/2}}{D_0^*}\right)(XY)^{1/2}(\log D^*)^2.$$

Note that we picked up a factor $\log D^*$ from the estimate (Theorem A.17)

$$\sum_{s<D^*}\frac{1}{\varphi(s)}\ll \log D^*.$$

Choose $B = 4A$ and $D_0^* = (\log Y)^A$. Since $X > (\log Y)^{2A}$ and $Y \gg (\log Y)^{2A}$, it follows that the left side of (10.16) is

$$\ll \frac{XY(\log D^*)^2}{(\log Y)^A} + \left(\frac{D^*}{(XY)^{1/2}} + \frac{1}{X^{1/2}} + \frac{1}{Y^{1/2}} + \frac{1}{D_0^*} \right) XY(\log D^*)^2$$

$$\ll \left(\frac{1}{X^{1/2}} + \frac{1}{Y^{1/2}} + \frac{1}{(\log Y)^A} \right) XY(\log XY)^2$$

$$\ll \frac{XY(\log XY)^2}{(\log Y)^A}.$$

This completes the proof.

We can now derive the upper bound (10.15) for the remainder term

$$R^{(\ell)} = \sum_{\substack{d < D \\ d \mid P(y)}} |r_d^{(\ell)}|,$$

where $z \le \ell < y$. From the definition (10.13) of the sets $\mathcal{B}^\ell$, we obtain the individual error terms

$$r_d^{(\ell)} = |B_d^{(\ell)}| - \frac{1}{\varphi(d)} |B^{(\ell)}|$$

$$= \sum_{\substack{z \le p_1 < y \le p_2 \le p_3 \\ \ell \le p_1 < (1+\varepsilon)\ell \\ \ell p_2 p_3 < N, (p_2 p_3, N) = 1 \\ p_1 p_2 p_3 \equiv N \pmod d}} 1 - \frac{1}{\varphi(d)} \sum_{\substack{z \le p_1 < y \le p_2 \le p_3 \\ \ell \le p_1 < (1+\varepsilon)\ell \\ \ell p_2 p_3 < N, (p_2 p_3, N) = 1}} 1.$$

We delete some numbers from the second sum by adding the condition that $(p_1 p_2 p_3, d) = 1$. This is equivalent to $(p_1, d) = 1$, since the condition $(p_2 p_3, d) = 1$ already follows from the fact that d divides $P(y)$. This additional condition decreases the second term by at most

$$\frac{1}{\varphi(d)} \sum_{\substack{p_1 p_2 p_3 < (1+\varepsilon)N \\ p_1 \mid d, p_1 \ge z}} 1 < \frac{(1+\varepsilon)N}{\varphi(d)} \sum_{\substack{p_1 \mid d, p_1 \ge z}} \frac{1}{p_1} \le \frac{(1+\varepsilon)N\omega(d)}{z\varphi(d)} \ll \frac{(N \log d)}{z\varphi(d)}.$$

Let $a(n)$ be the characteristic function of the set of numbers of the form $n = p_2 p_3$, where $y \le p_2 \le p_3$ and $(p_2 p_3, N) = 1$. Then we can write the error term in the form

$$r_d^{(\ell)} = \sum_{n < X} \sum_{\substack{Z \le p < Y \\ np \equiv a \pmod d}} a(n) - \frac{1}{\varphi(d)} \sum_{n < X} \sum_{\substack{Z \le p < Y \\ (np, d) = 1}} a(n) + O\left(\frac{(N \log d)}{z\varphi(d)} \right),$$

where

$$X = N/\ell$$

$$Y = \min(y, (1+\varepsilon)\ell)$$

$$Z = \max(z, \ell)$$

$$a = N.$$

Since $\ell < y$, we have

$$
\begin{aligned}
D^* &= \frac{(XY)^{1/2}}{(\log Y)^A} \\
&\geq \frac{N^{1/2}\min(y/\ell,(1+\varepsilon)^{1/2})}{(\log y)^A} \\
&> \frac{N^{1/2}}{(\log N)^A} \\
&= D.
\end{aligned}
$$

Similarly,

$$
D^* < (XY)^{1/2} \leq (Ny)^{1/2} < N.
$$

By Theorem 10.7,

$$
\begin{aligned}
R^{(\ell)} &= \sum_{\substack{d<D \\ d\mid P(y)}} |r_d^{(\ell)}| \\
&\leq \sum_{\substack{d<D^* \\ d\mid P(y)}} |r_d^{(\ell)}| \\
&= \sum_{\substack{d<D^* \\ d\mid P(y)}} \left| \sum_{n<X} \sum_{\substack{z\leq p<Y \\ np\equiv a \pmod d}} a(n) - \frac{1}{\varphi(d)} \sum_{n<X} \sum_{\substack{z\leq p<Y \\ (np,d)=1}} a(n) \right| \\
&\quad + \sum_{\substack{d<D^* \\ d\mid P(y)}} O\left(\frac{N\log d}{z\varphi(d)} \right) \\
&\ll \frac{XY(\log XY)^2}{(\log Y)^A} + \frac{N\log D^*}{z} \sum_{d<D^*} \frac{1}{\varphi(d)} \\
&\ll \frac{N}{(\log N)^{A-2}} + N^{7/8}(\log D^*)^2 \\
&\ll \frac{N}{(\log N)^4} + N^{7/8}(\log)^2 \\
&\ll \frac{N}{(\log N)^4}
\end{aligned}
$$

if we choose $A = 6$. This completes the proof.

10.8   Conclusion

We can now prove Theorem 10.1.

Proof. It follows from the formula for $V(z)$ in Theorem 10.3 that

$$
\frac{NV(z)}{\log N} = \mathfrak{S}(N)\frac{8e^{-\gamma}N}{(\log N)^2}\left(1 + O\left(\frac{1}{\log N}\right)\right).
$$

Theorem 10.2 gives a lower bound for $r(N)$ in terms of three sieving functions. Using the estimates for these sieving functions in Theorems 10.4, 10.5, and 10.6, we obtain

$$r(N) > S(A, \mathcal{P}, z) - \frac{1}{2} \sum_{z \le q < y} S(A_q, \mathcal{P}, z) - \frac{1}{2} S(B, \mathcal{P}, y) - 2N^{7/8} - N^{1/3}$$

$$> (2 \log 3 - \log 6 - c - O(\varepsilon)) \frac{e^{\gamma} N V(z)}{4 \log N}$$

$$+ O\left(\frac{\varepsilon^{-1} N}{(\log N)^3}\right) - 2N^{7/8} - N^{1/3}$$

$$> (2 \log 3 - \log 6 - c - O(\varepsilon)) \mathfrak{S}(N) \frac{2N}{(\log N)^2} \left(1 + O\left(\frac{1}{\log N}\right)\right)$$

$$+ O\left(\frac{\varepsilon^{-1} N}{(\log N)^3}\right) - 2N^{7/8} - N^{1/3}.$$

Since

$$2 \log 3 - \log 6 - c = 0.042 \ldots > 0,$$

we can choose ε such that $0 < \varepsilon < 1/200$ and

$$2 \log 3 - \log 6 - c - O(\varepsilon) > 0.$$

For this fixed value of ε, we have

$$O\left(\frac{\varepsilon^{-1} N}{(\log N)^3}\right) = O\left(\frac{N}{(\log N)^3}\right).$$

Then

$$r(N) \gg \mathfrak{S}(N) \frac{2N}{(\log N)^2}.$$

This completes the proof of Chen's theorem.

10.9   Notes

Chen [10, 11] announced his theorem in 1966 but did not publish the proof until 1973, apparently because of difficulties arising from the Cultural Revolution in China. An account of Chen's original proof appears in Halberstam and Richert's *Sieve Methods* [44]. The proof in this chapter is based on unpublished notes and lectures of Henryk Iwaniec [67]. The argument uses standard results from multiplicative number theory (Dirichlet characters, the large sieve, and the Siegel–Walfisz and Bombieri–Vinogradov theorems), all of which can be found in Davenport [19]. Other good references for these results are the monographs of Montgomery [83] and Bombieri [3]. For bilinear form inequalities, see Bombieri, Friedlander, and Iwaniec [4].

Part III

Appendix

Arithmetic functions

A.1   The ring of arithmetic functions

An *arithmetic function* is a complex-valued function whose domain is the set of all positive integers. Let f and g be arithmetic functions. The sum $f + g$ is the arithmetic function defined by

$$(f + g)(n) = f(n) + g(n).$$

Addition of arithmetic functions is clearly associative and commutative, and every arithmetic function f has an inverse $-f$ defined by $(-f)(n) = -f(n)$.

The *Dirichlet convolution* of the arithmetic functions f and g is defined by

$$(f * g)(n) = \sum_{d|n} f(d)g(n/d).$$

It is easy to see that Dirichlet convolution is commutative, that is, $f * g = g * f$, and distributes over addition in the following way:

$$f * (g + h) = f * g + f * h$$

The following theorem shows that Dirichlet convolution is also associative.

Theorem A.1 *If f, g, and h are arithmetic functions, then*

$$f * (g * h) = (f * g) * h.$$

Proof. For any $n \geq 1$,

$$((f * g) * h)(n) = \sum_{d \mid n} (f * g)(d) h \left(\frac{n}{d} \right)$$

$$= \sum_{dm=n} (f * g)(d) h(m)$$

$$= \sum_{dm=n} \sum_{k\ell=d} f(k) g(\ell) h(m)$$

$$= \sum_{k\ell m=n} f(k) g(\ell) h(m)$$

$$= \sum_{k \mid n} f(k) \sum_{\ell m = n/k} g(\ell) h(m)$$

$$= \sum_{k \mid n} f(k) \sum_{\ell \mid (n/k)} g(\ell) h \left(\frac{n}{k\ell} \right)$$

$$= \sum_{k \mid n} f(k)(g * h) \left(\frac{n}{k} \right)$$

$$= (f * (g * h))(n).$$

This completes the proof.

We define the arithmetic function $\delta(n)$ by

$$\delta(n) = \begin{cases} 1 & \text{if } n = 1, \\ 0 & \text{if } n \geq 2. \end{cases}$$

Then for any arithmetic function f we have

$$(f * \delta)(n) = \sum_{d \mid n} f(d) \delta \left(\frac{n}{d} \right) = f(n),$$

and so the set of complex-valued arithmetic functions forms a commutative ring with identity element $\delta(n)$. This ring is an integral domain (Exercise 3).

The product $f \cdot g$ of the arithmetic functions f and g is defined by

$$(f \cdot g)(n) = f(n) g(n).$$

Let L be the arithmetic function $L(n) = \log n$. Multiplication by L is a derivation on the ring of arithmetic functions, that is,

$$L \cdot (f * g) = (L \cdot f) * g + f * (L \cdot g)$$

(Exercise 11).

A.2   Sums and integrals

In number theory, we often need to establish asymptotic formulas or at least good estimates for sums of the form

$$M_f(x) = \sum_{n \le x} f(n),$$

where $f(n)$ is an arithmetic function. It is sometimes possible to estimate these "mean values" by integrals.

Theorem A.2 *Let a and b be integers with $a < b$, and let $f(t)$ be a monotonic function on the interval $[a, b]$. Then*

$$\min(f(a), f(b)) \le \sum_{k=a}^{b} f(k) - \int_{a}^{b} f(t)dt \le \max(f(a), f(b)).$$

Proof. If $f(t)$ is increasing on $[a, b]$, then

$$f(k) \le \int_{k}^{k+1} f(t)dt$$

for $k = a, a+1, \ldots, b-1$, and

$$f(k) \ge \int_{k-1}^{k} f(t)dt$$

for $k = a+1, \ldots, b$. It follows that

$$\sum_{k=a}^{b} f(k) = \sum_{k=a}^{b-1} f(k) + f(b) \le \int_{a}^{b} f(t)dt + f(b)$$

and

$$\sum_{k=a}^{b} f(k) = \sum_{k=a+1}^{b-1} f(k) + f(a) \ge \int_{a}^{b} f(t)dt + f(a).$$

Thus,

$$f(a) \le \sum_{k=a}^{b} f(k) - \int_{a}^{b} f(t)dt \le f(b).$$

Similarly, if $f(t)$ is decreasing, then

$$f(b) \le \sum_{k=a}^{b} f(k) - \int_{a}^{b} f(t)dt \le f(a).$$

This completes the proof.
    From this result, we get a useful estimate for $n!$.

Theorem A.3 *For any positive integer n, we have*

$$e \left(\frac{n}{e} \right)^n \le n! \le en \left(\frac{n}{e} \right)^n.$$

Proof. Since the function $f(t) = \log t$ is increasing on the interval $[1, n]$, it follows from Theorem A.2 that

$$\log n! = \sum_{k=1}^{n} \log n \le \int_1^n \log t \, dt + \log n = n \log n - n + 1 + \log n$$

and

$$\log n! \ge \int_1^n \log t \, dt = n \log n - n + 1.$$

The result follows from exponentiating these two inequalities.

Partial summation is another simple and powerful tool for computing sums in analysis and number theory.

Theorem A.4 (Partial summation)   *Let $u(n)$ and $f(n)$ be arithmetic functions. Define the sum function*

$$U(t) = \sum_{n \le t} u(n).$$

Let a and b be nonnegative integers with $a < b$. Then

$$\sum_{n=a+1}^{b} u(n) f(n) = U(b) f(b) - U(a) f(a + 1) - \sum_{n=a+1}^{b-1} U(n)(f(n + 1) - f(n)).$$

Let x and y be real numbers such that $0 \le y < x$. If $f(t)$ is a function with a continuous derivative on the interval $[y, x]$, then

$$\sum_{y < n \le x} u(n) f(n) = U(x) f(x) - U(y) f(y) - \int_y^x U(t) f'(t) \, dt.$$

In particular, if $f(t)$ has a continuous derivative on $[1, x]$, then

$$\sum_{n \le x} u(n) f(n) = U(x) f(x) - \int_1^x U(t) f'(t) \, dt.$$

Proof. This is a straightforward calculation.

$$\sum_{n=a+1}^{b} u(n) f(n)$$

$$= \sum_{n=a+1}^{b} (U(n) - U(n - 1)) f(n)$$

$$= \sum_{n=a+1}^{b} U(n) f(n) - \sum_{n=a}^{b-1} U(n) f(n + 1)$$

$$= U(b) f(b) - U(a) f(a + 1) - \sum_{n=a+1}^{b-1} U(n)(f(n + 1) - f(n)).$$

If the function $f(t)$ is continuously differentiable on $[y, x]$, then

$$f(n+1) - f(n) = \int_n^{n+1} f'(t)dt$$

and

$$U(n)(f(n+1) - f(n)) = \int_n^{n+1} U(t)f'(t)dt.$$

Let $a = [y]$ and $b = [x]$. Then

$$\sum_{y < n \leq x} u(n)f(n)$$

$$= \sum_{n=a+1}^{b} u(n)f(n)$$

$$= U(b)f(b) - U(a)f(a+1) - \sum_{n=a+1}^{b-1} U(n)(f(n+1) - f(n))$$

$$= U(x)f(b) - U(y)f(a+1) - \sum_{n=a+1}^{b-1} \int_n^{n+1} U(t)f'(t)dt$$

$$= U(x)f(x) - U(y)f(y) - U(x)(f(x) - f(b)) - U(y)(f(a+1) - f(y))$$

$$- \int_{a+1}^{b} U(t)f'(t)dt$$

$$= U(x)f(x) - U(y)f(y) - \int_y^x U(t)f'(t)dt.$$

If $f(t)$ is continuously differentiable on $[1, x]$, then

$$\sum_{n \leq x} u(n)f(n) = u(1)f(1) + \sum_{1 < n \leq x} u(n)f(n)$$

$$= u(1)f(1) + U(x)f(x) - U(1)f(1) - \int_1^x U(t)f'(t)dt$$

$$= U(x)f(x) - \int_1^x U(t)f'(t)dt.$$

This completes the proof.

Here is an application of partial summation. Recall that every real number x can be written in the form

$$x = [x] + \{x\},$$

where $[x]$ is the integer part of x and $\{x\}$ is the fractional part of x.

Theorem A.5   *Let*

$$\gamma = 1 - \int_1^\infty \frac{\{t\}}{t^2}dt.$$

Then $0 < \gamma < 1$ and

$$\sum_{n \le x} \frac{1}{n} = \log x + \gamma + O\left(\frac{1}{x}\right).$$

The real number γ is called Euler's constant .

Proof. Since $0 \le \{t\} < 1$ for all t, we have

$$0 < \int_1^\infty \frac{\{t\}}{t^2} dt < \int_1^\infty \frac{1}{t^2} dt = 1,$$

and so Euler's constant γ is a well-defined real number in the interval $(0, 1)$. We apply partial summation with $u(n) = 1$ for all n and $f(t) = 1/t$. Then

$$U(t) = [t] = t - \{t\}$$

and

$$\begin{aligned}
\sum_{n \le x} \frac{1}{n} &= \sum_{n \le x} u(n) f(n) \\
&= \frac{[x]}{x} + \int_1^x \frac{[t]}{t^2} dt \\
&= 1 - \frac{\{x\}}{x} + \int_1^x \frac{1}{t} dt - \int_1^x \frac{\{t\}}{t^2} dt \\
&= \log x + 1 - \int_1^\infty \frac{\{t\}}{t^2} dt + \int_x^\infty \frac{\{t\}}{t^2} dt - \frac{\{x\}}{x} \\
&= \log x + \gamma + O\left(\frac{1}{x}\right).
\end{aligned}$$

This completes the proof.

As another application of partial summation, we obtain the Euler sum formula.

Theorem A.6 (Euler sum formula) *Let $f(t)$ be a function with a continuous derivative on $[y, x]$. Then*

$$\sum_{y < n \le x} f(n) = \int_y^x f(t) dt + R,$$

where

$$R = \int_y^x \{t\} f'(t) dt + \{y\} f(y) - \{x\} f(x) = \int_y^x \theta(t) f'(t) dt + \theta(y) f(y) - \theta(x) f(x),$$

where

$$\theta(t) = \{t\} - \frac{1}{2}.$$

Proof. We apply partial summation with $a(n) = 1$ for all n. Then $A(t) = [t] = t - \{t\}$ and

$$\sum_{y < n \le x} f(n)$$

$$= [x]f(x) - [y]f(y) - \int_y^x [t]f'(t)dt$$

$$= [x]f(x) - [y]f(y) - \int_y^x tf'(t)dt + \int_y^x \{t\}f'(t)dt$$

$$= [x]f(x) - [y]f(y) - \left(xf(x) - yf(y) - \int_y^x f(t)dt\right) + \int_y^x \{t\}f'(t)dt$$

$$= \int_y^x f(t)dt + \int_y^x \{t\}f'(t)dt + \{y\}f(y) - \{x\}f(x).$$

This completes the proof.

There is a simple expression for partial summation in terms of Riemann–Stieltjes integrals. If f and g are bounded functions on $[y, x]$ and if $\int_y^x f dg$ exists, then $\int_y^x g df$ also exists and

$$\int_y^x f dg + \int_y^x g df = f(x)g(x) - f(y)g(y).$$

This lovely reciprocity law is called *integration by parts*. (See Apostol [1, chapter 9].) Let $u(n)$ be a nonnegative arithmetic function, and let

$$U(t) = \sum_{n \le t} u(n).$$

If f is continuous on $[y, x]$, then

$$\sum_{y < n \le x} u(n)f(n) = \int_y^x f(t)dU(t) = U(x)f(x) - U(y)f(y) - \int_y^x U(t)df(t).$$

If f has a continuous derivative on $[y, x]$, then

$$\int_y^x U(t)df(t) = \int_y^x U(t)f'(t)dt,$$

and we recover the formula for partial summation. Similarly, if we let

$$U(t) = \sum_{n < t} u(n)$$

and if f is continuous on $[y, x]$, then

$$\sum_{y \le n < x} u(n)f(n) = U(x)f(x) - U(y)f(y) - \int_y^x U(t)df(t). \qquad (A.1)$$

A.3   Multiplicative functions

An arithmetic function $f(n)$ is *multiplicative* if

$$f(mn) = f(m)f(n)$$

whenever m and n are relatively prime positive integers. Since $f(1) = f(1 \cdot 1) = f(1)^2$, we have $f(1) = 1$ or 0. If $f(1) = 0$, then $f(n) = f(n \cdot 1) = f(n)f(1) = 0$ for all $n \geq 1$. Therefore, if the multiplicative function f is not identically zero, then $f(1) = 1$.

If f and g are multiplicative functions, then the Dirichlet convolution $f * g$ is multiplicative (Exercise 2). An arithmetic function $f(n)$ is *completely multiplicative* if $f(mn) = f(m)f(n)$ for all positive integers m and n.

Theorem A.7 *Let f be a multiplicative function. Then*

$$f([m, n])f((m, n)) = f(m)f(n).$$

Proof. Let $p_1, \ldots, p_r$ be the prime numbers that divide m or n. Then

$$m = \prod_{i=1}^{r} p_i^{r_i}$$

and

$$n = \prod_{i=1}^{r} p_i^{n_i},$$

where $r_1, \ldots, r_r, s_1, \ldots, s_r$ are nonnegative integers. Moreover,

$$[m, n] = \prod_{i=1}^{r} p_i^{\max(r_i, s_i)}$$

and

$$(m, n) = \prod_{i=1}^{r} p_i^{\min(r_i, s_i)}.$$

Since

$$\{\max(r_i, s_i), \min(r_i, s_i)\} = \{r_i, s_i\}$$

and since f is multiplicative, it follows that

$$f([m, n])f((m, n)) = \prod_{i=1}^{r} f(p_i^{\max(r_i, s_i)}) \prod_{i=1}^{r} f(p_i^{\min(r_i, s_i)})$$

$$= \prod_{i=1}^{r} f(p_i^{r_i}) \prod_{i=1}^{r} f(p_i^{s_i})$$

$$= f(m)f(n).$$

This completes the proof.

The *Möbius function* $\mu(n)$ is defined by

$$\mu(n) = \begin{cases} 1 & \text{if } n = 1, \\ 0 & \text{if } n \text{ is divisible by the square of a prime,} \\ (-1)^r & \text{if } n \text{ is the product of } r \text{ distinct primes.} \end{cases}$$

Thus, $\mu(n) \neq 0$ if and only if n is square-free, and

$$\mu(n) = (-1)^{\omega(n)}$$

for square-free integers n, where $\omega(n)$ is the number of distinct prime divisors of n. It is easy to check that the arithmetic function $\mu(n)$ is multiplicative.

Theorem A.8 *Let f be a multiplicative function with $f(1) = 1$. Then*

$$\sum_{d|n} \mu(d) f(d) = \prod_{p|n} (1 - f(p)).$$

Proof. This is certainly true for $n = 1$. For $n > 1$, let n^* be the product of the distinct primes dividing n. Since $\mu(d) = 0$ if d is not square-free, it follows that

$$\sum_{d|n} \mu(d) f(d) = \sum_{d|n^*} \mu(d) f(d) = \prod_{p|n} (1 - f(p)).$$

This completes the proof.

Theorem A.9 *Let $f(n)$ be a multiplicative function. If*

$$\lim_{p^k \to \infty} f(p^k) = 0$$

as p^k runs through the sequence of all prime powers, then

$$\lim_{n \to \infty} f(n) = 0.$$

Proof. There exist only finitely many prime powers p^k such that $|f(p^k)| \geq 1$. Let

$$A = \prod_{|f(p^k)| \geq 1} |f(p^k)|.$$

Then $A \geq 1$. Let $0 < \varepsilon < A$. There exist only finitely many prime powers p^k such that $|f(p^k)| \geq \varepsilon/A$. It follows that there are only finitely many integers n such that

$$|f(p^k)| \geq \varepsilon/A$$

for every prime power p^k that exactly divides n. Therefore, if n is sufficiently large, then n is divisible by at least one prime power p^k such that $|f(p^k)| < \varepsilon/A$, and so n can be written in the form

$$n = \prod_{i=1}^{r} p_i^{k_i} \prod_{i=r+1}^{r+s} p_i^{k_i} \prod_{i=r+s+1}^{r+s+t} p_i^{k_i},$$

where $p_1, \ldots, p_{r+s+t}$ are pairwise distinct prime numbers such that

$$1 \leq |f(p_i^{k_i})| \quad \text{for } i = 1, \ldots, r,$$

$$\varepsilon/A \leq |f(p_i^{k_i})| < 1 \quad \text{for } i = r+1, \ldots, r+s,$$

$$|f(p_i^{k_i})| < \varepsilon/A \quad \text{for } i = r+s+1, \ldots, r+s+t,$$

and

$$t \geq 1.$$

Therefore,

$$|f(n)| = \prod_{i=1}^{r} |f(p_i^{k_i})| \prod_{i=r+1}^{r+s} |f(p_i^{k_i})| \prod_{i=r+s+1}^{r+s+t} |f(p_i^{k_i})| < A(\varepsilon/A)^t \leq \varepsilon.$$

This completes the proof.

A.4   The divisor function

The divisor function $d(n)$ counts the number of positive divisors of n. For example, $d(n) = 1$ if and only if $n = 1$, and $d(n) = 2$ if and only if n is prime.

Theorem A.10 *Let*

$$m = p_1^{k_1} \cdots p_r^{k_r}$$

be a positive integer, where $p_1, \ldots, p_r$ are distinct primes and $k_1, \ldots, k_r$ are nonnegative integers. Then

$$d(m) = (k_1 + 1) \cdots (k_r + 1)n.$$

If m and n are any positive integers, then

$$d(mn) \leq d(m)d(n).$$

If $(m, n) = 1$, then

$$d(mn) = d(m)d(n),$$

that is, the divisor function is multiplicative.

Proof. Every divisor d of m can be written uniquely in the form

$$d = p_1^{j_1} \cdots p_r^{j_r},$$

where

$$0 \leq j_i \leq k_i$$

for $i = 1, \ldots, r$. Since there are $k_i + 1$ choices of j_i for each $i = 1, \ldots, r$, it follows that

$$d(m) = \prod_{i=1}^{n} (k_i + 1).$$

Let n be a positive integer, and let

$$n = p_1^{\ell_1} \cdots p_r^{\ell_r},$$

where $\ell_1, \ldots, \ell_r$ are nonnegative integers. Then

$$d(n) = \prod_{i=1}^{r} (\ell_i + 1).$$

Since

$$mn = \prod_{i=1}^{r} p_i^{k_i + \ell_i},$$

and since

$$k_i + \ell_i + 1 \le (k_i + 1)(\ell_i + 1)$$

for all nonnegative numbers k_i and ℓ_i, it follows that

$$d(mn) = \prod_{i=1}^{r} (k_i + \ell_i + 1) \le \prod_{i=1}^{r} (k_i + 1)(\ell_i + 1) = d(m)d(n).$$

If $(m, n) = 1$, then $k_i = 0$ or $\ell_i = 0$ for each $i = 1, \ldots, r$. In this case,

$$k_i + \ell_i + 1 = (k_i + 1)(\ell_i + 1)$$

and

$$d(mn) = \prod_{i=1}^{r} (k_i + \ell_i + 1) = \prod_{\substack{i=1 \\ k_i \ne 0}}^{r} (k_i + 1) \prod_{\substack{i=1 \\ \ell_i \ne 0}}^{r} (\ell_i + 1) = d(m)d(n).$$

This completes the proof.

Theorem A.11

$$d(n) \ll_\varepsilon n^\varepsilon$$

for every $\varepsilon > 0$.

Proof. Let $f(n) = d(n)/n^\varepsilon$. We shall prove that $f(n) = o(1)$. Since the arithmetic functions $d(n)$ and n^ε are multiplicative, it follows that $f(n)$ is multiplicative, and so, by Theorem A.9, it suffices to prove that

$$\lim_{p^k \to \infty} f(p^k) = 0.$$

Since $(k + 1)/2^{k\varepsilon/2}$ is bounded for $k \geq 1$, we have

$$
\begin{aligned}
f(p^k) &= \frac{d(p^k)}{p^{k\varepsilon}} \\
&= \frac{k+1}{p^{k\varepsilon}} \\
&= \left(\frac{k+1}{p^{k\varepsilon/2}}\right)\left(\frac{1}{p^{k\varepsilon/2}}\right) \\
&\leq \left(\frac{k+1}{2^{k\varepsilon/2}}\right)\left(\frac{1}{p^{k\varepsilon/2}}\right) \\
&\ll \left(\frac{1}{p^k}\right)^{\varepsilon/2}.
\end{aligned}
$$

This completes the proof.

Theorem A.12

$$
D(x) = \sum_{n \leq x} d(n) = x \log x + (2\gamma - 1)x + O(\sqrt{x}).
$$

Proof. We can interpret the divisor function $d(n)$ and the sum function $D(x)$ geometrically. In the uv–plane,

$$
d(n) = \sum_{d \mid n} 1 = \sum_{n=uv} 1
$$

counts the number of lattice points (u, v) on the rectangular hyperbola $uv = n$ that lie in the quadrant $u > 0$, $v > 0$. Then $D(x)$ counts the number of lattice points in this quadrant that lie on or under the hyperbola $uv = x$, that is, the number of points (u, v) with positive integral coordinates such that $1 \leq u \leq x$ and $1 \leq v \leq x/u$. These lattice points can be divided into three pairwise disjoint classes:

$$
1 \leq u \leq \sqrt{x} \quad \text{and} \quad 1 \leq v \leq \sqrt{x},
$$

or

$$
1 \leq u \leq \sqrt{x} \quad \text{and} \quad \sqrt{x} < v \leq x/u,
$$

or

$$
\sqrt{x} < u \leq x \quad \text{and} \quad 1 \leq v \leq x/u.
$$

The last class consists of the lattice points (u, v) such that

$$
1 \leq v \leq \sqrt{x} \quad \text{and} \quad \sqrt{x} < u \leq x/v.
$$

It follows from Theorem A.5 that

$$
D(x) = \left[\sqrt{x}\right]^2 + \sum_{1 \leq u \leq \sqrt{x}} \left(\left[\frac{x}{u}\right] - \left[\sqrt{x}\right]\right) + \sum_{1 \leq v \leq \sqrt{x}} \left(\left[\frac{x}{v}\right] - \left[\sqrt{x}\right]\right)
$$

$$= \left[\sqrt{x}\right]^2 + 2 \sum_{1 \le u \le \sqrt{x}} \left(\left[\frac{x}{u}\right] - \left[\sqrt{x}\right]\right)$$

$$= 2 \sum_{1 \le u \le \sqrt{x}} \left[\frac{x}{u}\right] - \left[\sqrt{x}\right]^2$$

$$= 2 \sum_{1 \le u \le \sqrt{x}} \left(\frac{x}{u} - \left\{\frac{x}{u}\right\}\right) - \left(\sqrt{x} - \left\{\sqrt{x}\right\}\right)^2$$

$$= 2x \sum_{1 \le u \le \sqrt{x}} \frac{1}{u} - 2 \sum_{1 \le u \le \sqrt{x}} \left\{\frac{x}{u}\right\} - x + O(\sqrt{x})$$

$$= 2x \left(\log \sqrt{x} + \gamma + O\left(\frac{1}{\sqrt{x}}\right)\right) - x + O(\sqrt{x})$$

$$= x \log x + (2\gamma - 1)x + O(\sqrt{x}).$$

This completes the proof.

Theorem A.13

$$\sum_{n \le x} \frac{d(n)}{n} = \frac{1}{2}(\log x)^2 + O(\log x).$$

Proof. It follows from Theorem A.12 that

$$D(x) = \sum_{n \le x} d(n) = x \log x + O(x).$$

By partial summation, we obtain

$$\sum_{n \le x} \frac{d(n)}{n} = \frac{D(x)}{x} + \int_1^x \frac{D(t)}{t^2} dt$$

$$= \frac{x \log x + O(x)}{x} + \int_1^x \frac{t \log t + O(t)}{t^2} dt$$

$$= \log x + O(1) + \int_1^x \frac{\log t}{t} dt + O\left(\int_1^x \frac{1}{t} dt\right)$$

$$= \frac{1}{2}(\log x)^2 + O(\log x).$$

This completes the proof.

Theorem A.14

$$\sum_{n \le x} d(n)^2 \ll x (\log x)^3.$$

Proof. Since $d(ab) \le d(a)d(b)$ for all positive integers a and b, we have

$$\sum_{n \le x} d(n)^2 = \sum_{n \le x} d(n) \sum_{n=ab} 1$$

$$= \sum_{ab \le x} d(ab)$$

$$\le \sum_{ab \le x} d(a)d(b)$$

$$= \sum_{a \le x} d(a) \sum_{b \le x/a} d(b)$$

$$= \sum_{a \le x} d(a) \left(\left(\frac{x}{a} \right) \log \left(\frac{x}{a} \right) + O \left(\frac{x}{a} \right) \right)$$

$$\le x \log x \sum_{a \le x} \frac{d(a)}{a} + O \left(x \sum_{a \le x} \frac{d(a)}{a} \right)$$

$$\ll x (\log x)^3.$$

This completes the proof.

A.5   The Euler φ–function

Let $n \ge 1$. We denote by $\varphi(n)$ the number of positive integers $a \le n$ such that $(a, n) = 1$. If $a \equiv b \pmod{n}$, then $(a, n) = (b, n)$, and so $\varphi(n)$ also counts the number of congruence classes modulo n that are relatively prime to n. This is exactly the order of the multiplicative group of units in the ring $\mathbf{Z}/n\mathbf{Z}$.

Theorem A.15 *The arithmetic function $\varphi(n)$ is multiplicative, and*

$$\varphi(n) = n \prod_{p \mid n} \left(1 - \frac{1}{p} \right).$$

Proof. Let $(m, n) = 1$, and let $\varphi(m) = r$ and $\varphi(n) = s$. Let $a_1, \ldots, a_r$ and $b_1, \ldots, b_s$ be complete sets of representatives of the congruence classes relatively prime to m and n, respectively. We shall prove that the rs numbers $a_i n + b_j m$ for $i = 1, \ldots, r$ and $j = 1, \ldots, s$ form a complete set of representatives of the congruence classes relatively prime to mn. If

$$a_i n + b_j m \equiv a_k n + b_\ell n \pmod{mn},$$

then

$$a_i n + b_j m \equiv a_k n + b_\ell n \pmod{n}$$

and so

$$b_j m \equiv b_\ell m \pmod{n}.$$

Since $(m, n) = 1$, we have

$$b_j m \equiv b_\ell m \pmod{n}.$$

This implies that $j = \ell$. Similarly, we obtain $i = k$. Thus, the rs integers $a_i n + b_j m$ represent distinct congruence classes modulo mn. If $(a_i n + b_j m, mn) > 1$ for some i and j, then some prime p divides mn and $a_i n + b_j m$. Since $(m, n) = 1$, the prime p divides exactly one of m and n. If p divides m, then p divides $a_i n$, and so p divides a_i. This contradicts the fact that $(a_i, m) = 1$. Therefore, $(a_i n + b_j m, mn) = 1$ for all i and j.

We shall show that every congruence class relatively prime to mn is of this form. We note that $(m, n) = 1$ implies that the r integers $a_i n$ form a complete set of representatives of the congruence classes relatively prime to m, and the s integers $b_j m$ form a complete set of representatives of the congruence classes relatively prime to n. Let $(c, mn) = 1$. Then $(c, m) = 1$, and so

$$c \equiv a_i n \pmod{m}$$

for some i. Since

$$(c, n) = (c - a_i n, n) = 1,$$

it follows that

$$c - a_i n \equiv b_j m \pmod{n}$$

for some j. Therefore,

$$c \equiv a_i n + b_j m \pmod{n}$$

and

$$c \equiv a_i n + b_j m \pmod{m};$$

hence

$$c \equiv a_i n + b_j m \pmod{mn}.$$

Thus,

$$\varphi(mn) = rs = \varphi(m)\varphi(n).$$

This proves that φ is multiplicative. If p is prime and $k \geq 1$, the only integers not prime to p^k are multiples of p, and so

$$\varphi(p^k) = p^k - p^{k-1} = p^k \left(1 - \frac{1}{p}\right).$$

Therefore,

$$\varphi(n) = \prod_{\substack{p^k \| n \\ k \geq 1}} \varphi(p^k) = \prod_{\substack{p^k \| n \\ k \geq 1}} p^k \left(1 - \frac{1}{p}\right) = n \prod_{p | n} \left(1 - \frac{1}{p}\right).$$

This completes the proof.

Theorem A.16 *Let $\varepsilon > 0$. Then*

$$n^{1-\varepsilon} < \varphi(n) < n$$

for all sufficiently large n.

Proof. It is clear that $\varphi(n) < n$ for all $n > 1$. We shall prove that

$$\lim_{n \to \infty} \frac{n^{1-\varepsilon}}{\varphi(n)} = 0.$$

Since $p/(p-1) \leq 2$ for every prime number p, we have

$$\frac{p^{m(1-\varepsilon)}}{\varphi(p^m)} = \frac{p^{m(1-\varepsilon)}}{p^m - p^{m-1}} = \frac{p}{p-1} \frac{p^{m(1-\varepsilon)}}{p^m} \leq \frac{2}{p^{m\varepsilon}}.$$

Therefore,

$$\lim_{p^m \to \infty} \frac{p^{m(1-\varepsilon)}}{\varphi(p^m)} = 0.$$

Since the arithmetic function $n^{1-\varepsilon}/\varphi(n)$ is multiplicative, the result follows from Theorem A.9.

Theorem A.17

$$\sum_{n \leq x} \frac{1}{\varphi(n)} \ll \log x.$$

Proof. Let d^* denote the square-free part of d, that is,

$$d^* = \prod_{p \mid d} p.$$

Then

$$\frac{1}{\varphi(n)} = \frac{1}{n} \prod_{p \mid n} \left(1 - \frac{1}{p}\right)^{-1} = \frac{1}{n} \sum_{\substack{d=1 \\ d^* \mid n}}^{\infty} \frac{1}{d},$$

and so

$$\sum_{n \leq x} \frac{1}{\varphi(n)} = \sum_{n \leq x} \frac{1}{n} \sum_{\substack{d=1 \\ d^* \mid n}}^{\infty} \frac{1}{d}$$

$$= \sum_{d=1}^{\infty} \frac{1}{d} \sum_{\substack{n \leq x \\ d^* \mid n}} \frac{1}{n}$$

$$= \sum_{d=1}^{\infty} \frac{1}{d} \sum_{m \leq x/d^*} \frac{1}{d^* m}$$

$$\ll \sum_{d=1}^{\infty} \frac{1}{dd^*} \log x.$$

The integers of the form dd^* are precisely the integers that are *square-full* in the sense that if p divides d, then p^2 divides d for every prime p. We have

$$\sum_{d=1}^{\infty} \frac{1}{dd^*} = \prod_p \left(1 + \frac{1}{p^2} + \frac{1}{p^3} + \cdots\right)$$

$$= \prod_p \left(1 + \frac{1}{p^2}\left(1 - \frac{1}{p}\right)^{-1}\right)$$

$$= \prod_p \left(1 + \frac{1}{p(p-1)}\right)$$

$$= O(1).$$

This completes the proof.

A.6   The Möbius function

The fundamental property of the Möbius function is the following.

Theorem A.18

$$\sum_{d|n} \mu(d) = \delta(n) = \begin{cases} 1 & \text{if } n = 1, \\ 0 & \text{if } n \geq 2. \end{cases}$$

Proof. This is clearly true for $n = 1$. If $n \geq 2$, then

$$n = \prod_{i=1}^{k} p_i^{r_i},$$

where $k \geq 1$, $p_1, \ldots, p_k$ are distinct prime numbers, and $r_i \geq 1$ for $i = 1, \ldots, k$. Let $\sum'$ denote a sum over square-free integers. Then

$$\sum_{d|n} \mu(d) = \sum_{d|n}' \mu(d)$$

$$= \sum_{d|p_1 \cdots p_k} \mu(d)$$

$$= \sum_{d|p_1 \cdots p_k} (-1)^{\omega(d)}$$

$$= \sum_{\ell=0}^{k} \binom{k}{\ell}(-1)^\ell$$

$$= 0.$$

This completes the proof.

We define the arithmetic function

$$1(n) = 1$$

for all n. Then Theorem A.18 can be rewritten in the form

$$\mu * 1 = \delta.$$

A nonempty set $\mathcal{D}$ of positive integers is called *divisor-closed* if whenever $n \in \mathcal{D}$ and d divides n, then $d \in \mathcal{D}$. For example, the set N of all positive integers and the set of positive integers less than a fixed number z are examples of divisor-closed sets. The set of all divisors of a fixed positive integer is divisor-closed. If f and g are functions defined on a divisor-closed set $\mathcal{D}$, then their Dirichlet convolution $f * g$ is also defined on $\mathcal{D}$.

Theorem A.19 *Let $\mathcal{D}$ be a divisor-closed set, and let $f(n)$ be a function defined for all $n \in \mathcal{D}$. If g is the function defined on $\mathcal{D}$ by*

$$g(n) = \sum_{d \mid n} f(d),$$

then

$$f(n) = \sum_{d \mid n} \mu\left(\frac{n}{d}\right) g(d)$$

for all $n \in \mathcal{D}$.

Conversely, let g be a function defined on $\mathcal{D}$. If f is the function defined on $\mathcal{D}$ by

$$f(n) = \sum_{d \mid n} \mu\left(\frac{n}{d}\right) g(d),$$

then

$$g(n) = \sum_{d \mid n} f(d)$$

for all $n \in \mathcal{D}$.

Proof. If $n \in \mathcal{D}$ and $d \mid n$, then $d \in \mathcal{D}$, since $\mathcal{D}$ is divisor-closed. Let

$$g(n) = \sum_{d \mid n} f(d)$$

for $n \in \mathcal{D}$. Then

$$g = f * 1,$$

and so

$$\sum_{d \mid n} \mu\left(\frac{n}{d}\right) g(d) = (g * \mu)(n)$$

$$= ((f * 1) * \mu)(n)$$
$$= (f * (1 * \mu))(n)$$
$$= (f * \delta)(n)$$
$$= f(n).$$

Similarly, if

$$f(n) = \sum_{d \mid n} \mu\left(\frac{n}{d}\right) g(d) = (g * \mu)(n),$$

then

$$\sum_{d|n} f(d) = (f * 1)(n)$$

$$= ((g * \mu) * 1)(n)$$
$$= (g * (\mu * 1))(n)$$
$$= (g * \delta)(n)$$
$$= g(n).$$

This completes the proof.

Theorem A.20 *Let f and g be arithmetic functions. Then*

$$g(n) = \sum_{d|n} f(d)$$

if and only if

$$f(n) = \sum_{d|n} \mu\left(\frac{n}{d}\right) g(d).$$

Proof. This follows immediately from Theorem A.19 with the divisor-closed set $\mathcal{D}$ equal to the set **N** of all positive integers.

Theorem A.21 *Let $f(x)$ and $g(x)$ be functions defined for all real numbers $x \geq 1$. Then*

$$g(x) = \sum_{d \leq x} f(x/d)$$

if and only if

$$f(x) = \sum_{d \leq x} \mu(d) g(x/d).$$

Proof. Let f be a function defined for all $x \geq 1$. If

$$g(x) = \sum_{d \leq x} f(x/d),$$

then

$$\sum_{d \leq x} \mu(d) g(x/d) = \sum_{d \leq x} \mu(d) \sum_{d' \leq x/d} f(x/dd')$$

$$= \sum_{dd' \leq x} \mu(d) f(x/dd')$$

$$= \sum_{m \leq x} f(x/m) \sum_{d|m} \mu(d)$$

$$= f(x).$$

The proof in the opposite direction is similar.

Theorem A.22 *Let $\mathcal{D}$ be a finite divisor-closed set, and let f and g be functions defined on $\mathcal{D}$. If*

$$g(n) = \sum_{\substack{d\in\mathcal{D}\\ n|d}} f(d)$$

for all $n \in \mathcal{D}$, then

$$f(n) = \sum_{\substack{d\in\mathcal{D}\\ n|d}} \mu\left(\frac{d}{n}\right) g(d)$$

for all $n \in \mathcal{D}$. Conversely, if

$$f(n) = \sum_{\substack{d\in\mathcal{D}\\ n|d}} \mu\left(\frac{d}{n}\right) g(d)$$

for all $n \in \mathcal{D}$, then

$$g(n) = \sum_{\substack{d\in\mathcal{D}\\ n|d}} f(d)$$

for all $n \in \mathcal{D}$.

Proof. This is a straightforward computation:

$$\sum_{\substack{d\in\mathcal{D}\\ n|d}} \mu\left(\frac{d}{n}\right) g(d) = \sum_{\substack{d\in\mathcal{D}\\ n|d}} \mu\left(\frac{d}{n}\right) \sum_{\substack{k\in\mathcal{D}\\ d|k}} f(k)$$

$$= \sum_{\substack{nh\in\mathcal{D}}} \mu(h) \sum_{\substack{k\in\mathcal{D}\\ nh|k}} f(k)$$

$$= \sum_{\substack{nh\in\mathcal{D}}} \mu(h) \sum_{\substack{nh\ell\in\mathcal{D}}} f(nh\ell)$$

$$= \sum_{\substack{nr\in\mathcal{D}}} f(nr) \sum_{\substack{h\in\mathcal{D}\\ h|r}} \mu(h)$$

$$= \sum_{\substack{nr\in\mathcal{D}}} f(nr) \sum_{h|r} \mu(h)$$

$$= f(n).$$

The proof in the opposite direction is similar.

A.7   Ramanujan sums

Let q and n be integers with $q \geq 1$. The exponential sum

$$c_q(n) = \sum_{\substack{a=1\\ (a,q)=1}}^{q} e\left(\frac{an}{q}\right) \tag{A.2}$$

is called the *Ramanujan sum*. These sums play an important role in the proof of Vinogradov's theorem (Chapter 8).

Theorem A.23 *The Ramanujan sum $c_q(n)$ is a multiplicative function of q, that is, if $(q, q') = 1$, then*

$$c_{qq'}(n) = c_q(n)c_{q'}(n).$$

Proof. Since every congruence class relatively prime to qq' can be written uniquely in the form $aq'+a'q$ with $1 \le a \le q, 1 \le a' \le q'$, and $(a, q) = (a', q') = 1$, it follows that if $(q, q') = 1$, then

$$c_q(n)c_{q'}(n) = \sum_{\substack{a=1 \\ (a,q)=1}}^{q} e\left(\frac{an}{q}\right) \sum_{\substack{a'=1 \\ (a',q')=1}}^{q'} e\left(\frac{a'n}{q'}\right)$$

$$= \sum_{\substack{a=1 \\ (a,q)=1}}^{q} \sum_{\substack{a'=1 \\ (a',q')=1}}^{q'} e\left(\frac{(aq'+a'q)n}{qq'}\right)$$

$$= \sum_{\substack{a''=1 \\ (a'',qq')=1}}^{qq'} e\left(\frac{a''n}{qq'}\right)$$

$$= c_{qq'}(n).$$

Theorem A.24 *The Ramanujan sum can be expressed in the form*

$$c_q(n) = \sum_{d|(q,n)} \mu\left(\frac{q}{d}\right) d.$$

In particular, if $(q, n) = 1$, then

$$c_q(n) = \mu(q).$$

Proof. Since

$$f_d(n) = \sum_{\ell=1}^{d} e\left(\frac{\ell n}{d}\right) = \begin{cases} d & \text{if } d|n \\ 0 & \text{if } d \nmid n, \end{cases}$$

it follows that

$$c_q(n) = \sum_{\substack{k=1 \\ (k,q)=1}}^{q} e\left(\frac{kn}{q}\right)$$

$$= \sum_{k=1}^{q} e\left(\frac{kn}{q}\right) \sum_{d|(k,q)} \mu(d)$$

$$= \sum_{d|q} \mu(d) \sum_{\substack{k=1 \\ d|k}}^{q} e\left(\frac{kn}{q}\right)$$

$$= \sum_{d|q} \mu(d) \sum_{\ell=1}^{q/d} e\left(\frac{\ell n}{q/d}\right)$$

$$= \sum_{d|q} \mu(d) f_{q/d}(n)$$

$$= \sum_{d|q} \mu(q/d) f_d(n)$$

$$= \sum_{\substack{d|q \\ d|n}} \mu(q/d)d$$

$$= \sum_{d|(n,q)} \mu(q/d)d.$$

If $(q, n) = 1$, then $c_q(n) = \mu(q)$.

Theorem A.25 *The Ramanujan sum can be expressed in the form*

$$c_q(n) = \frac{\mu(q/(q, n))\varphi(q)}{\varphi(q/(q, n))}.$$

Proof. We define

$$q' = q/(q, n).$$

If the prime p divides q but not q', then $p|(q, n)$. It follows from Theorem A.15 that

$$\frac{\varphi(q)}{\varphi(q')} = \frac{q \prod_{p|q}(1 - 1/p)}{q' \prod_{p|q'}(1 - 1/p)}$$

$$= (q, n) \prod_{\substack{p|q \\ p\nmid q'}}(1 - 1/p)$$

$$= (q, n) \prod_{\substack{p|(q,n) \\ p\nmid q'}}(1 - 1/p).$$

Then

$$c_q(n) = \sum_{d|(q,n)} \mu\left(\frac{q}{d}\right)d$$

$$= \sum_{d|(q,n)} \mu\left(\frac{q}{(q, n)}\frac{(q, n)}{d}\right)d$$

$$= \sum_{cd=(q,n)} \mu\left(q'c\right)d$$

$$= \sum_{\substack{cd=(q,n) \\ (q',c)=1}} \mu\left(q'\right)\mu(c)d$$

$$= \sum_{\substack{cd=(q,n) \\ (q',c)=1}} \mu\left(q'\right)\frac{\mu(c)}{c}cd$$

$$= \mu\left(q'\right)(q,n) \sum_{\substack{c|(q,n)\\(q'.c)=1}} \frac{\mu(c)}{c}$$

$$= \mu\left(q'\right)(q,n) \prod_{\substack{p|(q,n)\\p|q'}} \left(1 - \frac{1}{p}\right)$$

$$= \frac{\mu(q')\varphi(q)}{\varphi(q')}.$$

This completes the proof.

A.8 Infinite products

This is a brief introduction to infinite products and Euler products.

Let $\alpha_1, \alpha_2, \ldots, \alpha_n, \ldots$ be a sequence of complex numbers. The *nth partial product* of this sequence is the number

$$p_n = \alpha_1 \cdots \alpha_n = \prod_{k=1}^{n} \alpha_k.$$

If as n tends to infinity, the sequence of nth partial products converges to a limit α different from zero, then we say that the infinite product $\prod_{k=1}^{\infty} \alpha_k$ converges and

$$\prod_{k=1}^{\infty} \alpha_k = \lim_{n \to \infty} p_n = \lim_{n \to \infty} \prod_{k=1}^{n} \alpha_k = \alpha.$$

We say that the infinite product diverges if either the limit of the sequence of partial products does not exist or the limit exists but is equal to zero. In the latter case, we say that the infinite product *diverges to zero*.

Let

$$\alpha_k = 1 + a_k.$$

If the infinite product $\prod_{k=1}^{\infty}(1 + a_k)$ converges, then $a_k \neq -1$ for all k. Moreover,

$$\lim_{k \to \infty} (1 + a_k) = \lim_{k \to \infty} \frac{p_k}{p_{k-1}} = 1,$$

and so

$$\lim_{k \to \infty} a_k = 0.$$

Theorem A.26 *Let $a_k \geq 0$ for all $k \geq 1$. The infinite product $\prod_{k=1}^{\infty}(1 + a_k)$ converges if and only if the infinite series $\sum_{k=1}^{\infty} a_k$ converges.*

Proof. Let $s_n = \sum_{k=1}^{n} a_k$ be the nth partial sum and let $p_n = \prod_{k=1}^{n}(1 + a_k)$ be the nth partial product. Since $a_n \geq 0$, the sequences $\{s_n\}$ and $\{p_n\}$ are both monotonically increasing, and $p_n \geq 1$ for all n. Since

$$1 + x \leq e^x$$

for all real numbers x, we have

$$0 \le \sum_{k=1}^{n} a_k < \prod_{k=1}^{n}(1 + a_k) \le \prod_{k=1}^{n} e^{a_k} = \exp\left(\sum_{k=1}^{n} a_k\right),$$

and so

$$0 \le s_n < p_n \le e^{s_n}.$$

This inequality implies that the sequence $\{p_n\}$ converges if and only if the sequence $\{s_n\}$ converges. This completes the proof.

We say that the infinite product $\prod_{n=1}^{\infty}(1 + a_n)$ *converges absolutely* if the infinite product

$$\prod_{n=1}^{\infty}(1 + |a_n|)$$

converges.

Theorem A.27 *If the infinite product $\prod_{n=1}^{\infty}(1 + a_n)$ converges absolutely, then it converges.*

Proof. Let

$$p_n = \prod_{k=1}^{n}(1 + a_k)$$

and let

$$P_n = \prod_{k=1}^{n}(1 + |a_k|).$$

If the infinite product converges absolutely, then the sequence of partial products $\{P_n\}$ converges and so the series

$$\sum_{n=2}^{\infty}(P_n - P_{n-1})$$

converges. Since

$$0 \le |p_n - p_{n-1}|$$
$$= |a_n p_{n-1}|$$
$$= \left|a_n \prod_{k=1}^{n-1}(1 + a_k)\right|$$
$$\le |a_n| \prod_{k=1}^{n-1}(1 + |a_k|)$$
$$= |a_n| P_{n-1}$$
$$= P_n - P_{n-1},$$

it follows that

$$\sum_{n=2}^{\infty} |p_n - p_{n-1}|$$

converges, and so

$$\sum_{n=2}^{\infty}(p_n - p_{n-1}) = \lim_{n\to\infty}\sum_{k=2}^{n}(p_k - p_{k-1}) = \lim_{n\to\infty}(p_n - p_1)$$

converges. Thus, the sequence of partial products $\{p_n\}$ converges to some finite limit.

We must prove that this limit is not zero. Since the infinite product $\prod_{k=1}^{\infty}(1 + a_k)$ converges absolutely, it follows from Theorem A.26 that the series $\sum_{k=1}^{\infty}|a_k|$ converges, and so the numbers a_k converge to zero. Therefore, for all sufficiently large integers k,

$$|1 + a_k| \geq 1/2$$

and

$$\left|\frac{-a_k}{1+a_k}\right| \leq 2|a_k|$$

It follows that the series

$$\sum_{k=1}^{\infty}\left|\frac{-a_k}{1+a_k}\right|$$

converges, and so the infinite product

$$\prod_{k=1}^{\infty}\left(1 - \frac{a_k}{1+a_k}\right)$$

converges absolutely. This implies that the sequence of nth partial products

$$\prod_{k=1}^{n}\left(1 - \frac{a_k}{1+a_k}\right) = \prod_{k=1}^{n}\frac{1}{1+a_k} = \frac{1}{\prod_{k=1}^{n}(1+a_k)} = \frac{1}{p_n}$$

converges to a finite limit, and so the limit of the sequence $\{p_n\}$ is nonzero. Therefore, the infinite product $\prod_{k=1}^{\infty}(1 + a_k)$ converges.

An *Euler product* is an infinite product over the prime numbers. We denote sums and products over the primes by $\sum_p$ and $\prod_p$, respectively.

Theorem A.28 *Let $f(n)$ be a multiplicative function that is not identically zero. If the series*

$$\sum_{n=1}^{\infty} f(n)$$

converges absolutely, then

$$\sum_{n=1}^{\infty} f(n) = \prod_{p}\left(1 + f(p) + f(p^2) + \cdots\right) = \prod_{p}\left(1 + \sum_{k=1}^{\infty} f(p^k)\right).$$

If $f(n)$ is completely multiplicative, then

$$\sum_{n=1}^{\infty} f(n) = \prod_{p} (1 - f(p))^{-1}.$$

Proof. If $\sum_{n=1}^{\infty} f(n)$ converges absolutely, then the series

$$a_p = \sum_{k=1}^{\infty} f(p^k)$$

converges absolutely for every prime p. Also, the series

$$\sum_{p} |a_p| = \sum_{p} \left| \sum_{k=1}^{\infty} f(p^k) \right|$$

$$\leq \sum_{p} \sum_{k=1}^{\infty} |f(p^k)|$$

$$< \sum_{n=1}^{\infty} |f(n)|$$

converges, and so the infinite product

$$\prod_{p} (1 + a_p) = \prod_{p} \left(1 + \sum_{k=1}^{\infty} f(p^k) \right)$$

converges absolutely. By Theorem A.27, this infinite product converges.

Let $\varepsilon > 0$, and choose an integer N_0 such that

$$\sum_{n > N_0} |f(n)| < \varepsilon.$$

For every positive integer n, let $P(n)$ denote the greatest prime factor of n. Then $\sum_{P(n) \leq N}$ denotes the sum over the integers all of whose prime factors are less than or equal to N, and $\sum_{P(n) > N}$ denotes the sum over the integers that have at least one prime factor strictly greater than N. Since the series $\sum_{k=0}^{\infty} f(p^k)$ converges absolutely for every prime number p, any finite number of these series can be multiplied together term by term. Let $N \geq N_0$. It follows from the unique factorization of integers as products of primes that

$$\prod_{p \leq N} \left(1 + \sum_{k=1}^{\infty} f(p^k) \right) = \sum_{P(n) \leq N} f(n)$$

and so

$$\left| \sum_{n=1}^{\infty} f(n) - \prod_{p \leq N} \left(1 + \sum_{k=1}^{\infty} f(p^k) \right) \right| = \left| \sum_{n=1}^{\infty} f(n) - \sum_{P(n) \leq N} f(n) \right|$$

$$= \left| \sum_{P(n)>N} f(n) \right|$$

$$\leq \sum_{P(n)>N} |f(n)|$$

$$\leq \sum_{n>N} |f(n)|$$

$$\leq \sum_{n>N_0} |f(n)|$$

$$< \varepsilon.$$

Therefore,

$$\sum_{n=1}^{\infty} f(n) = \lim_{N \to \infty} \prod_{p \leq N} \left(1 + \sum_{k=1}^{\infty} f(p^k) \right) = \prod_{p} \left(1 + \sum_{k=1}^{\infty} f(p^k) \right).$$

If $f(n)$ is completely multiplicative, then $f(p^k) = f(p)^k$ for all primes p and all nonnegative integers k. Since $f(p^k)$ tends to zero as k tends to infinity, it follows that $|f(p)| < 1$. Summing the geometric progression, we obtain

$$1 + \sum_{k=1}^{\infty} f(p^k) = 1 + \sum_{k=1}^{\infty} f(p)^k = \frac{1}{1 - f(p)},$$

and so

$$\prod_{p} \left(1 + \sum_{k=1}^{\infty} f(p^k) \right) = \prod_{p} (1 - f(p))^{-1}.$$

This completes the proof.

A.9   Notes

All of the material in this chapter is basic elementary number theory. Comprehensive standard references are the books of Hardy and Wright[51] and Hua [63]. Cashwell and Everett [8] proved that the ring of arithmetic functions is a unique factorization domain. Hardy's book *Ramanujan* [46] contains a chapter on Ramanujan's function $c_q(n)$ and its connection to the problem of representing numbers as sums of squares.

A.10   Exercises

1. Prove that

$$\sum_{k|n} \mu(k) d(n/k) = 1$$

for all $n \geq 1$.

2. Prove that if f and g are multiplicative functions, then the Dirichlet convolution $f * g$ is multiplicative.

3. Let f and g be arithmetic functions. Prove that if $f * g = 0$, then either $f = 0$ or $g = 0$. Thus, the ring of arithmetic functions is an integral domain.

4. An arithmetic function $f(n)$ is *additive* if $f(mn) = f(m) + f(n)$ for all positive integers m and n such that $(m, n) = 1$. An arithmetic function $f(n)$ is *completely additive* if $f(mn) = f(m) + f(n)$ for all positive integers m and n. Let $n = p_1^{r_1} \cdots p_k^{r_k}$. We define the arithmetic functions $\omega(n)$ and $\Omega(n)$ as follows. The arithmetic function $\omega(n)$ counts the number of distinct prime factors of n:

$$\omega(n) = k.$$

The arithmetic function $\Omega(n)$ counts the number of prime factors of n with multiplicities:

$$\Omega(n) = r_1 + \cdots + r_k.$$

Prove that $\omega(n)$ is additive but not completely additive. Prove that $\Omega(n)$ is completely additive.

5. Let $n = p_1^{r_1} \cdots p_k^{r_k}$. Liouville's function $\lambda(n)$ is defined by

$$\lambda(n) = (-1)^{\Omega(n)} = (-1)^{r_1 + \cdots + r_k}.$$

Prove that $\lambda(n)$ is completely additive.

6. Let $f(n)$ be an arithmetic function. There exists a unique completely multiplicative function $f_1(n)$ such that $f_1(p) = f(p)$ for all primes p. Show that $\mu_1(n) = \lambda(n)$.

7. Show that the functions $\mu(n)$, $\varphi(n)$, and $\sigma_\lambda(n)$ are not completely multiplicative.

8. Prove that

$$d(n) \le 2^{\Omega(n)} \le n$$

for every positive integer n. Prove that if n is square-free, then

$$d(n) = 2^{\omega(n)} = 2^{\Omega(n)}.$$

9. Prove that

$$\sum_{n \le x} (d(n))^2 \gg x(\log x)^2,$$

Hint: Apply the Cauchy–Schwarz inequality to $\sum_{n \le x} d(n)$.

10. Let f be an arithmetic function. Prove that f is invertible in the ring of arithmetic functions if and only if $f(1) = 1$.

11. Let f and g be arithmetic functions. Define the function L by

$$L(n) = \log n.$$

Prove that pointwise multiplication by $L(n)$ is a derivation on the ring of arithmetic functions, that is,

$$L \cdot (f * g) = (L \cdot f) * g + f * (L \cdot g).$$

12. Let f and g be arithmetic functions with Dirichlet generating functions $F(s)$ and $G(s)$, respectively. Prove that $F'(s)$ is the generating function for $L \cdot f$ and that $(F(s)G(s))'$ is the generating function for $L \cdot (f * g)$.

13. Prove that

$$f_q(n) = \sum_{a=1}^{q} e\left(\frac{an}{q}\right) = \sum_{d|q} c_d(n).$$

Use Möbius inversion to deduce Theorem A.24 from this identity.

14. Let

$$\sigma(n) = \sum_{d|n} d.$$

Prove that

$$n < \sigma(n) \le n \log n + O(n).$$

Hint: $\sigma(n) = \sum_{d|n} n/d$.

15. Let $\mu(n)$ be the Möbius function. Prove that

$$\sum_{n=1}^{\infty} \frac{\mu(n)}{n^s} = \prod_{p}\left(1 - \frac{1}{p^s}\right)$$

for all $s > 1$.

16. Prove that the Dirichlet convolution of arithmetic functions is associative, that is, if $f(n)$, $g(n)$, and $h(n)$ are arithmetic functions, then

$$(f * g) * h = f * (g * h).$$

17. Let $L(n) = \log n$ for all $n \ge 1$. For any arithmetic function f, define Lf by $Lf(n) = L(n)f(n)$. Prove that L is a derivation on the ring of arithmetic functions, that is,

$$L(f * g) = (Lf) * g + f * (Lg).$$

18. Let f, g, and h be arithmetic functions. Prove that

$$g(n) = \sum_{d|n} f(d)h(n/d)$$

    if and only if

$$f(n) = \sum_{d|n} \mu(d)g(n/d)h(d).$$

19. Compute

$$\prod_{k=2}^{\infty}\left(1 - \frac{2}{k(k+1)}\right).$$

20. Show that the infinite product

$$\prod_{k=2}^{\infty}\left(1 + \frac{(-1)^{k-1}}{k}\right)$$

    converges, but not absolutely.

21. Let $0 \le b_n < 1$ for all n. Prove that if $\sum_{n=1}^{\infty} b_n$ converges, then $\prod_{n=1}^{\infty}(1-b_n)$ converges.

22. Let $0 \le b_n < 1$ for all n. Prove that if $\sum_{n=1}^{\infty} b_n$ diverges, then $\prod_{n=1}^{\infty}(1-b_n)$ diverges to zero.

Bibliography

[1] T. M. Apostol. *Mathematical Analysis*. Addison-Wesley, Reading, Mass., 1957.

[2] R. Balasubramanian. On Waring's problem: $g(4) \leq 20$. *Hardy–Ramanujan J.*, 8:1–40, 1985.

[3] E. Bombieri. *Le grand crible dans la théorie analytique des nombres*. Number 18 in Astérisque. Société Mathématique de France, Paris, 1974.

[4] E. Bombieri, J. B. Friedlander, and H. Iwaniec. Primes in arithmetic progressions to large moduli. *Acta Math.*, 156:203–251, 1986.

[5] R. P. Brent. Irregularities in the distribution of primes and twin primes. *Math. Comput.*, 29:43–56, 1975.

[6] J. Brüdern. On Waring's problem for cubes. *Math. Proc. Cambridge Philos. Soc.*, 109:229–256, 1991.

[7] V. Brun. Le crible d'Eratosthène et le théorème de Goldbach. *Skrifter utgit av Videnskapsselskapet i Kristiania, I. Matematisk-Naturvidenskabelig Klasse*, 1(3):1–36, 1920.

[8] E. D. Cashwell and C. J. Everett. The ring of number-theoretic functions. *Pacific J. Math.*, 9:975–985, 1959.

[9] A. L. Cauchy. Démonstration du théorème général de fermat sur les nombres polygones. *Mém. Sci. Math. Phys. Inst. France*, 14(1):177–220, 1813–1815. Oeuvres(2), vol. 6, 320–353.

[10] J. Chen. On the representation of a large even integer as the sum of a prime and the product of at most two primes. *Kexue Tongbao*, 17:385–386, 1966.

[11] J. Chen. On the representation of a larger even integer as the sum of a prime and the product of at most two primes. *Sci. Sinica*, 16:157–176, 1973.

[12] S. L. G. Choi. Covering the set of integers by congruence classes of distinct moduli. *Math. Comput.*, 25:885–895, 1971.

[13] S. L. G. Choi, P. Erdős, and M. B. Nathanson. Lagrange's theorem with $N^{1/3}$ squares. *Proc. Am. Math. Soc.*, 79:203–205, 1980.

[14] N. G. Chudakov. On the density of the set of even integers which are not representable as a sum of two odd primes. *Izv. Akad. Nauk SSSR*, 2:25–40, 1938.

[15] B. Cipra. How number theory got the best of the pentium chip. *Science*, 267:175, 1995.

[16] R. Crocker. On the sum of a prime and two powers of two. *Pacific J. Math.*, 36:103–107, 1971.

[17] H. Davenport. On Waring's problem for cubes. *Acta Math.*, 71:123–143, 1939.

[18] H. Davenport. *Analytic Methods for Diophantine Equations and Diophantine Inequalities*. Ann Arbor Publishers, Ann Arbor, 1962.

[19] H. Davenport. *Multiplicative Number Theory*. Springer-Verlag, New York, 2nd edition, 1980.

[20] V. A. Dem'yanenko. On sums of four cubes. *Izv. Vyssh. Uchebn. Zaved. Mat.*, 54(5):64–69, 1966.

[21] J.-M. Deshouillers and F. Dress. Sums of 19 biquadrates: On the representation of large integers. *Annali Scuola Normale Super. Pisa*, 19:113–153, 1992.

[22] L. E. Dickson. *History of the Theory of Numbers*. Carnegie Institute of Washington, Washington, 1919; reprinted by Chelsea Publishing Company in 1971.

[23] L. E. Dickson. All positive integers are sums of values of a quadratic function of x. *Bull. Am. Math. Soc.*, 33:713–720, 1927.

[24] L. E. Dickson. All integers except 23 and 239 are sums of eight cubes. *Bull. Am. Math. Soc.*, 45:588–591, 1939.

[25] F. Dress. Théorie additive des nombres, probléme de waring et théorème de Hilbert. *Enseign. Math.*, 18:175–190, 301–302, 1972.

[26] H. B. Dwight. *Mathematical Tables*. Dover Publications, New York, 3rd edition, 1961.

[27] N. Elkies and I. Kaplansky. Problem 10426. *Am. Math. Monthly*, 102:70, 1995.

[28] W. J. Ellison. Waring's problem. *Am. Math. Monthly*, 78:10–36, 1971.

[29] W. J. Ellison and F. Ellison. *Prime Numbers*. John Wiley & Sons, New York, 1985.

[30] P. Erdős and P. Turán. Ein zahlentheoretischer Satz. *Izv. Inst. Math. Mech. Tomsk State Univ.*, 1:101–103, 1935.

[31] P. Erdős. On the integers of the form $x^k + y^k$. *J. London Math. Soc.*, 14:250–254, 1939.

[32] P. Erdős. On integers of the form $2^k + p$ and some related problems. *Summa Brasil. Math.*, 2:113–123, 1950.

[33] P. Erdős. On some problems of Bellman and a theorem of Romanoff. *J. Chinese Math. Soc. (N.S.)*, 1:409–421, 1951.

[34] P. Erdős. Some recent advances and current problems in number theory. In *Lectures on Modern Mathematics*, volume 3, pages 196–244. Wiley, New York, 1965.

[35] P. Erdős and K. Mahler. On the number of integers which can be represented by a binary form. *J. London Math. Soc.*, 13:134–139, 1938.

[36] P. Erdős and M. B. Nathanson. Lagrange's theorem and thin subsequences of squares. In J. Gani and V. K. Rohatgi, editors, *Contributions to Probability*, pages 3–9. Academic Press, New York, 1981.

[37] T. Estermann. On Goldbach's problem: Proof that almost all positive integers are sums of two primes. *Proc. London Math. Soc.*, 44:307–314, 1938.

[38] T. Estermann. *Introduction to Modern Prime Number Theory*. Campridge University Press, Cambridge, England, 1952.

[39] P. Fermat. *Oeuvres*. Gauthier-Villars et Fils, Paris, 1891.

[40] A. Fleck. Über die Darstellung ganzer Zahlen als Summen von sechsten Potenzen ganzer Zahlen. *Mat. Annalen*, 64:561, 1907.

[41] K. B. Ford. New estimates for mean values of Weyl sums. *Int. Math. Res. Not.*, (3):155–171, 1995.

[42] I. S. Gradshteyn and I. M. Ryzhik. *Table of Integrals, Series, and Products*. Academic Press, San Diego, 5th edition, 1994.

[43] E. Grosswald. *Representations of Integers as Sums of Squares*. Springer-Verlag, New York, 1985.

[44] H. Halberstam and H.-E. Richert. *Sieve Methods*. Academic Press, London, 1974.

[45] G. H. Hardy. On the representation of an integer as the sum of any number of squares, and in particular of five. *Trans. Am. Math. Soc.*, 21:255–284, 1920.

[46] G. H. Hardy. *Ramanujan: Twelve Lectures on Subjects Suggested by his Life and Work*. Chelsea Publishing Company, New York, 1959.

[47] G. H. Hardy and J. E. Littlewood. A new solution of Waring's problem. *Q. J. Math.*, 48:272–293, 1919.

[48] G. H. Hardy and J. E. Littlewood. Some problems of "Partitio Numerorum". A new solution of Waring's problem. *Göttingen Nach.*, pages 33–54, 1920.

[49] G. H. Hardy and J. E. Littlewood. Some problems of "Partitio Numerorum": VI. Further researches in Waring's problem. *Mat. Z.*, 23:1–37, 1925.

[50] G. H. Hardy and S. Ramanujan. Asymptotic formulae in combinatory analysis. *Proc. London Math. Soc.*, 17:75–115, 1918.

[51] G. H. Hardy and E. M. Wright. *An Introduction to the Theory of Numbers*. Oxford University Press, Oxford, 5th edition, 1979.

[52] F. Hausdorff. Zur Hilbertschen Lösung des Waringschen Problems. *Mat. Annalen*, 67:301–305, 1909.

[53] T. L. Heath. *Diophantus of Alexandria: A Study in the History of Greek Algebra*. Dover Publications, New York, 1964.

[54] D.R. Heath-Brown. Cubic forms in ten variables. *Proc. London Math. Soc.*, 47:225–257, 1983.

[55] D.R. Heath-Brown. Weyl's inequality, Hua's inequality, and Waring's problem. *J. London Math. Soc.*, 38:216–230, 1988.

[56] D. Hilbert. Beweis für die Darstellbarkeit der ganzen zahlen durch eine feste Anzahl n^{ter} Potenzen (Waringsches Problem). *Mat. Annalen*, 67:281–300, 1909.

[57] C. Hooley. On the representation of a number as a sum of two cubes. *Mat. Z.*, 82:259–266, 1963.

[58] C. Hooley. On the numbers that are representable as the sum of two cubes. *J. reine angew. Math.*, 314:146–173, 1980.

[59] C. Hooley. On nonary cubic forms. *J. reine angew. Math.*, 386:32–98, 1988.

[60] C. Hooley. On nonary cubic forms. *J. reine angew. Math.*, 415:95–165, 1991.

[61] C. Hooley. On nonary cubic forms. *J. reine angew. Math.*, 456:53–63, 1994.

[62] H. L. Hua. On Waring's problem. *Q. J. Math.*, 9:199–202, 1938.

[63] H. L. Hua. *Introduction to Number Theory*. Springer-Verlag, Berlin, 1982.

[64] L. K. Hua. *Additive Theory of Prime Numbers*, volume 13 of *Translations of Mathematical Monographs*. American Mathematical Society, Providence, R.I., 1965.

[65] A. Hurwitz. Über die Darstellung der ganzen Zahlen als Summen von n^{ten} Potenzen ganzer Zahlen. *Mat. Annalen*, 65:424–427, 1908.

[66] A. E. Ingham. *The Distribution of Prime Numbers*. Number 30 in Cambridge Tracts in Mathematics and Mathematical Physics. Cambridge University Press, Cambridge, 1932. Reprinted in 1992.

[67] H. Iwaniec. Introduction to the prime number theory. Unpublished lecture notes, 1994.

[68] H. Iwaniec. Sieve methods. Unpublished lecture notes, 1996.

[69] W. B. Jurkat and H.-E. Richert. An improvement of Selberg's sieve method. I. *Acta Arith.*, 11:207–216, 1965.

[70] A. Kempner. Bemerkungen zum Waringschen Problem. *Mat. Annalen*, 72:387–399, 1912.

[71] H. D. Kloosterman. On the representation of numbers in the form $ax^2+by^2+cz^2+dt^2$. *Acta Math.*, 49:407–464, 1925.

[72] H. D. Kloosterman. On the representation of numbers in the form $ax^2+by^2+cz^2+dt^2$. *Proc. London Math. Soc.*, 25:143–173, 1925.

[73] H. D. Kloosterman. Over het uitdrukken van geheele positieve getallen in den vorm $ax^2 + by^2 + cz^2 + dt^2$. *Verslag Amsterdam*, 34:1011–1015, 1925.

[74] M. I. Knopp. *Modular Functions in Analytic Number Theory*. Markham Publishing Co., Chicago, 1970; reprinted by Chelsea in 1994.

[75] E. Landau. Über eine Anwendung der Primzahltheorie auf das Waringsche Problem in der elementaren Zahlentheorie. *Mat. Annalen*, 66:102–106, 1909.

[76] E. Landau. Die Goldbachsche Vermutung und der Schnirelmannsche Satz. *Gottinger Nachrichten, Math. Phys. Klasse*, pages 255–276, 1930.

[77] E. Landau. *Über einige neuere Fortschritte der additiven Zahlentheorie*. Cambridge University Press, Cambridge, 1937.

[78] E. Landau. *Elementary Number Theory*. Chelsea Publishing Company, New York, 1966.

[79] V. A. Lebesgue. *Exercices d'Analyse Numl' erique*. Paris, 1859.

[80] A.-M. Legendre. *Théorie des Nombres*. Firmin-Didot, Paris, 3rd edition, 1830.

[81] Yu. V. Linnik. On the representation of large numbers as sums of seven cubes. *Mat. Sbornik NS*, 12:218–224, 1943.

[82] K. Mahler. Note on hypothesis K of Hardy and Littlewood. *J. London Math. Soc.*, 11:136–138, 1936.

[83] H. L. Montgomery. *Topics in Multiplicative Number Theory*. Number 227 in Lecture Notes in Mathematics. Springer-Verlag, Berlin, 1971.

[84] H. L. Montgomery and R. C. Vaughan. The exceptional set in Goldbach's problem. *Acta Arith.*, 27:353–370, 1975.

[85] L. J. Mordell. On the four integer cubes problem. *J. London Math. Soc.*, 11:208–218, 1936.

[86] L. J. Mordell. *Diophantine Equations*. Academic Press, London, 1969.

[87] Y. Motohashi. *Sieve Methods and Prime Number Theory*. Tata Institute for Fundamental Research, Bombay, India, 1983.

[88] M. B. Nathanson. Products of sums of powers. *Math. Mag.*, 48:112–113, 1975.

[89] M. B. Nathanson. Waring's problem for sets of density zero. In M. I. Knopp, editor, *Analytic Number Theory*, volume 899 of *Lecture Notes in Mathematics*, pages 301–310, Berlin, 1981. Springer-Verlag.

[90] M. B. Nathanson. A generalization of the Goldbach-Shnirel'man theorem. *Am. Math. Monthly*, 94:768–771, 1987.

[91] M. B. Nathanson. A short proof of Cauchy's polygonal number theorem. *Proc. Am. Math. Soc.*, 99:22–24, 1987.

[92] M. B. Nathanson. Sums of polygonal numbers. In A.C. Adolphson, J. B. Conrey, A. Ghosh, and R. I. Yager, editors, *Analytic Number Theory and Diophantine Problems*, volume 70 of *Progress in Mathematics*, pages 305–316, Boston, 1987. Birkhäuser.

[93] M. B. Nathanson. *Additive Number Theory: Inverse Problems and the Geometry of Sumsets*, volume 165 of *Graduate Texts in Mathematics*. Springer-Verlag, New York, 1996.

[94] A. Oppenheim. Hilbert's proof of Waring's problem. *Messenger Math.*, 58:153–158, 1928.

[95] T. Pepin. Démonstration du théoreme de Fermat sur les nombres polygones. *Atti Accad. Pont. Nuovi Lincei*, 46:119–131, 1892-93.

[96] H. Poincaré. Rapport sur le prix Bolyai. *Acta Math.*, 35:1–28, 1912.

[97] K. Prachar. *Primzahlverteilung*. Springer-Verlag, Berlin, 1957.

[98] H. Rademacher. *Topics in Analytic Number Theory*. Springer-Verlag, New York, 1973.

[99] D. Raikov. Über die Basen der natürlichen Zahlentreihe. *Mat. Sbornik N. S. 2*, 44:595–597, 1937.

[100] O. Ramaré. On šnirel'man's constant. Preprint, 1995.

[101] P. Revoy. Sur les sommes de quatre cubes. *Enseignement Math.*, 29:209–220, 1983.

[102] G. J. Rieger. Zur Hilbertschen Lösung des Waringschen Problems: Abschätzung von $g(n)$. *Archiv Math.*, 4:275–281, 1953.

[103] N. Romanov. Über einige Sätze der additiven Zahlentheorie. *Mat. Annalen*, 109:668–678, 1934.

[104] M. Rosen. A generalization of Mertens' theorem. Preprint, 1995.

[105] E. Schmidt. Zum Hilbertschen Beweise des Waringschen Theorems. *Mat. Annalen*, 74:271–274, 1913.

[106] W. M. Schmidt. *Analytische Methoden für Diophantische Gleichungen*. Birkhäuser Verlag, Basel, 1984.

[107] W. M. Schmidt. The density of integer points on homogeneous varieties. *Acta Math.*, 154:243–296, 1985.

[108] B. Scholz. Bemerkung zu einem Beweis von Wieferich. *Jahrber. Deutsch. Math. Ver.*, 58:45–48, 1955.

[109] A. Selberg. On an elementary method in the theory of primes. *Norske Vid. Selsk. Forh., Trondheim*, 19(18):64–67, 1947.

[110] A. Selberg. *Collected Papers, Volume I.* Springer-Verlag, Berlin, 1989.

[111] A. Selberg. *Collected Papers, Volume II.* Springer-Verlag, Berlin, 1991.

[112] D. Shanks and Jr. J. W. Wrench. Brun's constant. *Math. Comp.*, 28:293–299, 1183, 1974.

[113] L. G. Shnirel'man. On the additive properties of integers. *Izv. Donskovo Politekh. Inst. Novocherkasske*, 14:3–27, 1930.

[114] L. G. Shnirel'man. Über additive Eigenschaften von Zahlen. *Mat. Annalen*, 107:649–690, 1933.

[115] J. H. Silverman. Taxicabs and sums of two cubes. *Am. Math. Monthly*, 100:331–340, 1993.

[116] J. H. Silverman and J. Tate. *Rational Points on Elliptic Curves.* Springer-Verlag, New York, 1992.

[117] M. K. Sinisalo. Checking the Goldbach conjecture up to $4 \cdot 10^{11}$. *Math. Comput.*, 61:931–934, 1993.

[118] J. Spencer. Four squares with few squares. Preprint, 1990.

[119] A. Stöhr. Eine Basis h-Ordnung für die Menge aller natürlichen Zahlen. *Mat. Z.*, 42:739–743, 1937.

[120] E. Stridsberg. Sur la démonstration de M. Hilbert du théorème de Waring. *Mat. Annalen*, 72:145–152, 1912.

[121] G. Tenenbaum. *Introduction to Analytic and Probabilistic Number Theory.* Cambridge University Press, Cambridge, 1995.

[122] J. V. Uspensky and M. A. Heaslet. *Elementary Number Theory.* McGraw-Hill, New York, 1939.

[123] J. G. van der Corput. Sur l'hypothèse de Goldbach pour presque tous les nombres pair. *Acta Arith.*, 2:266–290, 1937.

[124] R. C. Vaughan. Sommes trigonométriques sur les nombres premiers. *C. R. Acad. Sci. Paris, Sér. A*, 285:981–983, 1977.

[125] R. C. Vaughan. *The Hardy–Littlewood Method.* Cambridge University Press, Cambridge, 1981.

[126] R. C. Vaughan. On Waring's problem for cubes. *J. reine angew. Math.*, 365:122–170, 1986.

[127] R. C. Vaughan. A new iterative method in Waring's problem. *Acta Math.*, 162:1–71, 1989.

[128] R. C. Vaughan. The use in additive number theory of numbers without large prime factors. *Philos. Trans. Royal Soc. London A*, 345:363–376, 1993.

[129] R. C. Vaughan and T. D. Wooley. On Waring's problem: some refinements. *Proc. London Math. Soc.*, 63:35–68, 1991.

[130] R. C. Vaughan and T. D. Wooley. Further improvements in Waring's problem. *Acta Math.*, 174:147–240, 1995.

[131] I. M. Vinogradov. On Waring's theorem. *Izv. Akad. Nauk SSSR, Otd. Fiz.-Mat. Nauk*, (4):393–400, 1928. English translation in *Selected Works*, pages 101–106, Springer-Verlag, Berlin, 1985.

[132] I. M. Vinogradov. Representation of an odd number as the sum of three primes. *Doklady Akad. Nauk SSSR*, 15(6–7):291–294, 1937. English translation in *Selected Works*, pages 129–132, Springer-Verlag, Berlin, 1985.

[133] I. M. Vinogradov. Some theorems concerning the theory of primes. *Mat. Sbornik*, 2(44):179–195, 1937.

[134] I. M. Vinogradov. *The Method of Trigonometric Sums in the Theory of Numbers*, volume 23. Trud. Mat. Inst. Steklov, Moscow, 1947. English translation published by Interscience, New York, 1954.

[135] I. M. Vinogradov. *The Method of Trigonometric Sums in Number Theory*. Nauka, Moscow, 1980. English translation in *Selected Works*, pages 181–295, Springer-Verlag, Berlin, 1985.

[136] R. D. von Sterneck. *Sitzungsber. Akad. Wiss. Wien (Math.)*, 112, IIa:1627–1666, 1903.

[137] Y. Wang. *Goldbach Conjecture*. World Scientific, Singapore, 1984.

[138] E. Waring. *Meditationes Algebraicae*. Cambridge University Press, Cambridge, 1770.

[139] G. L. Watson. A proof of the seven cube theorem. *J. London Math. Soc.*, 26:153–156, 1951.

[140] A. Weil. Sur les sommes de trois et quatre carrés. *Enseign. Math.*, 20:215–222, 1974.

[141] H. Weyl. Über die Gleichverteilung von Zahlen mod Eins. *Mat. Annalen*, 77:313–352, 1913.

[142] H. Weyl. A half-century of mathematics. *Am. Math. Monthly*, 58:523–553, 1942. Reprinted in *Gesammelte Abhandlungen*, volume IV, pages 464–494, Springer-Verlag, Berlin, 1968.

[143] H. Weyl. David Hilbert and his mathematical work. *Bull. Am. Math. Soc.*, 50:612–654, 1944. Reprinted in *Gesammelte Abhandlungen*, volume IV, pages 130–172, Springer-Verlag, Berlin, 1968.

[144] A. Wieferich. Beweis des Satzes, daß sich eine jede ganze Zahl als Summe von höchstens neun positiven Kuben darstellen läßt. *Mat. Annalen*, 66:95–101, 1909.

[145] E. Wirsing. Thin subbases. *Analysis*, 6:285–308, 1986.

[146] T. D. Wooley. Large improvements in Waring's problem. *Ann. Math.*, 135:131–164, 1992.

[147] T. D. Wooley. On Vinogradov's mean value theorem. *Mathematika*, 39:379–399, 1992.

[148] T. D. Wooley. Breaking classical convexity in Waring's problem: Sums of cubes and quasi-diagonal behavior. *Inventiones Math.*, 122:421–451, 1995.

[149] T. D. Wooley. Sums of two cubes. *Int. Math. Res. Not.*, (4):181–185, 1995.

[150] E. M. Wright. An easier Waring's problem. *J. London Math. Soc.*, 9:267–272, 1934.

[151] J. Zöllner. *Der Vier-Quadrate-Satz und ein Problem von Erdős und Nathanson*. PhD thesis, Johannes Gutenberg-Universität, Mainz, 1984.

[152] J. Zöllner. Über eine Vermutung von Choi, Erdős, und Nathanson. *Acta Arith.*, 45:211–213, 1985.

Index

Additive basis, 7
additive function, 328
adjoint equation, 262
almost prime, 271
asymptotic basis, 33

Basis, 7
basis of finite order, 192
binary quadratic form, 9
Brun's constant, 173
Brun's theorem, 173

Cauchy's lemma, 30
Cauchy's theorem, 31
Chebyshev functions, 154
Chen's theorem, 271
Choi–Erdős–Nathanson theorem, 24
circle method, 121
classical bases, 7
completely multiplicative function, 308
counting function, 191
covering congruences, 204

Difference operator, 99

Dirichlet convolution, 301
Dirichlet series, 151
discriminant of a form, 8
divisor-closed set, 318

Easier Waring's problem, 72, 102
equivalent matrices, 8
equivalent quadratic forms, 8
Erdős–Mahler theorem, 61
Euler products, 325
Euler sum formula, 306
Euler's constant, 306
exceptional set, 230

Goldbach conjecture, 177
Goldbach–Shnirel'man theorem, 197

Hardy–Littlewood asymptotic formula, 146
Hermite polynomial, 77
Hilbert–Waring theorem, 88
Hooley–Wooley theorem, 66
Hua's lemma, 116

Implied constant, xiii
inclusion-exclusion principle, 174
infinite product, 323

Jurkat–Richert theorem, 257

Lagrange's theorem, 5
large sieve inequality, 295
Legendre's formula, 232
linear sieve, 238
Linnik's theorem, 46
lower bound sieve, 234

Major arcs, 126, 213
minor arcs, 127, 213
multiplicative function, 308

Partial summation, 304
polygonal number theorem, 31
polygonal numbers, 4
positive-definite form, 9

Quadratic form, 8

Ramanujan sum, 321
Ramanujan's sum, 212
Ramaré's theorem, 208

Riemann zeta-function, 151

Selberg sieve, 180
seven cube theorem, 46
Shnirel'man density, 192
Shnirel'man's addition theorem, 193
Shnirel'man's constant, 208
Siegel–Walfisz theorem, 46, 216
sieve dimension, 238
sieving function, 232
sieving level, 232
sieving range, 232, 234
singular series, 137
sumset, 192
support level, 234
symmetric matrix, 7

Ternary quadratic form, 9

Upper bound sieve, 234

Vinogradov's theorem, 212

Waring's problem, 37
well approximated, 126
Weyl's inequality, 114
Wieferich–Kempner theorem, 41

Graduate Texts in Mathematics

continued from page ii

66 WATERHOUSE. Introduction to Affine Group Schemes.
67 SERRE. Local Fields.
68 WEIDMANN. Linear Operators in Hilbert Spaces.
69 LANG. Cyclotomic Fields II.
70 MASSEY. Singular Homology Theory.
71 FARKAS/KRA. Riemann Surfaces. 2nd ed.
72 STILLWELL. Classical Topology and Combinatorial Group Theory. 2nd ed.
73 HUNGERFORD. Algebra.
74 DAVENPORT. Multiplicative Number Theory. 2nd ed.
75 HOCHSCHILD. Basic Theory of Algebraic Groups and Lie Algebras.
76 IITAKA. Algebraic Geometry.
77 HECKE. Lectures on the Theory of Algebraic Numbers.
78 BURRIS/SANKAPPANAVAR. A Course in Universal Algebra.
79 WALTERS. An Introduction to Ergodic Theory.
80 ROBINSON. A Course in the Theory of Groups. 2nd ed.
81 FORSTER. Lectures on Riemann Surfaces.
82 BOTT/TU. Differential Forms in Algebraic Topology.
83 WASHINGTON. Introduction to Cyclotomic Fields.
84 IRELAND/ROSEN. A Classical Introduction to Modern Number Theory. 2nd ed.
85 EDWARDS. Fourier Series. Vol. II. 2nd ed.
86 VAN LINT. Introduction to Coding Theory. 2nd ed.
87 BROWN. Cohomology of Groups.
88 PIERCE. Associative Algebras.
89 LANG. Introduction to Algebraic and Abelian Functions. 2nd ed.
90 BRØNDSTED. An Introduction to Convex Polytopes.
91 BEARDON. On the Geometry of Discrete Groups.
92 DIESTEL. Sequences and Series in Banach Spaces.
93 DUBROVIN/FOMENKO/NOVIKOV. Modern Geometry-Methods and Applications. Part I. 2nd ed.
94 WARNER. Foundations of Differentiable Manifolds and Lie Groups.
95 SHIRYAEV. Probability. 2nd ed.

96 CONWAY. A Course in Functional Analysis. 2nd ed.
97 KOBLITZ. Introduction to Elliptic Curves and Modular Forms. 2nd ed.
98 BRÖCKER/TOM DIECK. Representations of Compact Lie Groups.
99 GROVE/BENSON. Finite Reflection Groups. 2nd ed.
100 BERG/CHRISTENSEN/RESSEL. Harmonic Analysis on Semigroups: Theory of Positive Definite and Related Functions.
101 EDWARDS. Galois Theory.
102 VARADARAJAN. Lie Groups, Lie Algebras and Their Representations.
103 LANG. Complex Analysis. 3rd ed.
104 DUBROVIN/FOMENKO/NOVIKOV. Modern Geometry-Methods and Applications. Part II.
105 LANG. $SL_2(\mathbf{R})$.
106 SILVERMAN. The Arithmetic of Elliptic Curves.
107 OLVER. Applications of Lie Groups to Differential Equations. 2nd ed.
108 RANGE. Holomorphic Functions and Integral Representations in Several Complex Variables.
109 LEHTO. Univalent Functions and Teichmüller Spaces.
110 LANG. Algebraic Number Theory.
111 HUSEMÖLLER. Elliptic Curves.
112 LANG. Elliptic Functions.
113 KARATZAS/SHREVE. Brownian Motion and Stochastic Calculus. 2nd ed.
114 KOBLITZ. A Course in Number Theory and Cryptography. 2nd ed.
115 BERGER/GOSTIAUX. Differential Geometry: Manifolds, Curves, and Surfaces.
116 KELLEY/SRINIVASAN. Measure and Integral. Vol. I.
117 SERRE. Algebraic Groups and Class Fields.
118 PEDERSEN. Analysis Now.
119 ROTMAN. An Introduction to Algebraic Topology.
120 ZIEMER. Weakly Differentiable Functions: Sobolev Spaces and Functions of Bounded Variation.
121 LANG. Cyclotomic Fields I and II. Combined 2nd ed.
122 REMMERT. Theory of Complex Functions. *Readings in Mathematics*

123 EBBINGHAUS/HERMES ET AL. Numbers.
    Readings in Mathematics
124 DUBROVIN/FOMENKO/NOVIKOV. Modern
    Geometry-Methods and Applications. Part
    III.
125 BERENSTEIN/GAY. Complex Variables: An
    Introduction.
126 BOREL. Linear Algebraic Groups.
127 MASSEY. A Basic Course in Algebraic Topology.
128 RAUCH. Partial Differential Equations.
129 FULTON/HARRIS. Representation Theory: A
    First Course.
    Readings in Mathematics
130 DODSON/POSTON. Tensor Geometry.
131 LAM. A First Course in Noncommutative Rings.
132 BEARDON. Iteration of Rational Functions.
133 HARRIS. Algebraic Geometry: A First Course.
134 ROMAN. Coding and Information Theory.
135 ROMAN. Advanced Linear Algebra.
136 ADKINS/WEINTRAUB. Algebra: An Approach via
    Module Theory.
137 AXLER/BOURDON/RAMEY. Harmonic Function
    Theory.
138 COHEN. A Course in Computational Algebraic
    Number Theory.
139 BREDON. Topology and Geometry.
140 AUBIN. Optima and Equilibria. An Introduction
    to Nonlinear Analysis.
141 BECKER/WEISPFENNING/KREDEL. Gröbner Bases.
    A Computational Approach to Commutative
    Algebra.
142 LANG. Real and Functional Analysis. 3rd ed.
143 DOOB. Measure Theory.
144 DENNIS/FARB. Noncommutative Algebra.
145 VICK. Homology Theory. An Introduction to
    Algebraic Topology. 2nd ed.

146 BRIDGES. Computability: A Mathematical
    Sketchbook.
147 ROSENBERG. Algebraic K-Theory and Its
    Applications.
148 ROTMAN. An Introduction to the Theory of
    Groups. 4th ed.
149 RATCLIFFE. Foundations of Hyperbolic
    Manifolds.
150 EISENBUD. Commutative Algebra with a View
    Toward Algebraic Geometry.
151 SILVERMAN. Advanced Topics in the Arithmetic
    of Elliptic Curves.
152 ZIEGLER. Lectures on Polytopes.
153 FULTON. Algebraic Topology: A First Course.
154 BROWN/PEARCY. An Introduction to Analysis.
155 KASSEL. Quantum Groups.
156 KECHRIS. Classical Descriptive Set Theory.
157 MALLIAVIN. Integration and Probability.
158 ROMAN. Field Theory.
159 CONWAY. Functions of One Complex Variable
    II.
160 LANG. Differential and Riemannian Manifolds.
161 BORWEIN/ERD,LYI. Polynomials and Polynomial
    Inequalities.
162 ALPERIN/BELL. Groups and Representations.
163 DIXON/MORTIMER. Permutation Groups.
164 NATHANSON. Additive Number Theory: The
    Classical Bases.
165 NATHANSON. Additive Number Theory: Inverse
    Problems and the Geometry of Sumsets.
166 SHARPE. Differential Geometry: Cartan's
    Generalization of Klein's Erlangen Programme.
167 MORANDI. Field and Galois Theory.